DARWIN'S HARVEST

DARWIN'S HARVEST

New Approaches to the Origins,
Evolution, and Conservation of Crops

Edited by
TIMOTHY J. MOTLEY,
NYREE ZEREGA,
and **HUGH CROSS**

Columbia University Press
NEW YORK

Columbia University Press
Publishers Since 1893
New York, Chichester, West Sussex
Copyright © 2006 Columbia University Press
All rights reserved

Library of Congress Cataloging-in-Publication Data

Darwin's harvest: new approaches to the origins, evolution,
and conservation of crops / edited by Timothy J. Motley, Nyree
Zerega, and Hugh Cross.
p. cm.
Includes bibliographical references and index.
ISBN 978-0-231-13316-6 (cloth : alk. paper)—ISBN 978-0-231-50809-4 (e-book)
1. Crops–Origin. 2. Crops–Evolution. 3. Plant conservation.
I. Motley, Timothy J.,
1965–II. Zerega, Nyree. III. Cross, Hugh (Hugh B.)

Printed in the United States of America

CONTENTS

Robin G. Allaby
Faculty of Life Sciences
Jacksons Mill
The University of Manchester
P.O. Box 88
Manchester, M60 1QD
United Kingdom

Mallikarjuna K. Aradhya
USDA National Clonal Germplasm
    Repository
University of California at Davis
One Shields Ave.
Davis, CA 95616

Giovanna Attene
Dipartimento di Scienze
Agronomiche e Genetica
Vegetale Agraria
Università degli Studi di Sassari
Via E. de Nicola, 07100, Sassari
Italy

Kenneth Birnbaum
Department of Biology
New York University
100 Washington Sq. East
1009 Main Building
New York, NY 10003

Terence A. Brown
Faculty of Life Sciences
Jacksons Mill
The University of
    Manchester

P.O. Box 88
Manchester, M60 1QD
United Kingdom

Edward S. Buckler IV
Cornell University
USDA–ARS Research Geneticist
Institute for Genomic Diversity
159 Biotechnology Building
Ithaca, NY 14853-2703

Luiz J. C. B. Carvalho
Brazilian Agricultural Research
    Corporation–EMBRAPA
SAIN Parque Rural Edificio Sede de
    EMPRAPA
Brasilia–DF, 70770-901
Brazil

Hugh Cross
Nationaal Herbarium Nederland
Universiteit Leiden Branch
Einsteinweg 2, P.O. Box 9514
2300 RA, Leiden
The Netherlands

Angélique D'Hont
Programme Canne à Sucre
CIRAD, TA 40/03
Avenue Agropolis
Montpellier, 34398–Cedex 5
France

Eve Emshwiller
Field Museum

1400 S. Lake Shore Dr.
Chicago, IL 60605-2496

Mary W. Eubanks
Department of Biology
Duke University
Box 90338
Durham, NC 27708

Jean-Christophe Glaszmann
Programme Canne à Sucre
CIRAD, TA 40/03
Avenue Agropolis
Montpellier, 34398–Cedex 5
France

Laurent Grivet
Programme Canne à Sucre
CIRAD, TA 40/03
Avenue Agropolis
Montpellier, 34398–Cedex 5
France

Abigail V. Harter
Department of Biology
Indiana University
Jordan Hall
Bloomington, IN 47405

Wilbert L. A. Hetterscheid
Botanical Gardens
Wageningen University
Gen. Foulkesweg 37
6703 BL Wageningen
The Netherlands

Vincent Lebot
Scientific Coordinator SPYN and
  TANSAO
CIRAD
P.O. Box 946
Port Vila
Vanuatu

Sarah Lindsay
Faculty of Life Sciences
Jacksons Mill
The University of Manchester
P.O. Box 88
Manchester, M60 1QD
United Kingdom

Rafael Lira Saade
Laboratorio de Recursos Naturales,
  UBIPRO
Facultad de Estudios Superiores
  Iztacala, UNAM
Av. de los Barrios I, Los Reyes
  Iztacal
Tl anepantla, CP 54090
México

Roger Malapa
VARTC
P.O. Box 231
Luganville, Santa
Vanuatu

Jean-Leu Marchand
CIRAD-Ca
TA 70/16
Montpellier, 34398–Cedex 5
France

Timothy J. Motley
The Lewis B. and Dorothy Cullman
  Program for Molecular Systematics
  Studies
The New York Botanical Garden
201st Street and Southern Blvd.
Bronx, NY 10458-5126

Laura Nanni
Dipartimento di Scienze degli
  Alimenti
Facoltà di Agraria
Università Politecnica delle
  Marche

Via Brecce Bianche
Ancona, 60131
Italy

Jean-Louis Noyer
CIRAD Biotrop
TA 40/03
Montpellier, 34398–Cedex 5
France

Kenneth M. Olsen
Department of Genetics
North Carolina State University
Raleigh, NC 27695-7614

Roberto Papa
Dipartimento di Scienze degli
  Alimenti
Facoltà di Agraria
Università Politecnica delle
  Marche
Via Brecce Bianche
Ancona, 60131
Italy

Daniel Potter
Department of Pomology
University of California at Davis
One Shields Ave.
Davis, CA 95616

Diane Ragone
The Breadfruit Institute
National Tropical Botanical
  Garden
3530 Papalina Road
Kalaheo, HI 96741

Domenico Rau
Dipartimento di Scienze degli
  Alimenti
Facoltà di Agraria
Università Politecnica delle
  Marche
Via Brecce Bianche

Ancona, 60131
Italy

Loren H. Rieseberg
Department of Biology
Indiana University
Jordan Hall
Bloomington, IN 47405

Barbara A. Schaal
Department of Biology
Evolutionary and Population/Plant
  Biology Programs
Washington University
1 Brookings Ave.
Campus Box 1137
St. Louis, MO 63130

Delphine Sicard
UMR de Génétique Végétale
INRA/UPS/CNRS/INA-PG
Ferme du Moulon, 91190
Gif-sur-Yvette
France

Charles J. Simon
USDA, Agricultural Research
  Service
Plant Genetic Resources Unit
Cornell University
Geneva, NY 14456-0462

David M. Spooner
USDA, Agricultural Research Service
Department of Horticulture
University of Wisconsin
1575 Linden Drive
Madison, WI 53706

Natalie M. Stevens
Maize Genetics Research
Institute for Genomic Diversity
Cornell University
175 Biotechnology Building
Ithaca, NY 14853-2703

Sarah M. Ward
Department of Soil and Crop Sciences
Department of Bioagricultural
  Sciences and Pest Management
Colorado State University
Fort Collins, CO 80523-1170

Nyree Zerega
Northwestern University and Chicago
  Botanic Garden
Program in Biological Sciences
2205 Tech Drive
Evanston, IL 60208

Crop Plants
Past, Present, and Future

Research on crop plants often has been at the forefront of revolutions in plant biology. Notable achievements include Charles Darwin's studies of variation of plants under domestication (Darwin, 1883), the work of Gregor Mendel on the garden pea and the principles of inheritance, and the Nobel Prize–winning research of Barbara McClintock and her discovery of transposable elements in maize (McClintock, 1950). More recently with the development of the polymerase chain reaction (PCR) and automated sequencing technology, novel DNA markers and gene regions often are first used by crop plant researchers before being used in other botanical disciplines. These techniques have enabled crop scientists to address questions that they previously could not answer, such as the effects of domestication and selection on the entire plant genome (Emshwiller, in press). Rice (*Oryza sativa*) was the second plant species, after the model plant species *Arabidopsis thaliana,* to have its entire genome sequenced (Goff et al., 2002; Yu et al., 2002). Current genome sequencing projects, such as those at the Institute for Genomics Research, are focusing on agronomically important groups, including the grass, legume, tomato, and cabbage families (see www.tigr.org).

Research on crop plant origins and evolution is relevant to researchers in many disciplines. Geneticists, agronomists, botanists, systematists, population biologists, archaeologists, anthropologists, economic botanists, conservation biologists, and the general public all have an interest in natural history and the cutting-edge methods that are shaping the future of science and the plants that sustain humankind. One reason for this interest in crop plants is that agriculture is a large industry, and as the world population continues to increase, resources become scarcer, and as environments and climates continue to change, new developments in crop plants will play an integral role in shaping the future.

Crop plant evolution is an enormous subject. The goal of this book is to provide a broad sample of current research on a diverse group of crop plants. The chapters use many methods and molecular markers to shed further light on the topics of plant origin and present new data on crop plant evolution. As in any field, however, there are philosophical differences, disagreements, and competition. For instance, there have been disagreements as to the origins of maize (Mangelsdorf, 1974; Beadle, 1977), and the same debates remain today (see chapters 4 and 5). Although the majority of maize researchers (Bennetzen et al., 2001) now accept the Beadle teosinte hypotheses, having the freedom to revisit alternative or unpopular hypotheses is an invaluable part of science. In order to ensure quality and impartial scrutiny of the data presented, each chapter in this book was subjected to anonymous peer review.

The contributors to this volume have a broad range of experience, some coming from agricultural backgrounds and others from the field of systematics. Some authors have experience in archaeological research and sequencing ancient DNA; others have experience in genetics and molecular biology. The contributions were selected to represent a broad range of major and minor crops. Some of the crops such as corn, beans, wheat, and potatoes have a long history of research, are cultivated around the world, and are among the most important staples of human civilization. Others, including sugarcane, yams, cassava, and breadfruit, are cultivated and used each day throughout tropical regions. Still others, such as oca and chayote, are lesser known outside their native regions. Sugarcane is an example of a crop used each day throughout the world and cultivated widely throughout tropical regions, yet its origins in Southeast Asia and the southwestern Pacific are obscure.

In keeping with the theme of this book, the crop species discussed exhibit a wide range of traits. Both temperate and tropical crops are included. Some species are cultivated by seed; others are vegetatively propagated by tubers, cuttings, or rhizomes. The crops also span the breadth of habit and lifecycle variation. The tree crops, such as breadfruit, walnuts, and avocado, have long lifespans. In the case of walnuts, the time to reach reproductive maturity is equal to a third of a human lifespan, making controlled studies difficult during an academic career. On the other hand, in the case of annuals (e.g., wheat, sunflower, and corn) researchers can easily set up breeding studies and experiments on progeny, perhaps getting three or more harvests per year in controlled environments. Further complicating studies of plant evolutionary history is the fact that plants, unlike animals, can more easily hybridize with closely related species, often leading to chromosome variants (polyploids, aneuploids) that are not detrimental but rather provide additional genetic variation.

The chapters of this book cover many themes, including plant origins, evolutionary relationships to wild species, crop plant nomenclature, tracing patterns of human-mediated crop dispersal, gene flow, and hybridization. Some chapters cover the genetic effects of cultivation practices and human selection, the identification of genetic pathways for beneficial traits, and germplasm conservation and collection.

It is the goal of this introductory chapter to review the origins, evolution, and conservation of crop plants. An entire volume could be dedicated to each of the topics, but in this chapter I have only scratched the surface in order to provide a few interesting case studies. In doing this I have tried to introduce the reader to the subject of crop plant research and identify some of the challenges and pitfalls that the authors of *Darwin's Harvest* faced during their research.

Beginnings of Agriculture

It has been postulated that agriculture is a necessary step in the advancement of civilizations because it allows larger and more stable populations to prosper (MacNeish, 1991). As resources became consistently available, a nomadic lifestyle was no longer necessary, and groups began settling in areas fit for cultivation. As the group became larger, division of labor occurred, creating more free time for development of other cultural activities such as mining, arts, education, philosophy, and laws. However, Diamond (1999) points out that with agricultural society also comes a higher incidence of disease, caused in part by high population densities and shifts from high-protein to high-carbohydrate diets. Most successful civilizations were built around farming, but there are examples of nomadic hunters and gatherers living at sustainable levels that are equal to or greater than (in terms of caloric intake and energy expended) the level in early agricultural societies

(Harlan, 1967), but these groups never were able to reach similar levels of cultural, scientific, industrial, or governmental development.

The earliest records for agriculture come from archaeological remains of stored seeds or tools and suggest, based on ^{14}C dating, that agriculture arose approximately 10,000 years ago (Lee and DeVore, 1968) in the Fertile Crescent, a region that wraps around the eastern edge of the Mediterranean Sea along the river valleys of the Nile, Tigris, and Euphrates east to the Persian Gulf. However, dates from agricultural sites in Asia (China: Chang, 1977; Sun et al., 1981; Thailand: Gorman, 1969) and Central America (Sauer, 1952; Smith, 1997) are nearly as old. It is possible that the arid conditions around the Mediterranean, more favorable for preservation of archaeological remains, may account for the earlier dates in the Fertile Crescent.

Several factors have been proposed that contributed to the rise of agriculture, including population pressures, climate changes, and co-evolution between plants and humans. The population growth hypothesis (Cohen, 1977) argues that growing human populations exhausted the regional resources, and this made the hunter and gatherer lifestyle inefficient (i.e., greater energy output was needed for caloric reward), thus forcing a shift to agriculture. Similarly, Childe's (1952) climatic change hypothesis suggests that after the Pleistocene ice age the regions around the southern and eastern Mediterranean became drier, forcing humans to congregate along water sources, and agriculture was needed to sustain the increasing population density. Rindos's (1984) hypothesis based on co-evolutionary dependence is the most thought-provoking. It asserts that a mutualistic dependence has developed over many generations between plants and humans, and they now rely on one another for survival. Crop plants provide a product we desire, and some depend on humans for cultivation. Examples of this dependence vary from sterile triploid crops (banana, taro, and breadfruit) that completely rely on humans for propagation to others such as corn that need humans for dispersal or have become bred for highly specialized monoculture communities that need weeding and pest control to outcompete more aggressive species. Pollan (2001) adds an unusual twist to this idea, looking at it from a plant's viewpoint, suggesting that plants have selected for humans.

Determining the events that lead to an agronomic society probably is never as simple as one single explanation but rather entails a combination of factors, independent of one another in each case of domestication. This is what Harlan (1992) calls the "no model" model. The same may be said about the origins and evolution of individual crop plants. Often no single cause can explain the origins of domesticated crops or their present distributions.

Crop Plants

The definition of a crop is not simple. Under domestication, selective pressures act heavily on certain phenotypic traits desirable for cultivation. The classic advantageous crop traits are nonshattering infructescences, fewer and larger fruits, loss of bitterness, reduced branching, self-pollination, increased seed set, loss of seed dormancy, quick germination, short growing season, and higher carbohydrate levels. These traits are called the domestication syndrome (Harlan et al., 1973; de Wet and Harlan, 1975; Harlan, 1992; Smith, 1998). Harlan (1992) defines a crop as anything that is harvested, and he further divides these plants into four categories: wild, tolerated, encouraged, and domesticated.

Anderson (1954) describes species that he calls camp followers. These plants did well in areas where humans altered the environment and thus could be the progenitors of crop plants (de Wet and Harlan 1975). These plants would be defined as weeds. In many cases domestic plants evolved from weedy species (e.g., rice, sorghum, and carrots) and do well in disturbed areas, such as tilled fields and middens (Harlan, 1992).

Some crops were once weeds in human settlements before the origins of agriculture; other crop progenitors were weeds in fields after the establishment of agriculture and often are considered secondary domesticates (de Wet and Harlan, 1975). For example, oats and rye were once weeds infesting fields of barley and wheat (Vavilov, 1926), and false flax (*Camelina sativa*, Brassicaceae) began as a weed in Russian flax fields (Zohary and Hopf, 1994). Other crops such as lettuce may have been domesticated the same way.

Some crops escape from cultivation and revert to weeds. The bitter melon (*Momordica charantia*), prized in Chinese and Filipino cooking, was introduced to the Hawaiian Islands in the 1930s. It later escaped from cultivation and is now a noxious weed. The naturalized plants have adapted back to the wild, where natural selection favors smaller fruits and less desirable flavor. The wild forms are called *M. charantia* var. *abbreviata* (Telford, 1990). This demonstrates the fine line between weeds and crops and how critical human preferences and intervention can be for the continuation of a crop.

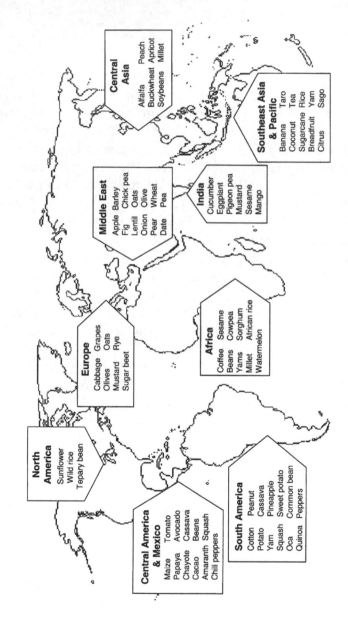

FIGURE 1.1 Areas of origin for crop plants according to recent scientific evidence.

Some crops have very local ranges; for example, tacaco (*Sechium tacaco*; Cucurbitaceae) is grown only in Costa Rica, whereas a related species, chayote (*Sechium edule*), has gained a wide acceptance beyond its native Mexico (chapter 8, this volume). What may be selected for in one area is not in another. Popular cultivars once valued and selected for their unique traits (heirloom varieties) may later vanish as popularity of alternative crops increases.

Many factors such as regional preferences, cultural bias, economics, and marketing may also play a role in a plant's use or disuse and determine whether it ultimately becomes a crop. When eating at an Italian restaurant it is difficult imagine that tomatoes were not a part of the cultural cuisine of Italy until just a few hundred years ago. Similarly, it is not easy to conceive of Ireland, Denmark, and Russia without potatoes. However, both tomatoes and potatoes are of New World origin (figure 1.1). At the time of their introduction into the Old World, Europeans did not immediately accept these crops because they were similar to local poisonous plants (deadly nightshades), they were thought to cause disease (under the *Doctrine of Signatures* the swollen tubers of potato were thought to cause leprosy), and they were associated with ethnic groups (eggplant and tomatoes were considered Jewish food; Davidson, 1992). Although we have overcome many prejudices and superstitions, today our crop preferences are being driven by economics and marketing. When most people think of a potato, they imagine the brown Irish potato, and outside the tropics most people envision a papaya as the pear-shaped solo variety, which packs and ships so nicely to consumers. Few new crops have been developed, and the world still relies on many of the staples it did in the past.

Today approximately 200 plant species have been domesticated worldwide (Harlan, 1992) out of approximately 250,000 known plant species (Heywood, 1993). However, fewer than 20 crops in eight plant families provide most of the world's food: wheat, rice, corn, beans, sugarcane, sugar beet, cassava, potato, sweet potato, banana, coconut, soybean, peanut, barley, and sorghum (Harlan, 1992). Only eight plant families stand between most humans and starvation, and 55 contain all our crop plants (Tippo and Stern, 1977).

Geographic Origins

Agriculture arose independently on several continents. If this were not the case and the knowledge of plant domestication were shared among the areas

Box 1.1

Russian scientist Nikolai I. Vavilov worked at the Bureau of Applied Botany (now VIR) in Leningrad from 1921 to 1940, where he laid down many of the foundations of modern crop plant research. Following advances in genetics in the early 19th century, Vavilov believed that improvement of Russian agriculture was best achieved through the collection of thousands of crop varieties from their areas of greatest diversity, followed by careful hybridization and selection of recombinant forms best adapted to local conditions. Vavilov's rival, Trofim D. Lysenko, did not agree with this method or the tenets of Darwinian–Mendelian genetics, favoring instead the Lamarckian model of inheritance whereby traits acquired in one generation are passed on to the progeny. Lysenko proposed that wheat and other crops could be induced to change by repeated exposure to harsh environments and would result in progeny better adapted to these conditions. For example, Lysenko subjected wheat seeds to cold treatment in the hope that they would result in cold-adapted progeny. Unfortunately, in the Soviet Union at this time scientific debate was not free from politics, and Lysenko's ideas (and his probably falsified field data) were favored by Stalin, and Lysenko eventually replaced Vavilov as president of the bureau. Soon after, while conducting fieldwork in the Ukraine, Vavilov was arrested for espionage. Vavilov died in a Soviet prison in 1943 (Popovsky, 1984).

BOX FIGURE 1.1 Monument outside VIR: Outstanding biologist and academician Nikolai Ivanovich Vavilov worked here from 1921 to 1940.

of agricultural origin, then at least some of the cultivated plant species would have changed hands as well. Almost certainly, different crops native to different regions of the world were domesticated separately in their respective regions, as seems to be the case of Old and New World crops.

In the 19th century de Candolle (1959) first put forth hypotheses for determining centers of origin for the various crop species using evidence from multiple disciplines (botany, geography, history, linguistics, and archaeology). de Candolle's multiple-discipline approach was primarily an intellectual effort. Vavilov (1992) greatly expanded de Candolle's ideas through the use of field research and breeding experiments. From this work, he developed his eight centers of origin theory, in which he proposed that the regions containing the highest genetic diversity of a crop species (species richness or number of varieties) probably were its area of origin. Vavilov's centers were broad (Tropical South Asiatic, East Asiatic, Southwestern Asiatic, Western Asiatic, Mediterranean, Abyssinian [Ethiopian], Central American, and Andean–South American), based on morphological similarities between wild species and crop plants or the number of cultivars or varieties of a crop species. Later he developed the idea of secondary centers to help explain crops that did not fit well into his defined centers of origin. Vavilov's work gave us a framework for studying the origins of crop plants, but perhaps his greatest contribution was his idea to collect the wild relatives of crop plants from these areas so they could be used in plant breeding programs for crop improvement (see Box 1.1 for a brief background on Vavilov's life).

Vavilov believed that a crop's center of diversity was also its center of origin. However, several researchers have shown that this is not always the case (see Smith, 1969). For example, the areas of greatest diversity of barley and rice are distant from their regions of domestication (Hancock, 2004). Furthermore, since Vavilov's work, new centers for crop origins have been proposed in North America (Heiser, 1990), and recent archaeological and paleontological records have been unearthed suggesting that New Guinea, a region outside Vavilov's Tropical South Asiatic center, is another region where agriculture arose independently, in this instance more than 6000 years ago (Denham et al., 2003).

Harlan (1971) redefined Vavilov's areas of crop origin with his "centers and noncenters" theory, in which he used archaeological evidence and the native ranges of crop progenitors to assign origins. He defined three centers of origin that he believed had never had contact with one another: the Near East (Fertile Crescent), North Chinese, and Mesoamerican. His noncenters were the African (central Africa), Southeast Asian and South Pacific, and South American. He suggested that noncenters were diffuse areas where origins could not be pinpointed and were perhaps influenced by other centers. Vavilov was also aware of these intermediate regions, which he called secondary centers. A common characteristic of every center is that a grain and a legume were always domesticated together (maize and common bean in the Americas, wheat and lentils in the Mediterranean, and rice and soybeans in Asia), providing complementary nutrition. Today researchers are using de Candolle's multidisciplinary approach by using advances in carbon dating and molecular techniques as well as archaeological (Kirch, 2000) and linguistic data (Diamond and Bellwood, 2003) and building on the hypotheses of Vavilov and Harlan to study crop origins and dispersal.

Based on our present knowledge, where are the centers of origin for our crop plants (figure 1.1)? In the New World sunflowers, tepary beans (*Phaseolus acutifolius* A. Gray) and wild rice (*Zizania aquatica*) appear to be of North American origin. Maize, papaya, cassava, cacao, avocado, beans (*Phaseolus* spp.), chayote, squash, cotton, and chili peppers have their origins in Mesoamerica. The Andes and rainforests of South America are centers for the domestication of potato, beans (*Phaseolus* spp.), sweet potato, quinoa, cotton, pineapple, yams, peppers, oca, cassava, and peanuts. In the Old World, African rice (*Oryza glaberrima*), coffee, beans (*Vigna* spp. and *Lablab niger*), pearl millet (*Pennisetum glaucum*), finger millet (*Eleusine coracana*), sorghum, watermelon, yams, and sesame are attributed to central Africa. In the Fertile Crescent of the Mediterranean, apples, barley, beans (*Vicia* spp.), lentils, olives, peas, pears, wheat, pomegranates, onions, grapes, figs, and dates were first brought into cultivation. Sugar beets, rye, mustard, oats, and cabbage are centered in southern Europe; cucumbers, eggplant, mustard, and sesame are from India; alfalfa, buckwheat, slender millet (*Panicum miliare*), and adzuki beans (*Vigna angularis*) are from central Asia; and bok choy, soybeans, peaches, broomcorn millet (*Panicum miliaceum*), and foxtail millet (*Setaria italica*) are from China. The tropical areas of Southeast Asia and the Pacific are the source areas for rice (*Oryza sativa*), taro, sugarcane, breadfruit, yams, citrus, and banana.

For some plants it is difficult to determine an exact locality of origin because the species disperse easily over long distances or human dispersal has clouded the issue. Various regions have been suggested as the area of origin for coconut, but the most favored are the western Pacific (Beccari, 1963; Corner, 1966; Moore, 1973; Harries, 1978) or the Neotropics (Guppy, 1906; Cook, 1910; Hahn, 2002). Fossil coconuts or coconut-like fruits dated to 38 mya in some cases are known from New Zealand (Berry, 1926; Couper, 1952; Campbell et al., 2000), Australia (Rigby, 1995), and India (Kaul, 1951; Patil and Upadhye, 1984), lending support to a western Pacific origin. However, phylogenetic evidence from molecular sequencing (Gunn, 2003; Hahn, 2002) does not provide enough resolution to determine the closest relatives of coconut. As data accumulate from different sources, the origin and historical dispersal of coconut may become clearer.

The origins and distribution of the sweet potato also have proved to be an enigma. Linguistic and genetic data suggest a South American origin (Yen, 1974; Shewry, 2003), but this does not explain its wide prehistoric distributions in the Pacific. The numerous Polynesian cultivars of sweet potato (Yen, 1974) make eastern Polynesia a classic example of a secondary center of diversity. Based on anthropological, archaeological, and botanical data (statues, similar myths, and sweet potato distribution), Thor Heyerdahl (1952) speculated that the Polynesians had originated in South America. To test this idea he organized the *Kon Tiki* expedition to prove that humans could have reached the islands of Polynesia in a balsa raft and introduced sweet potatoes to the Pacific before European contact. This theory has since been refuted by an overwhelming amount of evidence from linguistics, archaeology, anthropology, botany, and human genetics indicating that Polynesians are of Southeast Asian origin (Kirch, 2000; Hurles et al., 2003). Although it appears that the people of South America did not introduce sweet potatoes to the islands of the Pacific, the possibility remains that Polynesians voyaged to the coast of South America and brought back the sweet potato.

Research on Crop Plants

Most phylogenetic systematic studies of plants take place at or above the species level, examining the hierarchical relationships of species or groups of species. Crop plant researchers are interested not only in phylogenetic hierarchy but also in intraspecific variation. The varieties, cultivars, and races of crop plants often are as morphologically differentiated as genera are in the natural world. The high levels of morphological variation can occur when artificial selection is intense, resulting in rapid phenotypic differentiation over a few generations (Ungerer et al., 1998). In some cases, such as maize, the selective pressures affecting the phenotypic variation are offset by genetic recombination among alleles during the domestication process and help maintain genotypic variability (Wang et al., 1999). Alternatively, *Brassica oleracea* (cabbage,

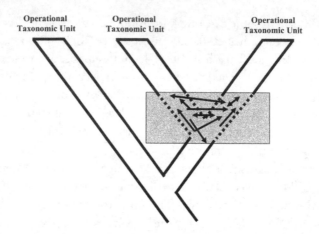

FIGURE 1.2   Phylogenetic tree. Gray box indicates region of interest in the evolutionary history of a plant lineage where crop scientists often focus their research efforts. Arrows indicate evolutionary events (e.g., hybridization, introgression, and polyploidy) that give rise the operational taxonomic units (species, varieties, cultivars).

broccoli, cauliflower, kohlrabi, Brussels sprouts, and its other cultivars) is an example of a plant complex that exhibits dramatic morphological variation but has low genetic variation (Kennard et al., 1994). In nature the same phenomenon occurs in the isolated habitats of island systems (Baldwin and Robichaux, 1995; Lindqvist et al., 2003). Furthermore, both agricultural and island populations undergo genetic bottlenecks (Ladizinsky, 1985) caused by either a founder event or genetic drift. Thus careful research and highly variable genetic markers are needed to achieve a clearer understanding of how this morphological variability is maintained in genetically similar crop plants.

Evolutionary events such as hybridization, introgression, and polyploidy can complicate crop plant research. Crop researchers must be concerned not only with a phylogenetic hierarchy (ancestral and sister relationships) but also with the plant's gene pool (figure 1.2). The ability of plants to survive polyploid events (although some level of sterility may occur), which usually are deleterious in animals, allows plants to overcome some of the limitations caused by genetic bottlenecks, founder effects, and selection. Allopolyploids result from the combination of two genetically different sets of chromosomes (through hybridization and incomplete meiotic division), whereas autopolyploids are the result of the multiplication of a set

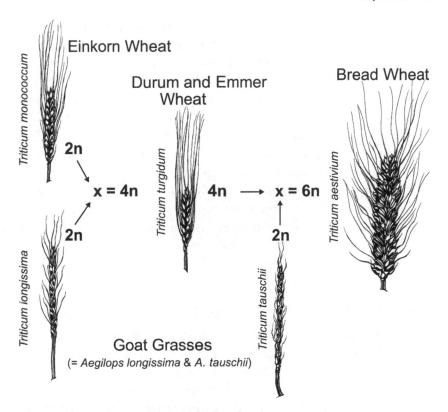

Einkorn Wheat

Triticum monococcum

Durum and Emmer Wheat

Bread Wheat

Triticum turgidum

Triticum aestivium

2n

x = 4n

2n

4n ⟶ x = 6n

2n

Triticum longissima

Triticum tauschii

Goat Grasses
(= *Aegilops longissima* & *A. tauschii*)

FIGURE 1.3 Evolutionary history of modern hexaploid bread wheat, showing two hybridization events leading to polyploid evolution and trigenomic accumulation.

of chromosomes from a single genome. These events can restore genetic variability and also produce desirable phenotypic results, but they also add another layer of complexity for the crop scientist to unravel.

Hybridization can occur when human dispersal of the crop brings it into contact with closely related species. The origin of our modern bread wheat may be one of the best-known and most complex examples of hybridization, allopolyploidy, and autopolyploidy in the evolution of crop plants (figure 1.3). Modern cultivated bread wheat incorporates three genomes. The early ancestor of wheat, *Triticum monococcum,* was diploid (2n = 14). Selection for shatterproof fruits and other desirable traits transformed the diploid ancestor into what we recognize as einkorn wheat. This wheat later hybridized with wild goat grass (*T. longissima*), producing sterile offspring.

Fertility was restored by the doubling of chromosomes (2n = 28), resulting in emmer and durum wheat (*T. turgidum* var. *dicoccum* and *T. turgidum* var. *durum,* respectively). Durum wheat was the variety prized for relaxed glumes at fruit maturity that allowed the fruit to be easily separated from the chaff. Later, a cross between the tetraploid (2n = 28) *T. turgidum* and another wild, diploid goat grass (*T. tauschii* [= *Aegilops squarrosa*]) resulted in modern hexaploid wheat (2n = 42), *T. aestivum* (see Feldman, 1976). This hexaploid and its high-protein varieties fill the breadbaskets of the world, although durum wheat is still cultivated today in dry regions for use in making products such as pasta and couscous. Similar cases of polyploidy and hybrid evolution are presented in other chapters of this book (e.g., oca, breadfruit, and corn), and Brown et al. (chapter 9, this volume) further explore the historical spread of wheat and its expansion into Europe.

Germplasm Collections and Maintenance

The establishment and maintenance of germplasm collections to preserve the genetic diversity of crop plants and their wild relatives are crucial but encounter many problems. Curators of these collections must deal with various lifecycles and ecological needs for each species (National Research Council, 1978; Gill, 1989), and this can raise costs. The more complicated the lifecycle needs or the more labor and land needed, the higher the financial costs of maintaining a collection. In general, it is easier to store seeds from temperate regions, such as cereals that undergo dormancy, than it is for tropical species that lack dormancy. Furthermore, it takes less space to maintain annual species whose seed is harvested and replanted each season rather than perennials or tree crops, which need large areas of land dedicated to preservation and perhaps more than 10 years for individuals to reach maturity. Another difficulty is the prevention of cross-pollination between plots to maintain the genetic purity of cultivar lines. Cryopreservation and tissue culture are alleviating some of these problems, but the long-term viability of these methods has not been fully tested (Razdan and Cocking, 1997a, 1997b).

In addition to biological challenges, political and economic difficulties also exist. Today, many museum collections and repositories face financial cutbacks and funding shortages. Each week it seems another notice is sent calling for scientists to help preserve collections that are in jeopardy (Miller et al., 2004). One germplasm collection and herbarium, the all-Russian

FIGURE 1.4 The Vavilov Institute of Plant Industry (VIR): **(A)** One of the two buildings housing the VIR, which are mirror images of one another across St. Isaac's Square in St. Petersburg; **(B)** seed germplasm collection; **(C)** herbarium collections of cotton cultivars; **(D)** maize varieties.

Vavilov Institute of Plant Industry (VIR), fought to survive both physical and financial threats (figure 1.4). This institute was established in 1890 to collect genetic resources from cultivated and wild plants. The collections were greatly enhanced by the expeditions of N. I. Vavilov (Vavilov, 1997) in the 1920s but were later threatened with destruction during the 900-day German siege of Leningrad (present-day St. Petersburg) during World War II. The collections include not only germplasm materials but also library and herbarium collections, some of which are rare or extinct cultivars. During the war Herculean efforts by the institute's staff saved the collections from the German bombs; they also prevented potato cultivars from freezing in winter temperatures that reached –40ºC, subdivided and shipped seed stock by military transport to alternative locations, propagated the seeds in plots near the front lines, and protected the most valuable cultivar accessions from the starving Leningrad population. Some of the researchers died of starvation surrounded by packets of rice and other food items that made up the collection (Alexanyan and Krivchenko, 1991). The staff realized the value of these collections, and some sacrificed everything.

Today the VIR is the second largest germplasm collection in the world, containing more than 320,000 plant accessions. Its main offices are in two large buildings that share a town square and prime piece of real estate with the gilded dome of St. Isaac's Cathedral in the heart of St. Petersburg. After the Soviet Union was dissolved, $5.5 million of Western funds (partially funded through a seed exchange program initiated by former U.S. Vice President Al Gore) were used to help renovate and update the germplasm storage facilities. However, as the new government in Russia adjusts to the new economy, the VIR is finding itself cut off from government funds, and the City Property Administration Committee (Webster, 2003) hopes to acquire the valuable real estate of the institute's buildings and relocate the collections. The financial and scientific costs of such a move would be tremendous. I had a chance to visit the VIR in 2002 and see the collections, and the integration of the library, herbarium, seed storage facility, and field stations is very impressive.

The costs associated with well-maintained germplasm facilities can be high and entail long-term commitments (Gill, 1989). However, it must be remembered that without these reservoirs of genetic diversity the costs could be far higher (Myers, 1988). In the early 1970s a fungal pathogen called southern corn blight (*Bipolaris maydis*) destroyed nearly $1 billion worth of the U.S. corn crop. Some states lost more than 50% of their yield. Southern corn blight (race T) was especially devastating to hybrid corn carrying Texas male sterile cytoplasm (Ullstrup, 1972). Male sterility was desirable for producing hybrid seed because it eliminated the need for the labor-intensive and costly detasseling process, and as a result much of the U.S. corn crop contained this cytoplasm (male-sterile plants act as the ovule donor or female in controlled crosses and cannot self-pollinate). However, because the majority of commercial hybrid seed had nearly identical maternal genotypes, vast expanses of uniform stands of corn were infected. Fortunately, gene banks were available to mitigate the effects of the blight. We may not always be as fortunate in the case of secondary crops for which fewer resources are available. In Mexico a similar epidemic now faces the monocultures of blue agaves used for tequila production (Valenzuela-Zapata and Nabhan, 2003). The recovery from this pathogen is ongoing. Germplasm collections are being set up from varieties collected in the wild, and cross-pollination and cultivation by seed, rather than by vegetative propagation, are being promoted to combat the agave pathogen.

Unfortunately, interest in germplasm collections wanes in times of abundance when there is no immediate need for new genetic resources. The need for germplasm repositories became clear during World War II, when Japan took over the extensive rubber plantations in eastern Asia, leaving the allies without a source of this strategic material. To combat the shortage, the U.S. government hired R. E. Schultes and other botanists (Davis, 1996) to establish a germplasm collection of rubber and related species in Costa Rica in the hopes of producing an alternative and genetically diverse source of rubber. Unfortunately, in a short-sighted move during a time of complacency after the war, fueled by the shift to cheaper petroleum-based synthetic rubber, the collection was abandoned and the investment lost. Today most natural rubber (still used in such items as airplane tires) comes from plantations that are resting on a narrow genetic base. A single pathogen similar to the southern corn blight could devastate the world's supply of natural rubber. In this volume similar struggles with germplasm conservation are described for chayote (chapter 8, this volume), and in the Pacific Ragone et al. (2001) have documented the loss of breadfruit collections or the corresponding records.

Gene banks and germplasm collections for preserving crop diversity are invaluable for researchers and plant breeders. Because of war and changes in the environment it is no longer possible to collect wild *Triticum* (figure 1.5) species in the mountains of Afghanistan, but because of past collecting efforts and preservation it is still possible to study them. However, without proper curation and accurate records, the value of the collection is diminished. A recent study exemplifies the value of accurate curation. Pope et al. (2001) describe the discovery of domesticated sunflower seeds from archaeological sites near Tabasco, Mexico. The results of this study led Lentz et al. (2001) to speculate that sunflower may have originated in Mexico rather than further north (Asch and Asch, 1985; Crites, 1993). Preliminary results from molecular data using ancient DNA and *Helianthus* accessions from the U.S. Department of Agriculture germplasm repository indicated the possibility of two separate origins for sunflower. However, closer examination of the material revealed that one accession was misidentified (D. Lentz, pers. comm., 2002). Luckily this misidentified collection was discovered by the researchers through careful scrutiny of the data, and new information has been brought forth on sunflowers further supporting a North American origin (chapter 2, this volume).

Sound systematics and careful recordkeeping are another important component of a well-maintained germplasm collection. Placing crop plants in taxonomic categories can be difficult. Differentiation between crop varieties often is slight and can be easily misinterpreted. Even potato experts have trouble recognizing and categorizing potato tubers from a single cultigen (chapter 13, this volume). To address the concerns of intraspecific classification, a new International Code of Nomenclature for Cultivated Plants (ICNCP: Trehane et al., 1995) was created (see chapter 13, this volume, for application of system). No matter which system of classification is used, sound systematics, well-vouchered collections, and continued genetic evaluation by researchers and breeders are all vital parts of an efficient and useful crop plant collection (Bernatsky and Tanksley, 1989).

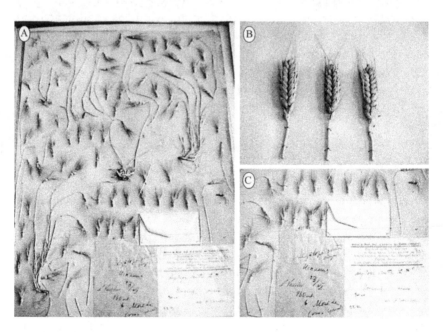

FIGURE 1.5   (A) Herbarium collection vouchering wild relatives of wheat collected by N. I. Vavilov on his expedition in the Fertile Crescent in 1926-1927; (B) close-up of specimen; (C) Vavilov collection label.

Molecular Studies of Crop Plants

New molecular techniques and applications are being developed continually. The sheer bulk of literature emerging with the rapid development of molecular techniques is evidence of the tremendous interest in genomic approaches to biology. Studies of crops at the molecular level have proliferated at an astounding rate since new methods, technologies, and tools have become available over the last 25 years (see Emshwiller, in press, for review). In the field of plant biology it is often crop researchers who embrace these tools first and show their usefulness in the study of evolutionary biology. The researcher must choose the levels of stringency, variability, and reproducibility needed for the question under consideration. Plants have three genomes from which scientists can draw information. The chloroplast and mitochondrial genomes typically are maternally inherited in most angiosperms. This makes both genomes good candidate regions for understanding parentage and lines of inheritance. The chloroplast genome typically evolves at a slower, more consistent rate and therefore is usually more useful at the generic and higher levels of the taxonomic hierarchy (Palmer, 1987). The chloroplast genome was more widely used in early molecular restriction fragment length polymorphism (RFLP) studies because it was easier to interpret the homology of markers. The mitochondrial genome of plants is more variable, exhibiting high levels of structural rearrangements, horizontal gene transfer, and lower levels of point mutations (Palmer and Herbon, 1988), making homology assessments of data more difficult (Doebley, 1992). Among closely related species, however, mitochondrial regions can provide more information and have only recently been used commonly in plant systematics (Cho et al., 1998). The nuclear genome typically is biparentally inherited and evolves at rates suitable for interspecific and in some cases intraspecific studies (Doebley, 1992).

DNA sequencing is a powerful tool for determining the closest relatives (and hence the wild progenitors) of crops. It is expected that the putative parental ancestors of crop plants would be on or near the same phylogenetic branch of the tree (Schilling et al., 1998). Zerega et al. (2004) used molecular sequence data to eliminate certain species of *Artocarpus* from consideration as putative ancestors of cultivated breadfruit and were able to narrow down the candidate ancestors to two other species that appeared to be more closely related. In some cases, such as soybean (*Glycine max*), no variability can be detected between the crop and the wild species, *G. soja* (Doyle and Beachy, 1985; Doyle, 1988), lending support to *G. soja* being the wild ancestor. When polyploidy or hybridization has played a role in the evolutionary history (e.g., wheat and maize) the answer is not so apparent. Unfortunately, the variability of most commonly sequenced gene regions typically is not sufficient to reveal intraspecific variation. At these subspecific levels, genome-wide approaches and fingerprinting techniques become useful.

Each type of molecular marker has strengths and weaknesses (see Vienne et al., 2003; appendix I, this volume) for crop plant studies (Gepts, 1993). Isozymes and allozymes were one of the first widely used methods. They generated reasonable amounts of data at low cost and allowed detection of genotypic differences and levels of heterozygosity (Hamrick and Godt, 1990). These enzymatic techniques provided early evidence of multiple origins of the common bean (Koenig and Gepts, 1989).

Because of the conservative nature of chloroplast DNA and hence the ease of making homology assessments between bands, the region often was targeted for polymorphic sites using RFLP (Botstein et al., 1980). This technique provided more variation than isozymes and allozymes but was costly and time-consuming. Studies using RFLPs have provided evidence for the hybrid origin and parentage of *Citrus* cultivars (Green et al., 1986) and revealed that papaya (*Carica papaya*) diverged early from all wild species of the genus in South America and evolved in isolation from its nearest relatives, probably in Central America (Aradhya et al., 1999).

Crop researchers are always seeking more variable markers and less costly and faster techniques. Therefore methods such as randomly amplified length polymorphisms (Williams et al., 1990) and amplified fragment length polymorphisms (Vos et al., 1995), which survey the entire genome, provide numerous polymorphic markers, and require no prior knowledge of the genome, became very popular, but they could not be used to assess levels of heterozygosity. These fingerprinting methods have been used to determine genetic differences between varieties of lentils (Ford et al., 1997), assess parentage for hybrid sugarcane (Lima et al., 2002) and corn cultivars (Welsh et al., 1991), genotype gooseberry cultivars (Lanham and Brennan, 1999), and screen for pathogen-resistant tomato lines (Martin et al., 1991). Microsatellite technology (Tautz, 1989) surveys hypervariable sequences in plants (Toth et al., 2000) and requires primers designed for each group of related organisms. It is quickly becoming more common as published primer pairs become more available. Microsatellites have been useful for fingerprinting germplasm accessions of grape species (Lamboy and Alpha, 1998) and for genotyping taro varieties and determining genetic and biogeographic relationships of Pacific island cultivars (Godwin et al., 2001). These marker technologies are also applied to construct linkage maps (giving a specific location of a gene on a chromosome by assigning distances between genes) and determine quantitative trait loci. Quantitative trait loci map measurable phenotypic traits (e.g., plant height) that are measured on a linear scale and allow the researcher to determine the genetic contribution that gene provides to the phenotypic trait.

Conclusions

Darwin, Mendel, McClintock, and many others have used domesticated species to study evolution in plants. This trend continues as the genomics wave sweeps through the scientific community. Recently, public concern about genetically modified organisms has brought crop studies into the headlines. The recent outcries against genetically modified crops are based on the fact that genes from organisms in other biological kingdoms, such as bacteria, are incorporated into the genome of plants. For centuries crop breeders have introduced beneficial alleles from closely related species into crops through hybridization and selection. The main difference between these traditional practices and genetically modified organisms is that crop scientists are no longer limited to the genetic material within the crop's gene pool, yet wild relatives of crop species remain a vital resource for crop improvement. Unfortunately, conservation of cultural or heirloom varieties is difficult, and habitats of wild species are being destroyed before the full utility of these resources can be realized. The time is ripe for taking another look at recent molecular studies of the origins, evolution, and conservation of crop plants.

Molecular techniques provide powerful tools to crop scientists at a time when it is possible to study entire crop genomes. As new questions arise, many crop researchers are revisiting classic evolutionary inquiries into crop plant evolution. What are the geographic origins of crop species? What are a crop's closest wild ancestors? What are the levels of genetic variation between species, varieties, and cultivars? What is the genetic influence of selection for agronomic traits? What is the most economic way to establish a germplasm repository that reflects the genetic diversity of a crop?

This is an exciting time for the evolutionary biologist as new technology gives hope that the long-sought answers to these questions will be found. In fact, there have been many new discoveries. Researchers using new molecular techniques in combination with data from multiple disciplines have revealed that some crops have multiple origins or new centers of origin. They have identified genetic pathways for desirable traits, genotyped germplasm collections to make maintenance more efficient and economical, gained a better understanding of the genetic effects of selection, and mounted new expeditions to collect the wild ancestors of crop species.

Although advances are being made at a rapid pace, crop evolution through human selection is not a straightforward or parsimonious process, and many questions remain unanswered. The chapters of this book will present just a few of the findings that have been made in recent years and give us a view into what the future holds for crop plant research.

Acknowledgments

I would like to thank Robbin Moran, Eve Emshwiller, and two anonymous reviewers for helpful comments on the manuscript. I also thank Jeremy Motley for contributing the illustrations of wheat. I am indebted to Gavrilova Vera, Boris Makarov, and Tamara Smekalova at the VIR, who opened up their facilities (research laboratories, herbarium, and seed storage units) and gave freely of their time and expertise while I conducted research in St. Petersburg. I am grateful to Olga Voronova and Tatyana Lobova for providing logistical and linguistic support in Russia. Financial support for this project was provided by the Torrey Botanical Society and the Lewis B. and Dorothy Cullman Foundation.

References

Alexanyan, S. M. and V. I. Krivchenko. 1991. Vavilov Institute scientists heroically preserve World Genetic Resources Collection during World War II siege of Leningrad. *Diversity* 7: 10–13.

Anderson, E. 1954. *Plants, Man, and Life*. A. Melrose, London, UK.

Aradhya, M. K., R. M. Manshardt, F. Zee, and C. W. Morden. 1999. A phylogenetic analysis of the genus *Carica* L. (Caricaceae) based on restriction fragment length variation in a cpDNA intergenic spacer region. *Genetic Resources and Crop Evolution* 46: 579–586.

Asch, D. L. and N. B. Asch. 1985. Prehistoric plant cultivation in west-central Illinois. In R. I. Ford (ed.), *Prehistoric Food Production in North America*, 149–203. Museum of Anthropology, Anthropological Paper No. 75. University of Michigan, Ann Arbor, MI, USA.

Baldwin, B. G. and R. H. Robichaux. 1995. Historical biogeography and ecology of the Hawaiian silversword alliance (Asteraceae): New molecular phylogenetic perspectives. In W. L. Wagner and V. A. Funk (eds.), *Hawaiian Biogeography: Evolution on a Hot Spot Archipelago*, 259–287. Smithsonian Institution Press, Washington, DC, USA.

Beadle, G. W. 1977. The origin of *Zea mays*. In C. A. Reed (ed.), *Origins of Agriculture*, 615–635. Mouton Press, The Hague.

Beccari, O. 1963. The origin and dispersal of *Cocos nucifera*. *Principes* 7: 57–69.

Bennetzen, J., E. Buckler, V. Chandler, J. Doebley, J. Dorweiler, B. Gaut, M. Freeling, S. Hake, E. Kellogg, R. S. Poethig, V. Walbot, and S. Wessler. 2001. Genetic evidence and the origin of maize. *Latin American Antiquity* 12: 84–86.

Bernatsky, R. and S. D. Tanksley. 1989. Restriction fragments as molecular markers for germplasm evaluation and utilization. In A. D. H. Brown, O. H. Frankil, D. R. Marshall, and J. T. Williams (eds.), *The Use of Plant Genetic Resources,* 353–362. Cambridge University Press, Cambridge, UK.

Berry, E. W. 1926. *Cocos* and *Phymatocaryon* in the Pliocene of New Zealand. *American Journal of Science* 12: 181–184.

Botstein, D., R. L. White, M. Skolnick, and R. W. Davis. 1980. Construction of a genetic linkage map in man using restriction fragment length polymorphisms. *American Journal of Human Genetics* 32: 314–331.

Campbell, J., R. Fordyce, A. Grebneff, and P. Maxwell. 2000. *Fossil coconuts from Mid-Cenozoic shallow marine sediments in southern New Zealand* (abstract). Otago University, Dunedin, NZ.

Chang, K. C. 1977. *The Archeology of Ancient China,* 3rd ed. Yale University Press, New Haven, CT, USA.

Childe, V. G. 1952. *New Light on the Most Ancient East.* Routledge and Kegan Paul, London, UK.

Cho, Y., Y.-L. Qiu, P. Kuhlman, and J. D. Palmer. 1998. Explosive invasion of plant mitochondria by a group I intron. *Proceedings of the National Academy of Science, USA* 95: 14244–14249.

Cohen, M. N. 1977. *The Food Crisis in Prehistory: Overpopulation and the Origins of Agriculture.* Yale University Press, New Haven, CT, USA.

Cook, O. F. 1910. History of the coconut palm in America. *Contributions to the U.S. National Herbarium* 14: 271–342.

Corner, E. J. H. 1966. *The Natural History of Palms.* University of California Press, Berkley, CA, USA.

Couper, R. 1952. The spore and pollen flora of the *Cocos*-bearing beds, Mangonui, North Auckland. *Transactions and Proceedings of the Royal Society of New Zealand* 79: 340–348.

Crites, G. D. 1993. Domesticated sunflower in fifth millennium B.P. temporal context: New evidence from middle Tennessee. *American Antiquity* 58: 146–148.

Darwin, C. 1883. *The Variation of Animals and Plants under Domestication,* 2nd ed. D. Appleton, New York, NY, USA.

Davidson, A. 1992. Europeans' wary encounter with tomatoes, potatoes, and other New World foods. In N. Foster and L. S. Cordell (eds.), *Chilies to Chocolate: Food the Americas Gave the World,* 1–14. University of Arizona Press, Tucson, AZ, USA.

Davis, W. 1996. The betrayal of the dream 1944–54. In *One River: Explorations and Discoveries in the Amazon Rain Forest,* 330–371. Simon & Schuster, New York, NY, USA.

de Candolle, A. 1959. *Origin of Cultivated Plants,* 2nd ed. (translated from the 1886 version). Hafner, New York, NY, USA.

Denham, T. P., S. G. Haberle, C. Lentfer, R. Fullgar, J. Field, M. Therin, N. Porch, and B. Winsborough. 2003. Origins of agriculture at Kuk swamp in the highlands of New Guinea. *Science* 301: 189–193.

de Wet, J. M. J. and J. R. Harlan. 1975. Weeds and domesticates: Evolution in the man-made habitat. *Economic Botany* 29: 99–107.

Diamond, J. 1999. *Guns, Germs, and Steel: The Fates of Human Societies.* W. W. Norton, New York, NY, USA.

Diamond, J. and P. Bellwood. 2003. Farmers and their languages: The first expansions. *Science* 300: 597–602.

Doebley, J. 1992. Molecular systematics and crop evolution. In P. S. Soltis, D. E. Soltis, and J. J. Doyle (eds.), *Molecular Systematics of Plants,* 202–220. Chapman Hall, New York, NY, USA.

Doyle, J.J. 1988. 5S ribosomal gene variation in the soybean and its progenitor. *Theoretical and Applied Genetics* 70: 369–376.

Doyle, J.J. and R.N. Beachy. 1985. Ribosomal gene variation in the soybean (*Glycine*) and its relatives. *Theoretical and Applied Genetics* 75: 621–624.

Emshwiller, E. In press. Genetic data and plant domestication. In M.A. Zeder, D. Decker Walters, D. Bradley, B. Smith, and E. Emshwiller (eds.), *Documenting Domestication: New Genetic and Archaeological Paradigms.* University of California Press, Berkeley, CA, USA.

Feldman, M. 1976. Wheats: *Triticum* spp. (Gramineae–Triticinae). In N.W. Simmonds (ed.), *Evolution of Crop Plants,* 120–127. Longman, New York, NY, USA.

Ford, R., E.C.K. Pand, and P.W.J. Taylor. 1997. Diversity analysis and species identification in *Lens* using PCR generated markers. *Euphytica* 96: 247–255.

Gepts, P. 1993. The use of molecular and biochemical markers in crop evolution studies. *Evolutionary Biology* 27: 51–94.

Gill, K.S. 1989. Germplasm collection and the public plant breeder. In A.D.H. Brown, O.H. Frankil, D.R. Marshall, and J.T. Williams (eds.), *The Use of Plant Genetic Resources,* 3–31. Cambridge University Press, Cambridge, UK.

Godwin, I.D., E.S. Mace, and Nurzuhairawaty. 2001. Genotyping Pacific island taro (*Colocasia esculenta* (L.) Schott) germplasm. In R.J. Henry (ed.), *Plant Genotyping: The DNA Fingerprinting of Plants,* 109–128. CABI Publishing, Oxfordshire, UK.

Goff, S.A., et al. 2002. A draft sequence of the rice genome (*Oryza sativa* L. ssp. *japonica*). *Science* 296: 92–100.

Gorman, C. 1969. Hoabinhian: A pebble-tool complex with early plant associations in Southeast Asia. *Science* 163: 671–673.

Green, R.M., A. Vardi, and E. Galun. 1986. The plastone of *Citrus.* Physical map variation among *Citrus* cultivars and species and comparison with related genera. *Theoretical and Applied Genetics* 72: 170–177.

Gunn, B.F. 2003. *The Phylogeny of the Cocoeae (Arecaceae) with Emphasis on* Cocos nucifera. M.S. thesis, University of St. Louis, MO, USA.

Guppy, H.B. 1906. *Observations of a Naturalist in the Pacific Between 1896 and 1899. Plant-dispersal,* Vol. 2. Macmillan, London, UK.

Hahn, W.J. 2002. A phylogenetic analysis of the arecoid line of palms based on plastid DNA sequence data. *Molecular Phylogenetics and Evolution* 23: 189–204.

Hamrick, J.L. and M.J.W. Godt. 1990. Allozyme diversity in plant species. In A.H.D. Brown, M.T. Clegg, A.L. Kahler, and B.S. Weir (eds.), *Plant Population Genetics, Breeding and Genetic Resources,* 43–63. Sinauer, Sunderland, MA, USA.

Hancock, J.F. 2004. *Plant Evolution and the Origin of Crop Species,* 2nd ed. CABI Publishing, Oxfordshire, UK.

Harlan, J.R. 1967. A wild wheat harvest in Turkey. *Archeology* 20:197–201.

Harlan, J.R. 1971. Agricultural origins: Centers and noncenters. *Science* 174: 468–474.

Harlan, J.R. 1992. *Crops and Man,* 2nd ed. American Society of Agronomy Inc. and Crop Science Society of America, Madison, WI, USA.

Harlan, J.R., J.M.J. de Wet, and E.G. Price. 1973. Comparative evolution of cereals. *Evolution* 27: 311–325.

Harries, H. 1978. The evolution, dissemination and classification of *Cocos nucifera* L. *Botanical Review* 44: 265–319.

Heiser, C. B. 1990. New perspectives on the origin and evolution of New World domesticated plants: Summary. *Economic Botany* 44 (supplement): 111–116.

Heyerdahl, T. 1952. *American Indians in the Pacific.* Victor Petersons Bokindustriaktiebolag, Stockholm, Sweden.

Heywood, V. H. 1993. *Flowering Plants of the World.* Oxford University Press, New York, NY, USA.

Hurles, M. E., E. Maund, J. Nicholoson, E. Bosch, C. Renfrew, B. C. Sykes, and M. A. Jobling. 2003. Native American Y chromosomes in Polynesia: The genetic impact of the Polynesian slave trade. *American Journal of Human Genetics* 72: 1282–1287.

Kaul, K. N. 1951. A palm fruit from Kapurdi (Jodhpur, Rajasthan Desert) *Cocos sahnii* sp. nov. *Current Science* 20: 138.

Kennard, W. C., M. K. Slocum, S. S. Figdore, and T. C. Osborn. 1994. Genetic analysis of morphological variation in *Brassica oleracea* using molecular markers. *Theoretical and Applied Genetics* 87: 721–732.

Kirch, P. V. 2000. *On the Road of the Winds.* University of California Press, Berkeley, CA, USA.

Koenig, R. and P. Gepts. 1989. Allozyme diversity in wild *Phaseolus vulgaris:* Further evidence for two major centers of diversity. *Theoretical and Applied Genetics* 78: 809–817.

Ladizinsky, G. 1985. Founder effect in crop evolution. *Economic Botany* 39: 191–199.

Lamboy, W. F. and C. G. Alpha. 1998. Using simple sequence repeats (SSRs) for DNA fingerprinting germplasm accessions of grape (*Vitis* L.) species. *Journal of the American Society of Horticultural Science* 123: 182–188.

Lanham, P. G. and R. M. Brennan. 1999. Genetic characterization of gooseberry (*Ribes grossularia* subgenus *Grossularia*) germplasm using RAPD, ISSR and AFLP markers. *Journal of Horticultural Science and Biotechnology* 74: 361–366.

Lee, R. B. and I. DeVore. 1968. Problems in the study of hunters and gatherers. In R. B. Lee and I. DeVore (eds.), *Man the Hunter,* 3–12. Aldine, Chicago, IL, USA.

Lentz, D. L., M. E. O. Pohl, K. O. Pope, and A. R. Wyatt. 2001. Prehistoric sunflower (*Helianthus annuus* L.) domestication in Mexico. *Economic Botany* 55: 370–376.

Lima, M. L. A., A. A. F. Garcia, K. M. Oliveira, S. Matsuoka, H. Arizono, C. L. de Souza Jr., and A. P. de Souza. 2002. Analysis of genetic similarity detected by AFLP and coefficient of parentage among genotypes of sugar cane (*Saccharum* spp.). *Theoretical and Applied Genetics* 104: 30–38.

Lindqvist, C., T. J. Motley, J. J. Jeffery, and V. A. Albert. 2003. Cladogenesis and reticulation in the Hawaiian endemic mints (Lamiaceae). *Cladistics* 19: 480–495.

MacNeish, R. S. 1991. *The Origins of Agriculture and the Settled Life.* University of Oklahoma Press, Norman, OK, USA.

Mangelsdorf, P. C. 1974. *Corn: Its Origin Evolution and Improvement.* Harvard University Press, Cambridge, MA, USA.

Martin, G. B., J. G. K. Williams, and S. D. Tanksley. 1991. Rapid identification of markers linked to a *Pseudomonas* resistance gene in tomato by using random primers and near-isogenic lines. *Proceedings of the National Academy of Sciences, USA* 88: 2336–2340.

McClintock, B. 1950. The origin and behavior of mutable loci in maize. *Proceedings of the National Academy of Sciences* 36: 344–355.

Miller, S. E., W. J. Kress, and K. Samper. 2004. Crisis for biodiversity collections. *Science* 303: 310.

Moore, H. E. Jr. 1973. Palms in the tropical ecosystems of Africa and South America. In B. J. Meggers, E. S. Ayensu, and W. D. Duckworth (eds.), *Tropical Forest Ecosystems in Africa and South America: A Comparative Review,* 63–88. Smithsonian Institute, Washington, DC, USA.

Myers, N. 1988. Draining the gene pool: The causes, course, and genetic consequences of genetic erosion. In J. R. Kloppenburg Jr. (ed.), *Seed and Sovereignty: The Use and Control of Plant Genetic Resources,* 90–113. Duke University Press, Durham, NC, USA.

National Research Council (U.S.), Committee on Germplasm Resources. 1978. *Conservation of Germplasm Resources: An Imperative.* National Academy of Sciences, Washington, DC, USA.

Palmer, J. D. 1987. Chloroplast DNA evolution and biosystematic uses of chloroplast DNA variation. *American Naturalist* 130: S6–S29.

Palmer, J. D. and L. A. Herbon. 1988. Plant mitochondrial DNA evolves rapidly in structure, but slowly in sequence. *Journal of Molecular Evolution* 28: 87–97.

Patil, G. V. and E. V. Upadhye. 1984. *Cocos*-like fruit from Mahgaonkalan Intertrappean beds. In A. K. Sharma (ed.), *Evolutionary Botany and Biostratigraphy,* 541–554. A. K. Gosh Commencement Volume, Birbal Sahni Institute, Lucknow, India.

Pollan, M. 2001. *The Botany of Desire: A Plant's Eye View of the World.* Random House, New York, NY, USA.

Popovsky, M. 1984. *The Vavilov Affair.* Archon Books, Hamden, CT, USA.

Pope, K. O., M. E. O. Pohl, J. G. Jones, D. L. Lentz, C. van Nagy, F. J. Vega, and I. R. Quitmyer. 2001. Origin and environmental setting of ancient agriculture in the lowlands of Mesoamerica. *Science* 1370–1373.

Ragone, D., D. H. Lorence, and T. Flynn. 2001. History of plant introductions to Pohnpei, Micronesia and the role of the Pohnpei Agricultural Station. *Economic Botany* 55: 290–303.

Razdan, M. K. and E. C. Cocking. 1997a. *Conservation of Plant Genetic Resources In Vitro.* Volume I: *General Aspects.* Agritech Publications, Shrub Oak, NY, USA.

Razdan, M. K. and E. C. Cocking. 1997b. *Conservation of Plant Genetic Resources In Vitro.* Volume II: *Applications and Limitations.* Agritech Publications, Shrub Oak, NY, USA.

Rigby, J. F. 1995. A fossil *Cocos nucifera* L. fruit from the latest Pliocene of Queensland, Australia. In D. D. Pant (ed.), *Birbal Sahni Institute, Centennial Volume,* 379–381. Allahabad University, Allahabad, India.

Rindos, D. 1984. *The Origins of Agriculture: An Evolutionary Prospective.* Academic Press, Orlando, FL, USA.

Sauer, C. O. 1952. *Agricultural Origins and Dispersals.* MIT Press, Cambridge, MA, USA.

Schilling, E. E., C. R. Linder, R. Noyes, and L. H. Rieseberg. 1998. Phylogenetic relationships in *Helianthus* (Asteraceae) based on nuclear ribosomal DNA internal transcribed spacer region sequence data. *Systematic Botany* 23: 177–188.

Shewry, P. R. 2003. Tuber storage proteins. *Annals of Botany* 91: 755–769.

Smith, B. D. 1997. The initial domestication of *Cucurbita pepo* in the Americas 10,000 years ago. *Science* 276: 932–934.

Smith, B. D. 1998. *The Emergence of Agriculture.* Scientific American Library Series No. 54, W. H. Freeman, New York, NY, USA.

Smith, C. E. Jr. 1969. From Vavilov to the present: A review. *Economic Botany* 23: 2–19.

Sun, W. J., N. G. Du, and M. H. Chen. 1981. The paleovegetation and paleoclimate during the time of Homudu people. *Acta Botanica Sinica* 23: 146–151.

Tautz, D. 1989. Hypervariability of simple sequences as a general source for polymorphic DNA markers. *Nucleic Acids Research* 17: 6463–6471.

Telford, I. R. H. 1990. Cucurbitaceae: Gourd family. In W. L. Wagner, D. R. Herbst, and S. H. Sohmer (eds.), *Manual of the Flowering Plants of Hawaii,* pp. 568–581. Bishop Museum Press/University of Hawaii Press, Honolulu, HI, USA.

Tippo, O. and W. L. Stern. 1977. *Humanistic Botany*. W. W. Norton, New York, NY, USA.

Toth, G., Z. Gaspari, and J. Juka. 2000. Microsatellites in different eukaryotic genomes: Survey and analysis. *Genome Research* 10: 967–981.

Trehane, P., C. D. Brickell, B. R. Baum, W. L. A. Hetterscheid, A. C. Leslie, J. McNeill, S. A. Spongberg, and F. Vrugtman. 1995. International code of nomenclature of cultivated plants. *Regnum Vegetabile* 133: 1–175.

Ullstrup, A. J. 1972. The impact of the southern corn leaf blight epidemics of 1970–71. *Annual Review of Phytopathology* 10: 37–50.

Ungerer, M. C., S. Baird, J. Pan, and L. H. Rieseberg. 1998. Rapid hybrid speciation in wild sunflowers. *Proceedings of the National Academy of Sciences, USA* 95: 11757–11762.

Valenzuela-Zapata, A. G. and G. P. Nabhan. 2003. When the epidemic hit the king of clones. In A. G. Valenzuela-Zapata and G. P. Nabhan (eds.), *Tequila: A Natural and Cultural History*, 57–61. University of Arizona Press, Tucson, AZ, USA.

Vavilov, N. I. 1926. Studies on the origins of cultivated plants. *Bulletin of Applied Botany, Genetics, and Plant Breeding* 16: 1–245.

Vavilov, N. I. 1992. *Origin and Geography of Cultivated Plants*. University Press, Cambridge, UK.

Vavilov, N. I. 1997. *Five Continents*. International Plant Genetic Resources Institute, Rome, Italy.

Vienne, D. de, S. Santoni, and M. Falque. 2003. Principal sources of molecular markers. In D. de Vienne (ed.), *Molecular Markers in Plant Genetics and Biotechnology*, 3–46. Science Publishers, Enfield, NH, USA.

Vos, P., R. Hogers, M. Bleeker, M. Rijans, T. Van de Lee, M. Hornes, A. Frijters, J. Pot, J. Peleman, M. Kuiper, and M. Zabeau. 1995. AFLP: A new technique for DNA fingerprinting. *Nucleic Acids Research* 23: 4407–4414.

Wang, R.-L., A. Stec, J. Hey, L. Lukens, and J. Doebley. 1999. The limits of selection during maize domestication. *Nature* 398: 236–239.

Webster, P. 2003. Prestigious plant institute in jeopardy. *Science* 299: 641.

Welsh, J., R. J. Honeycutt, M. McClelland, and B. W. S. Sobral. 1991. Parentage determination in maize hybrids using the arbitrarily primed polymerase chain reaction (AP-PCR). *Theoretical and Applied Genetics* 82: 473–476.

Williams, J. G. K., A. R. Kubelik, K. J. Livak, J. A. Rafalski, and S. V. Tingey. 1990. DNA polymorphism amplified by arbitrary primers are useful as genetic markers. *Nucleic Acids Research* 18: 6531–6535.

Yen, D. E. 1974. The sweet potato and Oceania. *Bernice P. Bishop Museum Bulletin, 236*.

Yu, J., et al. 2002. A draft sequence of the rice genome (*Oryza sativa* L. ssp. *indica*). *Science* 296: 79–92.

Zerega, N. J. C., D. Ragone, and T. J. Motley. 2004. Complex origins of breadfruit: Implications for human migrations in Oceania. *American Journal of Botany* 91: 760–766.

Zohary, D. and M. Hopf. 1994. *Domestication of Plants in the Old World*. Oxford Scientific Publications, Oxford University Press, Oxford, UK.

PART 1

GENETICS AND ORIGIN OF CROPS

Evolution and Domestication

Loren H. Rieseberg and Abigail V. Harter   CHAPTER 2

Molecular Evidence and the Evolutionary History of the Domesticated Sunflower

The domestication of plants and animals by prehistoric humans was perhaps the most far-reaching cultural development in human history. Not only were domesticated organisms crucial to the rise of modern civilization, but their widespread use has dramatically altered the ecology and evolutionary history of numerous other species (Diamond, 2002). As a consequence, there is great interest in determining the geographic origins and timing of domestication (Sauer, 1952; Harlan, 1971). Although seemingly straightforward, this task is complicated by poor preservation of plant remains, particularly in tropical regions, and by the difficulty of discriminating between wholly independent origins of domestication and the secondary introduction of crop plants from a core region (Cowan and Watson, 1992; Denham et al., 2003; Neumann, 2003).

In the New World, these complications have led to conflicting interpretations of archaeological and paleobotanical evidence regarding the relationship between Mesoamerica and other regions where evidence of food production is found. One interpretation holds that Mesoamerica served as a primary center of domestication from which domesticated plant lineages and food production practices spread to areas of secondary innovation (Harlan, 1971; Lentz et al., 2001). In this view, the midlatitude woodland region of eastern North America is considered to be one of these secondary areas, and the domestication of indigenous North American plant species is hypothesized to have been triggered by the introduction of major crops from Mesoamerica (Lentz et al., 2001). The alternative and more widely accepted interpretation is that agriculture in eastern North America arose wholly independently (Smith, 1989; Cowan and Watson, 1992, Neumann, 2003).

Evidence of an independent origin of agriculture in eastern North America derives primarily from the archaeobotanical record of four indigenous crops: thick-walled cucurbit or squash (*Cucurbita pepo* ssp. *ovifera*), sumpweed (*Iva annua*), goosefoot (*Chenopodium berlandieri*), and sunflower (*Helianthus annuus*). All exhibit morphological changes in reproductive propagules that are associated with domestication (Asch and Asch, 1985; Smith, 1989). The transition to fully domesticated forms occurred between 4000 and 3000 years BP (Smith, 1989), which substantially predates the introduction of maize *circa* 1800 years BP (Chapman and Crites, 1987); note that maize is thought to be the first tropical crop to be introduced into eastern North America (Smith, 1989). In addition, knotweed (*Polygonum erectum*), maygrass (*Phalaris caroliniana*), and little barley (*Hordeum pusillum*) were used as minor seed crops before the introduction of maize (Cowan, 1978; Asch and Asch, 1985), but there is insufficient evidence to establish strong cases for their domesticated status.

Despite strong archaeobotanical support, the eastern North American origin of three of the four main indigenous domesticates (thick-walled cucurbit, goosefoot, and sunflower) has been questioned. For example, a recent mitochondrial DNA study (Sanjur et al., 2002) was consistent with an origin for *C. pepo* ssp. *ovifera* from wild gourds in either northeastern Mexico (*C. pepo* ssp. *fraterna*) or eastern North America (*C. pepo* ssp. *ovifera* var. *ozarkana*). However, a possible progenitor role for *C. pepo* ssp. *fraterna* was quickly ruled out by random amplified polymorphic DNA (RAPD) data (Decker-Walters et al., 2002), which places the domesticate with *C. pepo* ssp. *ovifera* var. *ozarkana* as originally proposed (Decker-Walters et al., 1993). Likewise, Wilson (1990) postulates that goosefoot might have a Mexican origin because of its close resemblance to the Mexican cultivar *Chenopodium berlandieri* ssp. *nutalliae*.

The most serious challenge to the eastern North American domestication hypothesis derives from the discovery of a sunflower achene and seed at the San Andrés site in Tabasco, Mexico, that date to 4130 ± 40 years BP and 4085 ± 50 years BP, respectively (accelerator mass spectrometry [AMS] determined) (Lentz et al., 2001; Pope et al., 2001). The achene and seed clearly represent the domesticated form, and their age rivals that of the earliest domesticated achenes from eastern North America, which are from the Hayes site in Tennessee and date to 4265 ± 60 years BP (AMS determined; Crites, 1993). However, Lentz et al. (2001) questions the shrinkage factors used to correct carbonized achene sizes at sites from eastern North America (Yarnell, 1978) and argues that the achenes from the Hayes site and other early finds actually represent wild material (but see Smith, 2003). If the Lentz et al. arguments were valid, then the earliest domesticated sunflower remains in eastern North America would derive from the Higgs site in eastern Tennessee (2850 ± 85 years BP, AMS determined; Brewer, 1973) and the Marble Bluff Rockshelter in northwest Arkansas (2842 ± 44 years BP, AMS determined; Fritz, 1997).

So far, molecular evidence has had little impact on the debate over the geographic origins of the domesticated sunflower, although it has been interpreted as supporting both sides of the debate (Heiser, 2001; Lentz et al., 2001). Given disagreements regarding the interpretation of earlier molecular studies and the recent completion of a comprehensive microsatellite survey of sunflower origins (Harter et al., 2004), it seemed worthwhile to provide a critical review of molecular data relating to sunflower domestication. We will show that although sunflower appears to be easily domesticated, molecular evidence indicates that all extant domesticated sunflowers had a single origin in eastern North America.

Systematics and Biogeography of *H. annuus*

Helianthus comprises approximately 50 species of sunflower, all of which are native to North America (Schilling and Heiser, 1981; Seiler and Rieseberg, 1997). The genus is monophyletic (Schilling et al., 1994) and includes diploids (n = 17), tetraploids, and hexaploids. Although most species are perennial, section *Helianthus* (formerly section *Annui*) includes 11 or 12 species, most of which are self-incompatible, diploid annuals. Molecular phylogenetic studies indicate that the section is monophyletic and consistently place *H. annuus* in a clade with three other species: *H. argophyllus*, *H. bolanderi*, and *H. exilis* (Rieseberg, 1991; Rieseberg et al., 1991; Schilling, 1997; Schilling et al., 1998). In all trees, *H. argophyllus*, a silver-leaved sunflower from southern Texas, is sister to *H. annuus*. The two species do hybridize in areas of contact in southern Texas but retain their distinctive morphology and karyotype, presumably because of divergent ecological selection and a fairly strong chromosomal sterility barrier (Heiser, 1951a).

The domesticated sunflower is clearly derived from the wild form of *H. annuus,* or common sunflower (Heiser 1951b, 1954). Hybrids between wild and domesticated *H. annuus* are fully fertile (Heiser, 1954), and molecular studies all confirm the predicted progenitor-derivative relationship (e.g., Rieseberg and Seiler, 1990; Cronn et al., 1997; Tang and Knapp, 2003). Heiser (1954) gave formal recognition to four different forms of the common sunflower: *H. annuus* ssp. *lenticularis* (the western North American subspecies), *H. annuus* ssp. *texanus* (a form of *H. annuus* from Texas that has converged toward a local species, *H. debilis,* with which it hybridizes), *H. annuus* ssp. *annuus* (the midwestern and more weedy form of the species), and *H. annuus* ssp. *annuus* var. *macrocarpus* (the domesticated sunflower). Heiser later recognized the inadequacy of this classification because of extensive intergradation between forms, so he adopted a less formal treatment in his monograph of the genus (Heiser et al., 1969). However, in later discussions, Heiser (1976, 1978) once again used subspecific nomenclature but restricted the definition of ssp. *annuus* to the urban weed form of *H. annuus.* Molecular evidence indicates that there is significant structuring among populations of *H. annuus,* but it more closely tracks geography (i.e., isolation by distance) than subspecific categories (Harter et al., 2004).

Wild *H. annuus* currently occurs throughout the continental United States, southern Canada, and northern Mexico (Heiser et al., 1969; González-Elizondo and Gómez-Sánchez, 1992), but its prehistoric distribution is poorly understood. Heiser (1951b) speculated that the species was restricted to the southwestern United States before the arrival of *Homo sapiens* into the Americas. Native Americans used wild *H. annuus* for food, so Heiser (1951b) proposed that it became a camp-following weed and was thereby introduced into the central and eastern United States, where it was domesticated. However, it seems more likely that buffalo was the primary dispersal agent (Asch, 1993) and that wild *H. annuus* was widely distributed throughout the Great Plains, western United States, and northern Mexico before the colonization of North America by humans.

Previous Molecular Studies

The first comprehensive molecular analysis of sunflower domestication assayed chloroplast DNA (cpDNA) and allozyme variation in 5 Native American varieties, 3 modern cultivars, 15 old landraces, and 12 wild populations from throughout the continental United States (Rieseberg and Seiler, 1990). All 23 cultivars had the same chloroplast DNA haplotype,

implying a single origin for extant domesticated sunflowers. This haplotype was also found in wild populations from Missouri, New Mexico, and
California, so no conclusions could be made regarding the geographic origin
of the domesticates.

Wild and domesticated sunflowers were very similar at allozyme loci as
well. Twenty-nine of 30 alleles found in the domesticates also occurred in
wild populations, with an average genetic identity (I) between wild and
domesticated populations of 0.93, a value only slightly lower than that for
comparisons between wild populations (I = 0.96). Because of these very
similar high genetic identities, the question of geographic origins could
not be addressed. Nonetheless, high levels of allozyme variability in wild
plants and virtual monomorphism in cultivated lines reinforced the cpDNA
results: Extant domesticated sunflowers had a single origin from a very
limited gene pool (Rieseberg and Seiler, 1990).

Shortly after this initial study, Arias and Rieseberg (1995) attempted
to locate the geographic center of domestication for sunflower using RAPD
markers. However, the high RAPD identity between wild populations and
domesticated *H. annuus* (I = 0.976 to I = 0.997) once again precluded
determination of geographic origin. In fact, Arias and Rieseberg were
skeptical that molecular evidence could ever solve this problem, suggesting
that the weedy, human-dispersed nature of wild *H. annuus* populations
probably had erased evidence of geographic structure. Fortunately, as will
be discussed later in this chapter, we were unnecessarily pessimistic.

In 1997, another attempt was made to use allozyme variation to ascer-
tain the geographic origin of the domesticated sunflower (Cronn et al.,
1997). This study differed from that of Rieseberg and Seiler (1990) in
its inclusion of four additional allozyme loci, increased sampling of both
wild and cultivated accessions, the use of clusters of related populations as
operational taxonomic units in genetic distance trees, and the inclusion of
related wild species for rooting the trees. This improved method led to the
discovery of limited geographic structure among wild populations. More
significantly, they found that the domesticated sunflower was slightly more
similar genetically to wild populations from the Great Plains than from
the Southwest or California. However, support for this relationship was
very weak.

Recently, the development of microsatellite loci for sunflower has greatly
enhanced our ability to analyze genetic relationships between domesticated
and wild accessions (Whitton et al., 1997; Tang et al., 2002). In previous
work, cpDNA haplotypes and RAPD and allozyme allele frequencies were

not sufficiently differentiated between geographic locations to determine likely source populations for domesticated sunflower. However, microsatellites have proved superior to these markers for the study of domestication because there is more intraspecific genetic variation at these loci, making it feasible to dissect relationships between recently divergent populations.

In sunflower, microsatellites were first used for this purpose by Tang and Knapp (2003). With the exception of a wild accession from North Dakota, which appears to be the product of crop–wild hybridization, their study provided the first strong statistical support for the genetic separation of cultivated from wild material. Unfortunately, there was insufficient resolution between the wild populations and inadequate geographic coverage to determine the geographic origin of the domesticated sunflower. However, it is noteworthy that a wild population from the Great Plains (Oklahoma) clustered most closely with the domesticates, and the single wild population from Mexico was most distant.

The most intriguing result of Tang and Knapp (2003) was the large genetic distances observed between two of the Native American varieties (Hopi and Havusupai) and other domesticated sunflowers (0.714 to 0.798). Tang and Knapp interpreted the large distances as evidence that the domesticated sunflower might have multiple origins. This interpretation was consistent with earlier observations by Heiser (1976) on the morphological distinctness of the Hopi and Havusupai varieties, the discovery of domesticated sunflower remains at archaeological sites in both Mexico (Lentz et al., 2001) and eastern North America (Yarnell, 1978), and quantitative trait locus studies of domestication traits (Burke et al., 2002) indicating that sunflower was easily domesticated (domestication entailed few major genetic changes, and wild populations contain numerous alleles with effects in the direction of the cultivar).

There are two weaknesses with the multiple-origin hypothesis. First, the genetic distances reported by Tang and Knapp (2003) are exaggerated because only a single sample was analyzed per accession. Second, all sampled domesticated sunflowers appear to form a monophyletic lineage that derives from within the pool of wild variation. Note that this is not immediately apparent in figure 6 of Tang and Knapp (2003) because the consensus unweighted pair group method with arithmetic mean tree was rooted with a highly derived cultivar lineage rather than a primitive wild form, and they included the hybrid North Dakota population in the tree. If there were multiple origins of the domesticates, we would expect independently derived cultivar lineages to be placed sister to the wild progenitor populations from which they were derived, and this is not the case. On the other hand, given the lack of extensive sampling from Mexico, it was perfectly reasonable for Tang and Knapp to assume that probable progenitor populations for at least one of the origins had not been sampled.

Recent Work

A second microsatellite survey was recently completed by Harter et al. (2004). This study differed from that of Tang and Knapp (2003) in that there was complete geographic coverage of the prehistoric range of sunflower, including Mexico. Also, all wild populations were collected by the authors and attempts were made to choose large populations from natural sites that were far from cultivated fields to minimize the potential for crop–wild gene flow. Finally, in addition to standard tree-building methods, sophisticated model-based clustering approaches were used that are more appropriate and powerful for assigning domesticates to wild populations and for reconstructing the pattern of genetic drift between wild populations and domesticated strains arising from the domestication process.

Individuals from 21 geographically diverse populations of wild *H. annuus* from North America and Mexico and 10 domesticated lineages including 2 commercial lines and 8 Native American–developed landraces (figure 2.1) were genotyped for 18 microsatellite loci (Harter et al., 2004). The resultant data set was analyzed in three ways. Pairwise genetic distances between populations were calculated and used to construct a neighbor-joining (NJ) tree (figure 2.2). Second, a model-based clustering approach was implemented with the software program STRUCTURE (Pritchard et al., 2000; Falush et al., 2003) to infer population structure in wild *H. annuus* and then to assign the domesticates to inferred populations. Third, the STRUCTURE program (Pritchard et al., 2000; Falush et al., 2003) was used to make inferences about ancestral allele frequencies in the common ancestor of wild and domesticated sunflower and the degree of drift away from the ancestral genomic composition in each population.

Neighbor-Joining Tree

The topology of an NJ tree based on pairwise genetic distances between populations closely follows their geographic distribution, although some nodes are not well supported (figure 2.2). The Mexico plus Arizona grouping is supported by a high bootstrap value of 90% and includes

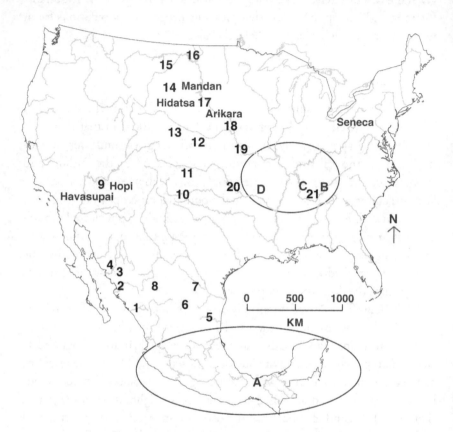

FIGURE 2.1 Map of sampling locations used by Harter et al. (2004), archaeological sites and Native American groups. Shaded areas = centers of domestication, with eastern North America to the north and Mesoamerica to the south; numbers = sampling locations of wild populations, where 1 = Sinaloa, 2 = Sonora5, 3 = Sonora4, 4 = Sonora6, 5 = Tamaulipas, 6 = Zacatecas, 7 = Nuevo León, 8 = Chihuahua, 9 = Arizona, 10 = Texas, 11 = Oklahoma2, 12 = Kansas, 13 = Colorado, 14 = Montana1, 15 = Montana2, 16 = North Dakota, 17 = South Dakota, 18 = Iowa, 19 = Missouri, 20 = Oklahoma1, 21 = Tennessee; names = historical locations of Native American groups; and letters = archaeological sites with oldest remains of domesticated sunflower, where A = San Andres, Tabasco, MX (4130 ± 40 BP), B = Higgs, TN, USA (2850 ± 85 BP), C = Hayes, TN, USA (4265 ± 60 BP) and D = Marble Bluff, AR, USA (2843 ± 44 BP). Identities of indigenous groups associated with Maíz de Tejas and Maíz Negro are unknown. USDA and Mammoth are modern cultivars derived from Russian stock. Therefore these strains do not appear on the map.

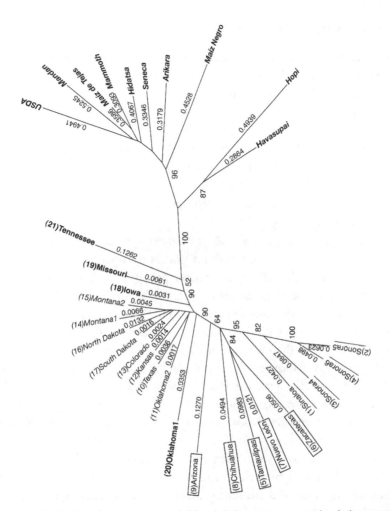

FIGURE 2.2 Majority rule consensus neighbor-joining tree summarized the genetic distances, D_A (Nei et al., 1983) between groups. West Mexico populations are underlined, east-central Mexico populations are in boxes, U.S. Great Plains populations are in italics, east-central U.S. populations are in bold, and cultivars are in bold and italics. Numbers in parentheses correspond to sampling locations of wild *H. annuus* populations, as shown in figure 2.1. Numbers along branches are mean drift values; the value for each domesticated strain is the average across all comparisons with wild populations, and the value for each wild population is the average across all comparisons with domesticated strains. Numbers at nodes indicate bootstrap values greater than 50% (1000 replicates). Because of space considerations, the 74% bootstrap value for the node subtending Colorado, the 54% bootstrap for the node subtending Seneca, and the 62% bootstrap for the node subtending Mammoth–Maíz de Tejas do not appear on the tree.

two clusters that correspond to the western coastal plain (Sinaloa, Sonora4, Sonora6, Sonora5) and northeastern Mexico (Tamaulipas, Nuevo León, Zacatecas), plus more interior populations (Arizona and Chihuahua) basal to them. The U.S. cluster has lower bootstrap values, but the Great Plains populations (Montana2, Montana1, North Dakota, South Dakota, Colorado, Kansas, Texas, and Oklahoma2) form a discrete group within which the branching order reflects geographic relationships. Populations to the east of the Great Plains (Tennessee, Missouri, Iowa, and Oklahoma1) do not form a distinct group. Instead, each is sister to the Great Plains group.

All cultivars belong to a single, strongly supported group (bootstrap = 100%) in the NJ tree. Although Hopi and Havasupai form a distinct and well-supported clade within the cultivar group, genetic distances (0.436 to 0.696) are not as large as those reported by Tang et al. (2003). Wild populations from the east-central United States, especially Tennessee, Missouri, and Iowa, which represent the eastern wild form (*H. annuus* ssp. *annuus*), have the closest genetic relationship with all the domesticated accessions. More broadly, the Great Plains populations, as a whole, cluster more closely with the domesticates (bootstrap = 90%) than do populations from Mexico. These results suggest a single origin of extant domesticated sunflowers from the east-central United States as originally hypothesized by Heiser (1951b). Note that this result is not inconsistent with genetic data suggesting that sunflower is readily domesticated (Burke et al., 2002) because domestication of even the most amenable wild taxon is a long and arduous process when compared with the spread of an already domesticated form.

Model-Based Clustering

The admixture model included in the STRUCTURE program was used to define genetic populations or clusters in wild *H. annuus* based on allele frequencies and then to assign domesticated genotypes probabilistically to these defined clusters. Genetic populations were defined at both a regional and a local scale. Note that the admixture model allows individuals to originate from more than one source population.

At the regional scale, two genetic populations or clusters of wild *H. annuus* were consistently found by the STRUCTURE program. One cluster comprised all Mexican populations plus Arizona, whereas all central U.S. populations (i.e., populations from the Great Plains and east-central United States) formed a second cluster (figure 2.3a). Assignment of domesticated individuals to these two clusters revealed that all extant domesticates had central U.S. ancestry (figure 2.4). Indeed, the average estimated ancestry for each domesticated strain was at least 0.997!

The regional clusters (figure 2.3a) were subjected to further independent analyses to identify local genetic populations. Tests for population structure on the Mexican subsample identified two clusters that correspond to distinct geographic regions: west Mexico and east-central Mexico (figure 2.3b). Likewise, the North America subsample could be subdivided genetically into a U.S. Great Plains and east-central U.S. cluster (figure 2.3b). Assignment of domesticated lineages to these local clusters revealed that, as predicted by the NJ tree, all domesticates were assigned to the east-central United States, with average estimated membership of at least 0.994 for all domesticates (figure 2.4). Thus both regional and local clustering analyses indicate that domesticated sunflowers are most similar to wild *H. annuus* from the central United States, particularly the easternmost populations.

Patterns of Genetic Drift

All previous studies of genetic variation in wild and domesticated sunflowers have reported much lower levels of variability in domesticated than in wild sunflowers (Rieseberg and Seiler, 1990; Cronn et al., 1997; Tang and Knapp, 2003), as would be predicted if there were a genetic bottleneck associated with domestication. Using the F model of the STRUCTURE program, Harter et al. (2004) investigated the pattern of genetic drift between wild and domesticated sunflowers in order to determine whether this pattern was consistent with domesticates arising via genetic drift from wild U.S. populations or from wild Mexican populations. The F model assumes that populations have independently drifted from the allele frequencies found in their common ancestor and uses a Bayesian approach to make inferences about ancestral allele frequencies and the rate of drift away from the ancestor. Wild populations that are most similar in allele frequency to the common ancestor of wild and domesticated *H. annuus* should exhibit little evidence of drift (i.e., have low drift values). Likewise, if domestication is associated with a strong selective bottleneck, domesticated lines should have much larger drift values than wild populations.

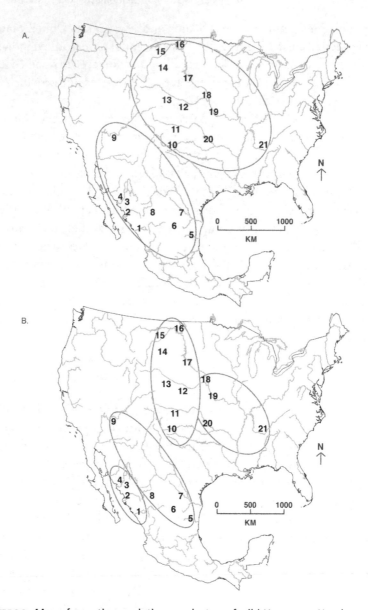

FIGURE 2.3 Map of genetic populations or clusters of wild *H. annuus*. Numbers = sampling locations of wild populations, where 1 = Sinaloa, 2 = Sonora5, 3 = Sonora4, 4 = Sonora6, 5 = Tamaulipas, 6 = Zacatecas, 7 = Nuevo León, 8 = Chihuahua, 9 = Arizona, 10 = Texas, 11 = Oklahoma2, 12 = Kansas, 13 = Colorado, 14 = Montana1, 15 = Montana2, 16 = North Dakota, 17 = South Dakota, 18 = Iowa, 19 = Missouri, 20 = Oklahoma1, 21 = Tennessee. **(A)** Regional clusters of wild *H. annuus*: Mexico plus Arizona and central United States. **(B)** Local clusters of wild *H. annuus*: west Mexico, east-central Mexico, Great Plains, and east-central United States.

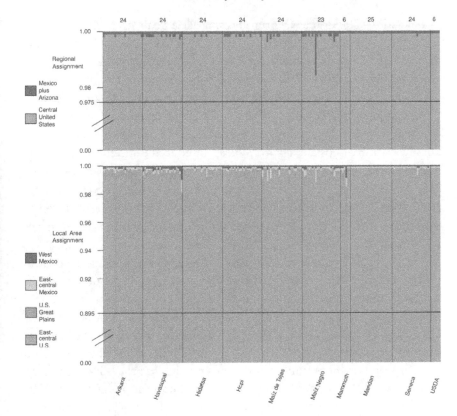

FIGURE 2.4 Results of the domesticated *H. annuus* genotypic cluster assignment. Each domesticated individual's genome is represented by a thin vertical line that is partitioned into colored segments in proportion to the estimated membership in each of the wild source clusters. Cultivars are separated with black lines, with names below and sample sizes above.

As predicted, domesticated populations had much higher drift values than wild populations (figure 2.2). The lowest mean drift value in a domesticate was more than 200 times that of the lowest mean value in a wild population, indicative of a strong genetic bottleneck associated with domestication (Harter et al., 2004).

Consistent with the cluster analyses, drift values in the wild populations place the ancestry of the domesticated sunflower in the central United States but fail to differentiate between a Great Plains or an east-central origin (figure 2.2). The nine lowest drift values are from central U.S. populations, and several populations actually have 90% credibility regions around the estimated drift value that includes 0.000 drift away from ancestral allele frequencies: Kansas, South Dakota, Oklahoma2 (all Great Plains), and Iowa (east-central United States). This is a remarkable result indicating that these contemporary wild populations are essentially identical in allele frequency to the wild ancestor of domesticated sunflowers. Several other populations from the Great Plains (North Dakota) or east-central United States (Oklahoma1 and Tennessee) have intermediate or high drift values, however, indicating that allele frequencies in these populations have drifted substantially from those found in the ancestral population. These populations cluster genetically with other wild populations from these areas (figure 2.2), a result consistent with strong localized drift events (Harter et al., 2004), perhaps because of recent founding events. This explanation is particularly likely for Tennessee, which is a roadside population far from the native range of *H. annuus*.

The combined results from the genetic distance tree, model-based clustering, and drift analyses indicate that the progenitor of the domesticated sunflower was genetically most similar to wild populations in the central plains of the United States. A more precise geographic location is difficult (and perhaps nonsensical) to infer because of high levels of gene flow between populations in this area. Nonetheless, of the sampled populations, Iowa probably is most similar to the ancestor in that the 90% credibility region around the estimated drift value for this population includes 0.000, it is placed very close to the cultivars in the NJ tree, and it belongs to the east-central U.S. genetic population to which the domesticates were assigned in the model-based cluster analysis.

Conclusions

Several inferences can be made from molecular genetic studies of the domesticated sunflower. First, all studies agree that domestication was associated with a strong genetic bottleneck. As a consequence, allele frequencies have changed more than 50 times faster in domesticated lineages than in wild populations since their divergence from a common ancestor. Second, molecular evidence is most consistent with a single origin of all extant domesticated sunflowers. All domesticates share the same chloroplast DNA haplotype, and in the most recent and convincing molecular studies, all extant domesticates are placed in a monophyletic group that is well separated from all wild populations. Third, genetic distance trees, model-based clustering, and drift analyses of microsatellite data all indicate that the domesticated sunflower was derived from wild sunflowers in the central plains of the United States. Because of extensive gene flow between populations in this area, it is not possible to assign the domesticated sunflower to a single wild population or local geographic area. However, of the sampled localities, a population from Iowa was most similar to the wild ancestor of the domesticated sunflower.

Based on these data, a likely scenario for domestication is that wild sunflowers from the central plains colonized adjacent regions to the east (i.e., Tennessee, Kentucky, Illinois, Missouri, Arkansas, and Ohio), perhaps because of human activities in the middle Holocene (Heiser, 1951b). The wild sunflowers were subsequently brought under cultivation and were domesticated over a period from approximately 4000 BP to 3000 BP (Smith, 1989). More generally, molecular evidence of a U.S. ancestry of extant domesticated sunflowers supports an origin in eastern North America independent of Mesoamerican domestication. However, the provenance of the domesticated achenes from the San Andrés site in Mexico remains a mystery. Possibly, there was an earlier and independent domestication in Mexico, but it does not appear to have influenced domestication in eastern North America. Alternatively, achenes may have been carried to San Andrés from the north. However, as far as we are aware, there is no evidence of long-distance trade at this time. Additional archaeobotanical work in Mexico is needed to establish the authenticity of the Mexican find by estimating the timing and duration of the Mexican domestication event (if it existed) and determining the date of extinction and its cause.

References

Arias, D.M. and L.H. Rieseberg. 1995. Genetic relationships among domesticated and wild sunflowers (*Helianthus annuus*, Asteraceae). *Economic Botany* 49: 239–248.

Asch, D.L. 1993. Common sunflower (*Helianthus annuus* L.): The pathway toward its domestication. In *Proceedings of the 58th Annual Meeting 17 May 1993*. Society of American Archaeology, St. Louis, MO, USA.

Asch, D.L. and N.B. Asch. 1985. Prehistoric plant cultivation in west-central Illinois. In R.I. Ford (ed.), *Prehistoric Food Production in North America*, 149–203. Anthropological Paper No. 75, Museum of Anthropology, University of Michigan, Ann Arbor, MI, USA.

Brewer, A.J. 1973. Analysis of floral remains from the Higgs site (40L045). In C.R. McCollough and Charles H. Faulkner (eds.), *Excavation of the Higgs and Doughty Sites: I-75 Salvage Archaeology*, 141–144. Miscellaneous Paper No. 12, Tennessee Archaeological Society, Department of Anthropology, University of Tennessee, Knoxville, TN, USA.

Burke, J. M., S. Tang, S. J. Knapp, and L. H. Rieseberg. 2002. Genetic analysis of sunflower domestication. *Genetics* 161: 1257–1267.

Chapman, J. and G. D. Crites. 1987. Evidence for early maize (*Zea mays*) from the ice-house bottom site, Tennessee. *American Antiquity* 52: 352–354.

Cowan, C. W. 1978. The prehistoric use and distribution of maygrass in eastern North America: Cultural and phytogeographical implications. In R. I. Ford (ed.), *The Nature and Status of Ethnobotany,* 263–288. Anthropological Paper No. 67, Museum of Anthropology, University of Michigan, Ann Arbor, MI, USA.

Cowan, C. W. and P. J. Watson (eds.). 1992. *The Origins of Agriculture: An International Perspective.* Smithsonian Institution Press, Washington, DC, USA.

Crites, G. D. 1993. Domesticated sunflower in fifth millennium B.P. temporal context: New evidence from middle Tennessee. *American Antiquity* 58: 146–148.

Cronn, R., M. Brothers, K. Klier, P. K. Bretting, and J. F. Wendel. 1997. Allozyme variation in domesticated annual sunflower and its wild relatives. *Theoretical Applied Genetics* 95: 532–545.

Decker-Walters, D. S., J. E. Staub, S. M. Chung, E. Nakata, and H. D. Quemada. 2002. Diversity in free-living populations of *Cucurbita pepo* (Cucurbitaceae) as assessed by random amplified polymorphic DNA. *Systematic Botany* 27: 19–28.

Decker-Walters, D. S., T. Walters, C. W. Cowan, and B. D. Smith. 1993. Isozymic characterization of wild populations of *Cucurbita pepo. Journal of Ethnobiology.* 13: 55–72.

Denham, T. P., S. G. Haberle, C. Lentfer, R. Fullagar, J. Field, M. Therin, N. Porch, and B. Winsborough. 2003. Origins of agriculture at Kuk Swamp in the highlands of New Guinea. *Science* 301: 189–193.

Diamond, J. 2002. Evolution, consequences and future of plant and animal domestication. *Nature* 418: 700–707.

Falush, D., M. Stephens, and P. Donnelly. 2003. Inference of population structure using multilocus genotype data: Linked loci and correlated allele frequencies. *Genetics* 164: 1567–1587.

Fritz, G. J. 1997. A three-thousand-year-old cache of seed crops from Marble Bluff, Arkansas. In K. J. Gremillion (ed.), *People, Plants, and Landscapes: Studies in Paleoethnobotany,* 42–62. University of Alabama Press, Tuscaloosa, AL, USA.

González-Elizondo, M. S. and D. Gómez-Sánchez. 1992. Notes on *Helianthus* (Compositae–Heliantheae) from Mexico. *Phytologia* 72: 63–70.

Harlan, J. R. 1971. Agricultural origins: Centers and noncenters. *Science* 174: 468–474.

Harter, A. V., K. A. Gardner, D. Falush, D. L. Lentz, R. Bye, and L. H. Rieseberg. 2004. Single origin of extant domesticated sunflowers in eastern North America. *Nature* 430: 201–205.

Heiser, C. B. 1951a. Hybridization in the annual sunflowers: *Helianthus annuus* × *H. argophyllus. American Naturalist* 85: 64–72.

Heiser, C. B. 1951b. The sunflower among the North American Indians. *Proceedings of the American Philosophical Society* 95: 432–448.

Heiser, C. B. 1954. Variation and subspeciation in the common sunflower, *Helianthus annuus. American Midland Naturalist* 51: 287–305.

Heiser, C. B. 1976. *The Sunflower.* University of Oklahoma Press, Norman, OK, USA.

Heiser, C. B. 1978. Taxonomy of *Helianthus* and origin of domesticated sunflower. In J. F. Carter (ed.), *Sunflower Science and Technology,* 31–53. American Society of Agronomy, Crop Science Society of America, Madison, WI, USA.

Heiser, C. B. 2001. About sunflowers. *Economic Botany* 55: 470–473.

Heiser, C. B., D. M. Smith, S. B. Clevenger, and W. C. Martin, Jr. 1969. The North American sunflowers (*Helianthus*). *Memoirs of the Torrey Botanical Club* 22: 1–218.

Lentz, D. L., M. E. D. Pohl, K. O. Pope, and A. R. Wyatt. 2001. Prehistoric sunflower (*Helianthus annuus* L.) domestication in Mexico. *Economic Botany* 55: 370–376.

Nei, M., F. Tajima, and Y. Tateno. 1983. Accuracy of estimated phylogenetic trees from molecular data. *Journal of Molecular Evolution* 19: 153–170.

Neumann, K. 2003. New Guinea: A cradle of agriculture. *Science* 301: 180–181.

Pope, K. O., M. E. D. Pohl, J. G. Jones, D. L. Lentz, C. von Nagy, F. J. Vega, and I. R. Quitmyer. 2001. Origin and environmental setting of ancient agriculture in the lowlands of Mesoamerica. *Science* 292: 1370–1373.

Pritchard, J. K., M. Stephens, and P. Donnelly. 2000. Inference of population structure using multilocus genotype data. *Genetics* 155: 945–959.

Rieseberg, L. H. 1991. Homoploid reticulate evolution in *Helianthus:* Evidence from ribosomal genes. *American Journal of Botany* 78: 1218–1237.

Rieseberg, L. H., S. Beckstrom-Sternberg, A. Liston, and D. Arias. 1991. Phylogenetic and systematic inferences from chloroplast DNA and isozyme variation in *Helianthus* sect. *Helianthus*. *Systematic Botany* 16: 50–76.

Rieseberg, L. H. and G. Seiler. 1990. Molecular evidence and the origin and development of the domesticated sunflower (*Helianthus annuus*). *Economic Botany* 44S: 79–91.

Sanjur, O. I., D. R. Piperno, T. C. Andres, and L. Wessel-Beaver. 2002. Phylogenetic relationships among domesticated and wild species of *Cucurbita* (Cucurbitaceae) inferred from a mitochondrial gene: Implications for crop plant evolution and areas of origin. *Proceedings of the National Academy of Sciences, USA* 99: 535–540.

Sauer, C. O. 1952. *Agricultural Origins and Dispersals*. American Geographical Society, New York, NY, USA.

Schilling, E. E. 1997. Phylogenetic analysis of *Helianthus* (Asteraceae) based on chloroplast DNA restriction site data. *Theoretical and Applied Genetics* 94: 925–933.

Schilling, E. E. and C. B. Heiser. 1981. Infrageneric classification of *Helianthus* (Compositae). *Taxon* 30: 393–403.

Schilling, E. E., C. R. Linder, R. Noyes, and L. H. Rieseberg. 1998. Phylogenetic relationships in *Helianthus* (Asteraceae) based on nuclear ribosomal DNA internal transcribed spacer region sequence data. *Systematic Botany* 23: 177–188.

Schilling, E. E, J. L. Panero, and U. H. Eliasson. 1994. Evidence from chloroplast DNA restriction site analysis on the relationships of *Scalesia* (Asteraceae, Heliantheae). *American Journal of Botany* 81: 248–254.

Seiler, G. J. and L. H. Rieseberg. 1997. Systematics, origin, and germplasm resources of the wild and domesticated sunflower. In A. A. Schneiter (ed.), *Sunflower Science and Technology*, 21–66. American Society of Agronomy, Crop Science Society of America, Madison, WI, USA.

Smith, B. D. 1989. Origins of agriculture in eastern North America. *Science* 246: 1566–1571.

Smith, B. D. 2003. *Rivers of Change: Essays on Early Agriculture in Eastern North America*. Smithsonian Institution Press, Washington, DC, USA.

Tang, S. X. and S. J. Knapp. 2003. Microsatellites uncover extraordinary diversity in native American land races and wild populations of cultivated sunflower. *Theoretical and Applied Genetics* 106: 990–1003.

Tang, S., J. K. Yu, M. B. Slabaugh, D. K. Shintani, and S. J. Knapp. 2002. Simple sequence repeat map of the sunflower genome. *Theoretical and Applied Genetics* 105: 1124–1136.

Whitton, J., L. H. Rieseberg, and M. C. Ungerer. 1997. Microsatellite loci are not conserved across the Asteraceae. *Molecular Biology and Evolution* 14: 204–209.

Wilson, H. D. 1990. *Quinua* and relatives (*Chenopodium* sect. *Chenopodium* subsect. *Cellulata*). *Economic Botany* 44S: 92–110.

Yarnell, R. 1978. Domestication of sunflower and sumpweed in eastern North America. In R. I. Ford (ed.), *The Nature and Status of Ethnobotany*, 289–299. Anthropological Paper No. 67, Museum of Anthropology, University of Michigan, Ann Arbor, MI, USA.

Laurent Grivet, Jean-Christophe
Glaszmann, and Angélique D'Hont

CHAPTER 3

Molecular Evidence of Sugarcane
Evolution and Domestication

Sugarcane is an important industrial plant in subtropical and tropical regions of the world, and almost 20 million ha is cultivated for its sucrose-rich stalks. Most of the crop is processed in specialized mills to extract sucrose. The primary use of sucrose is for human consumption, but in Brazil it is also used to produce ethanol, a renewable substitute for fossil fuels.

Sugarcane prehistory evidently occurred in a vast area covering India to Polynesia. As with many tropical plants that are consumed for their vegetative organs, few remnants of sugarcane have been reported from archeological records (Daniels and Daniels, 1993; Bayliss-Smith, 1996). As a consequence, most theories on sugarcane domestication have come from living wild and cultivated plants.

The art of making sugar from sugarcane was first reported from India and China (Daniels and Daniels, 1976). From these regions, the knowledge then disseminated to the west and southeast. Dissemination to the west began when Greeks reached the Indus Valley in the 4th century BC. Since then historians have documented the extension of sugarcane and the sugar industry toward the Middle East, North Africa, southern Europe, and America (Deerr, 1949). Sugar manufacture probably came to insular Southeast Asia through Buddhist influence from India (Daniels and

Daniels, 1976). It did not penetrate into Melanesia or further west into Polynesia. However, sugarcane is abundant in village gardens throughout these two regions and is simply consumed by chewing. In Melanesia, the plant is deeply rooted in the local culture, although traditions and knowledge about clone names and their specific uses are disappearing fast (Buzacott and Hughes, 1951; Warner, 1962).

The primary domestication of sugarcane probably occurred in New Guinea from the wild species *Saccharum robustum* and resulted in a series of sweet clones identified by botanists as *S. officinarum*. These cultivars were transported by humans to continental Asia, where they hybridized with a wild species, *S. spontaneum,* giving rise to a new series of cultivars better adapted to subtropical environments and to sugar manufacture. They are identified as *S. barberi* for cultivars from India and as *S. sinense* for cultivars from China. This scenario, popular among sugarcane specialists and first established by E. W. Brandes (1956) 50 years ago, is one of the many that have been hypothesized in the development of historical and botanical knowledge of this crop. Since the end of the 1980s, DNA-based markers have been used to estimate genetic relationships between individuals or populations of plants because of their reliability. They offer a unique opportunity to investigate the origin of sugarcane. This chapter summarizes the information they have provided.

Taxonomy and Distribution of Traditional Sugarcane Cultivars

Sugarcane cultivars are clones propagated by stem cuttings. Traditional cultivars have been described as species by botanists and have been given Latin binomials. Today they make a marginal contribution to the sugar industry because they have been replaced by interspecific hybrids developed by artificial breeding. However, they are essential to our understanding of the domestication of sugarcane.

S. officinarum L.

These cultivars are encountered in subsistence gardens throughout Melanesia. The highest morphological diversity is encountered in western New Guinea. They have brightly colored, thick stalks, rich in sugar. They generally have a chromosome complement of 2n = 80. The first Dutch breeders in Java used the term *Noble* to refer to their flamboyant colors and large size (Brandes, 1956).

S. barberi Jew. and *S. sinense* Roxb.

These cultivars were traditionally associated with sugar manufacture in India and China until the beginning of the 20th century. These clones generally have thinner stalks and leaves, flatter colors and lower sugar content than Nobles, a chromosome number greater than 80 (2n = 81 to 124), and a better adaptation to subtropical environments. They were formerly cultivated in mainland Asia, especially northern India and southern China, the probable birthplace of the sugar industry. Today they are confined to germplasm collections. Five morphocytological groups have been described: Mungo, Saretha, Nargori, Sunnabile, and Pansahi (Barber, 1922). *S. barberi* usually includes the first four groups, all reported from India. The fifth group is either included in *S. barberi* or called *S. sinense*. It was reported from China and was introduced to India at the close of the 18th century.

S. edule Hassk.

S. edule Hassk. is cultivated in subsistence gardens from New Guinea to Fiji for its edible, aborted inflorescence. Its large, thick-stalked canes contain no sugar. Chromosome number is in the range of 2n = 60 to 122, with multiples of 10 most common (Roach, 1972).

Taxonomy and Distribution of Wild Species Related to Sugarcane

Wild taxa related to sugarcane include two species from the genus *Saccharum* and several species from related genera. The status of a third *Saccharum* wild species remains ambiguous because it may derive from an intergeneric hybridization.

S. spontaneum L.

S. spontaneum generally has thin stalks with no or very low sugar content. It generally grows spontaneously in the vicinity of water resources. Its chromosome complement varies between 2n = 40 and 128. For 80% of individuals the complement is a multiple of eight, indicating a polyploid series with frequent aneuploidy (Panje and Babu, 1960; figure 3.1).

The species covers a huge geographic distribution. Panje and Babu (1960) published a map based on prospecting records, which extend from

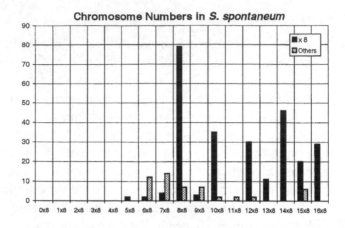

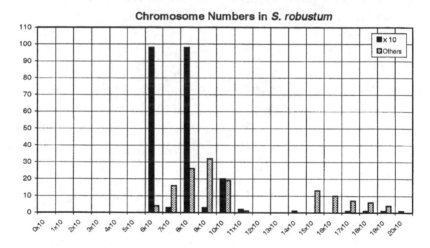

FIGURE 3.1   Chromosome complements in wild *Saccharum* species. The upper graph gives the frequency of chromosome numbers observed in *S. spontaneum* accessions collected worldwide (data from Panje and Babu, 1960). Multiples of 8 are figured in black; other numbers are in gray. The lower diagram gives the frequency of chromosome numbers in *S. robustum* accessions collected over the range distribution of the species. Data are from Price (1965). Multiples of 10 are figured in black, and others are in gray.

Africa to Southeast Asia. The continental Asian origin of *S. spontaneum* is in little doubt because of the high morphological, cytological, and ecological diversity encountered there (Panje and Babu, 1960; Chen et al., 1981). In Kalimantan, the species is abundant in the wild and shows morphological variability (Berding and Koike, 1980), indicating that it is probably indigenous (figure 3.2). In Sulawesi, *S. spontaneum* is abundant in natural

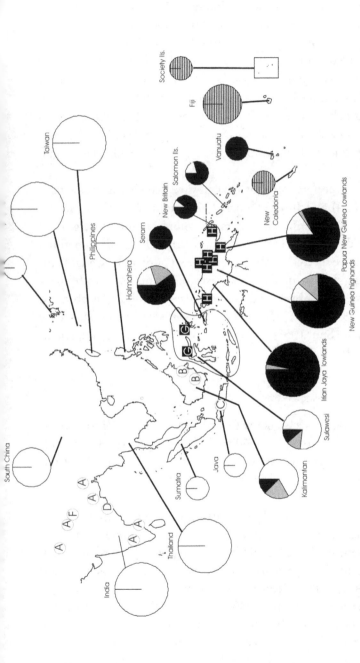

FIGURE 3.2 Frequency of wild *Saccharum* species collected during prospecting expeditions. Data are from Berding and Koike (1980), Buzacott and Hugues (1951), Chen et al. (1981), Coleman (1971), Daniels (1977), Engle et al. (1979), Grassl (1946), Krishnamurthi and Koike (1977), Lennox (1939), Lo and Sun (1969), Panje and Babu (1960), Price (1965), Price and Daniels (1968), Nagatomi et al. (1984) in Berding and Roach (1987), Sreenivasan et al. (1982, 1985), Sreenivasan and Sadakorn (1983) in Berding and Roach (1987), Tew et al. (1991), and Warner and Grassl (1958). Colors in pies are white for *S. spontaneum*, gray for *S. robustum* collected in anthropic environment (fence, garden), black for *S. robustum* collected in wild environment or when no precision is available, and striped for *S. maximum*. Size of the pie is small if total sample is <10, medium if sample is ≥10 and <100, and large if sample is ≥100. Molecular cytotypes identified in D'Hont et al. (1993) for *S. robustum* and *S. spontaneum* accessions are figured on the map when collection sites are known. A black square is for *S. robustum*, and a white circle is for *S. spontaneum*. Wallacea, the floristic transition zone between Southeast Asia and Melanesia, is circled.

habitats (Berding and Koike, 1980; Tew et al., 1991), but the morphological diversity seems limited. The general thinking is that *S. spontaneum* is not indigenous east of Kalimantan, although it is now abundant. In Irian Jaya, observations are scarce, favoring a recent introduction (Berding and Koike, 1980). In Papua New Guinea, *S. spontaneum* is locally abundant in wet depressions of extensive savanna grasslands, which probably have a human origin (Henty, 1982). In the Bismarck and Solomon archipelagos the species is locally abundant (Burcham, 1948; Warner and Grassl, 1958). It is also present in many tropical Pacific islands (Whistler, 1995; Welsh, 1998) and probably recently extended to Central America (Pohl, 1983; Hammond, 1999).

S. *spontaneum* has been reported as an aggressive weed in sugarcane fields in Java (Baker, 1874) and India (Barber, 1920). It is an efficient pioneer species, as shown, for example, by the rapid colonization of bare ground on Krakatau islands after the 1883 eruption (Turner, 1992). It sometimes behaves as an invasive species, as observed in the vicinity of the Canal Point sugarcane breeding station in Florida (Westbrooks and Miller, 1993) or on lands that have been subject to slash-and-burn agriculture, as observed in Panama (Hammond, 1999). In Taiwan, *S. spontaneum* is used for fence construction (Lo and Sun, 1969) and in West Africa for archery and thatching (Poilecot, 1999).

S. robustum Brandes & Jeswiet ex Grassl.

S. *robustum* has long, thick stalks with little or no sugar. Chromosome numbers vary generally from 2n = 60 to 110. Multiples of 10 are common (70%), and two cytotypes predominate: 2n = 60 and 2n = 80. A few clones have chromosomes numbers between 140 and 200 (Price, 1965; figure 3.1).

S. *robustum* has been reported as occurring in natural populations in the islands of Kalimantan, Sulawesi, Maluku, and New Guinea and in the Bismarck, Solomon, and Vanuatu archipelagos (figure 3.2). In Kalimantan, it is reported mostly from gardens, where it is used as a medicinal plant (Berding and Koike, 1980). Reports from the wild are limited to a single observation along a river in the 1930s (Price, 1965). In Sulawesi, it is abundant in natural habitats (Berding and Koike, 1980; Tew et al., 1991). The diversity is reduced to the Tannange type, commonly used by farmers for fencing. The highest morphological diversity is clearly encountered in New Guinea (Price, 1965; Berding and Koike, 1980).

Typical habitat corresponds to mud banks along watercourses, but it is also encountered on humid slopes or along roadside ditches. Where it occurs in the wild, *S. robustum* is often planted in native gardens, either for medicinal purposes or as a material for house or fence building.

S. robustum seems most likely to be native to regions southeast of Sulawesi and *S. spontaneum* to regions northwest of Sulawesi. It is therefore possible that before plant dispersal by humans, *S. spontaneum* and *S. robustum* had allopatric distributions on each side of this island. It is interesting to note that this threshold fits with a major floristic transition zone, Wallacea, between Southeast Asia and Melanesia (Steenis, 1950).

Genera Other Than *Saccharum*

The contribution to the emergence of sugarcane from various genera other than *Saccharum,* particularly *Erianthus, Sclerostachya, Narenga,* and *Miscanthus,* has been hypothesized by several sugarcane specialists. Their detailed description is given in Daniels and Roach (1987).

The genus *Erianthus* has a wide distribution in the Old Word from the Mediterranean Basin to New Guinea. Seven species have been described: three diploids with 2n = 20 and four with chromosome numbers between 2n = 20 and 2n = 60. Two species have large sugarcane-like stalks, the others have thin stems.

The genus *Miscanthus* is distributed in South Pacific and Southeast Asia up to Siberia (Daniels and Roach, 1987). This genus is currently divided into four sections and 12 species. The most common chromosome numbers reported are 2n = 38 and 2n = 76, except for section *Diantra,* for which it is 2n = 40.

The genera *Sclerostachya* and *Narenga* are closely related. *Sclerostachya* includes two or three species distributed from northern India to the Malay Peninsula and China. Chromosome numbers of 2n = 30 and 2n = 34 have been reported. *Narenga* includes two species distributed from North India to Vietnam (Mukherjee, 1957).

S. maximum (Brongn.) Trin.

S. maximum (syn. *Erianthus maximus* Brongn.) have large, thick, free-thrashing, brightly colored stalks with little sugar. Lennox (1939) included it in *S. robustum* and later, based on floral morphology, proposed it to be a *Saccharum* × *Miscanthus* hybrid (Daniels, 1967). Clones examined for chromosome number are in the range of 87–100 (Price and Daniels, 1968). *S. maximum* is reported as wild populations in New Caledonia, Fiji, and the Cook, Society, Marquesas, and Austral islands (Grassl, 1946; Daniels, 1967; Smith, 1979; Welsh, 1998). This distribution is contiguous to but does not overlap with the distribution of *S. robustum*.

S. maximum grows preferentially on slopes where rainfall is high but is also encountered along edges of small rivers in areas where rainfall is low (Daniels, 1967). A dispersion of *S. maximum* by humans throughout Polynesia is possible, and the species may have been locally cultivated (Lepovsky, 2003). In Fiji, Daniels (1967) reports that *S. maximum* is similar to *S. edule* in general appearance and overall dimensions and that natural populations are found in the same habitat type.

Hypothesis for Sugarcane Domestication

Daniels and Roach (1987) give a synopsis of hypotheses for the domestication of sugarcane. At the end of the 19th century, *S. maximum* was proposed to be the wild ancestor of *S. officinarum* because most cultivars had been encountered in the southern Pacific by European explorers, a region where the wild cane was also growing. Later, the exploration of New Guinea revealed a more spectacular diversity of *S. officinarum* (Lennox, 1939) and permitted the discovery of *S. robustum,* another thick-stalked wild relative. This species was then proposed as the wild ancestor of *S. officinarum,* but a contribution of *S. maximum* was not denied (Brandes, 1956). Brandes (1956) further proposed that *S. officinarum* may have been transported by humans east to Polynesia and to the northwest in subtropical continental Asia. In India and China it would have hybridized with *S. spontaneum* and given rise to *S. barberi* and *S. sinense,* respectively.

Other scenarios have also been proposed for the origin of *S. barberi* and *S. sinense.* A direct selection from *S. spontaneum,* in India, has been hypothesized for *S. barberi* based on the occurrence of an ancient sugar-making industry and the abundance and diversity of *S. spontaneum* in that area (Barber, 1920; Purseglove, 1976). A hybrid origin between *S. officinarum* and *Miscanthus sacchariflorus* has been proposed for *S. sinense* (Grassl, 1946). A direct emergence of *S. sinense* from a still undiscovered Chinese wild species has also been hypothesized (Daniels and Daniels, 1993).

In the 1950s and later, botanical inventories and prospecting efforts increased the number of wild sugarcane relatives in Asia and Pacific. Interest was raised about the genus *Saccharum* and related genera. It appeared that the Himalayan foothills contained an exceptional species diversity (Panje, 1953). Based on morphological evidence, Mukherjee (1957) proposed an origin of all traditional cultivars in that region through intercrossing between species of genera *Saccharum, Erianthus, Sclerostachya,* and *Narenga* that he called the *Saccharum* complex. Daniels et al. (1975) added genus *Miscanthus* to this group of species, based on analyses of morphology and leaf flavonoid pigments.

Molecular Differentiation of Wild Species

A different structural organization of the monoploid genome for *S. spontaneum* and *S. robustum* is suggested by the polyploid series based on multiples of eight and ten, respectively. Physical mapping of ribosomal RNAs *45S* and *5S* by fluorescent in situ hybridization (FISH) confirmed this difference and established basic chromosome numbers of x = 8 for *S. spontaneum* and x = 10 for *S. robustum* (D'Hont et al., 1998).

Allopatric populations of *S. spontaneum* and *S. robustum* are clearly differentiated at the DNA level. Indeed, *S. spontaneum* samples from Kalimantan and Sumatra and *S. robustum* from New Guinea and Halmahera are strongly differentiated by nuclear low copy restriction fragment length polymorphism (RFLP; Burnquist et al., 1992; Lu et al., 1994), mitochondrial DNA probes (D'Hont et al., 1993), and random amplified polymorphic DNA markers (Nair et al., 1999).

Data addressing relationships between sympatric populations of *S. spontaneum* and *S. robustum* are still sparse. In New Guinea, all *S. spontaneum* individuals observed have the same cytotype, 2n = 80. D'Hont et al. (1998) showed that this cytotype is decaploid, with a typical *S. spontaneum* basic chromosome number of x = 8. However, field observations show a morphological continuum between extreme types, and some individuals presenting intermediate morphological characteristics between *S. spontaneum* and *S. robustum* are difficult to classify (Henty, 1969). Moreover, a small sample of *S. spontaneum* individuals collected in New Guinea appear more closely related to *S. robustum* than to any other *S. spontaneum* based on RFLP with nuclear low copy RFLP markers (Besse et al., 1997) and on the hybridization signal intensity of a repeated satellite sequence *SoCIR1* (Alix et al., 1998). This suggests that *S. spontaneum* populations from New Guinea are genetically closer to *S. robustum* than *S. spontaneum* populations west of Sulawesi. A simple interpretation may be that genetic exchange does occur between the two species.

An *S. maximum* clone named Raitea has been positively identified as a *Saccharum–Miscanthus* hybrid, based on genomic in situ hybridization (GISH). It contained 80 *Saccharum* chromosomes and 19 *Miscanthus* chromosomes (unpublished data). Two other clones of *S. maximum*, Fiji15 and NC100, showed very little homology with the *Miscanthus* dispersed repeated specific probe (Alix, 1998). These sparse data suggest that *S. maximum* might be a heterogeneous group of populations with a different level of introgression between *Saccharum* and *Miscanthus*.

Origin of Traditional Cultivars Based on Molecular Data

S. officinarum

Multiple lines of molecular evidence support a direct descent of Noble clones from the wild species *S. robustum*. A single mitochondrial haplotype, H, was detected among a series of *S. officinarum* clones (D'Hont et al., 1993). This haplotype is the most common haplotype detected in a collection of *S. robustum* clones from New Guinea and New Britain (table 3.1; figure 3.2). It is different from haplotype G, detected in two *S. robustum* accessions from Sulawesi and Halmahera (islands located west of New Guinea). It is also different from the six haplotypes revealed in a collection of *S. spontaneum* individuals sampled over a large geographic range.

RFLP analysis of nuclear single copy DNA placed Noble cultivars very close to *S. robustum*. The average similarity between a Noble clone and an *S. robustum* clone is about the same as the average similarity between two *S. robustum* clones (table 3.1; Lu et al., 1994). Noble cultivars have a basic chromosome number of x = 10, as does *S. robustum* (D'Hont et al., 1998), and are octoploids like the most common cytotype (2n = 80) in the *S. robustum* wild species.

S. barberi and S. sinense

RFLP with low copy nuclear DNA (Lu et al., 1994) and GISH (D'Hont et al., 2002) clearly show that *S. barberi* and *S. sinense* cultivars are the result of interspecific hybridizations between representatives of the two genetic groups of the *Saccharum* genus, *S. spontaneum* on one side and *S. officinarum* or *S. robustum* on the other side. Because *S. barberi* and *S. sinense* clones have sweet stalks and because the region where they were formerly cultivated is outside the natural distribution range of *S. robustum*, the scenario of Brandes (1956) provides the simplest explanation for their

Table 3.1 Relationship Between *S. spontaneum* and *S. robustum* wild Accessions and
S. officinarum (Noble) Cultivars Based on Cytoplasmic (D'Hont et al., 1993) and Nuclear
RFLP Probes (Lu et al., 1994)

	Mitochondrial Haplotypes								Average Similarities with Nuclear RFLP		
	A	B	C	D	E	F	G	H	spo	rob	nob
S. spontaneum (spo)	11	2	1	2	1	1			31%	18%	20%
S. robustum (rob)							2	13		38%	37%
S. officinarum (nob)								15			66%

The first part of the table lists the frequency of 8 different mitochondrial haplotypes in the 3 groups
of accessions. The second part of the table lists the mean similarity between the 3 groups of accessions
based on nuclear RFLP data.

origins: *S. officinarum* cultivars probably were transported by humans to
mainland Asia, they then naturally crossed with local *S. spontaneum* and
gave rise to *S. barberi* and *S. sinense* in India and China, respectively.

It is likely that these clones are early-generation hybrids because no, or
very few, interspecific chromosome exchanges have been detected with GISH
(D'Hont et al., 2002). This is in contrast with the observations of higher
levels of interspecific chromosome exchange in modern cultivars. These
are known to be derived from six to eight meiotic events since the founding *S. officinarum–S. spontaneum* artificial interspecific crosses from which
they derive (D'Hont et al., 1996). *S. barberi* and *S. sinense* cultivars that
were tested have the mitochondrial haplotype H of *S. officinarum*, indicating that this species was the maternal parent and wild *S. spontaneum* the
paternal parent in the founding crosses (D'Hont et al., 1993). Low copy
nuclear RFLP suggests that each morphocytogenetic group represents a set
of somatic mutants derived from a single founding interspecific hybrid
event (D'Hont et al., 2002). The Pansahi group, alias *S. sinense,* is not particularly distinct from the other groups according to nuclear RFLPs. The
S. barberi and *S. sinense* cultivars thus are all derived from similar processes involving an interspecific hybridization event followed by morphological and genetic radiation through mutation, which may have occurred
in different geographic regions of continental Asia.

S. edule

Few molecular data are available for tracing the origin of *S. edule*. The
mitochondrial haplotype has been established for a single clone. It was H,
the same as *S. officinarum, S. barberi,* and *S. sinense* cultivars and most of

S. robustum (D'Hont et al., 1993). An independent investigation based on chloroplast RFLP markers of another clone led to a similar conclusion (Sobral et al., 1994). Recent sequence analysis of nuclear genes by single nucleotide polymorphism showed that *S. edule* clones are closely related to *S. robustum* (unpublished results). These sparse data support the hypothesis that *S. edule* correspond to a series of mutant clones identified in *S. robustum* populations and were preserved by humans.

Contribution of Genera Other Than *Saccharum* to Sugarcane

Molecular data do not support a contribution from *Erianthus* or *Miscanthus* to the genome of sugarcane cultivars. Current extant species of the genera *Saccharum, Erianthus,* and *Miscanthus* are clearly distinct according to isozyme and nuclear and cytoplasmic RFLP data (Glaszmann et al., 1989, 1990; Lu et al., 1994; D'Hont et al., 1993, 1995). However, these results were established by comparison with very few representatives from the genera *Erianthus* and *Miscanthus.* Another approach, relying less on the number of representatives used, yielded similar results: Fast-evolving sequence repeats with multiple dispersed loci in the genome were cloned in *Miscanthus* and *Erianthus* and were hybridized on DNA of representatives of traditional cultivars and wild *Saccharum.* No trace of these *Miscanthus* or *Erianthus* specific sequences were found in any tested individuals (Alix et al., 1998, 1999). Finally, an extensive survey of diversity in *Erianthus* was carried out with nuclear low copy DNA sequences. This showed that *Erianthus* probably is monophyletic and highly divergent from the genus *Saccharum* (Besse et al., 1997).

These data support the view of genus *Saccharum* as a well-defined lineage that includes cultivated sugarcanes plus two wild species, *S. spontaneum* and *S. robustum.* This lineage has diverged over a long period of evolution from the lineages leading to the genera *Erianthus* and *Miscanthus.* Thus, cultivated sugarcanes probably emerged from wild *Saccharum* species, and secondary introgressions with other genera are not likely pathways.

However, this does not mean that natural intergeneric hybridizations are impossible and may not account for some local peculiarities. An *S. maximum* clone has already clearly been identified as a *Saccharum–Miscanthus* hybrid by use of molecular markers. It is also possible that the giant clones of *S. robustum* used for fencing in the New Guinea highlands with high chromosome numbers and some clones classified as *S. edule* may be derived to some extent from intergeneric hybridization. This could be checked easily with molecular markers.

Restriction fragment analysis of the chloroplast genome (Sobral et al., 1994) and analysis of nuclear repeated sequences (Alix et al., 1998, 1999) suggested that *Saccharum* is more closely related to *Miscanthus* than to *Erianthus*. The concept of a *Saccharum* complex may have contributed to an overestimation of the contribution of other genera to the emergence of cultivated sugarcane. This concept was first developed based on geobotanical considerations (Mukherjee, 1957; Daniels et al., 1975) and later received apparent support from leaf flavonoid data. However, morphological traits and flavonoids can be misleading when they are used as single diagnostic markers, especially in polyploid species. They provide few independent tests of genetic variation, and their genetic controls may be polygenic and complexly regulated. For example, the flavone C-glycoside compound F_{13}, which is assumed to be diagnostic for *Erianthus*, occasionally appears in the progenies of crosses between *S. officinarum* and *S. spontaneum*, although it is not present in the parents (Williams et al., 1974). Such markers should be used cautiously because introgression may not always be distinguishable from homoplasty or artifacts. Similar caution should be taken with morphological characters when they are used as diagnostic markers. In the past, morphology has often been misleading in validating artificial intergeneric progenies, especially those involving *Saccharum* and *Erianthus* (D'Hont et al., 1995; Piperidis et al., 2000).

Origin of Modern Cultivars

The origin of modern cultivars is well known. They are derived from several artificial interspecific hybridizations between *S. officinarum* used as the female and *S. spontaneum* and, to a lesser extent, *S. barberi* as the pollen donor. F_1 hybrids were then backcrossed to *S. officinarum* to recover a high–sugar-producing type. These crosses and backcrosses were performed at the end of the 19th century in Java and India. All present-day cultivars are derived from interbreeding of these first interspecific hybrids. By tracing the genealogy of a series of modern cultivars it has been estimated that the elite gene pool derives from about 19 *S. officinarum* clones (four with high frequency), a few *S. spontaneum* (two with high frequency) clones, and one *S. barberi* clone (Arceneaux, 1967). Artificial interspecific hybridization has provided a major breakthrough in sugarcane breeding, solving some disease problems and also increasing yield and adaptability. Molecular cytogenetic data show that around 15–25% of the genomes of these cultivars are derived from *S. spontaneum* (D'Hont et al., 1996 and unpublished data).

Conclusions

Molecular marker evidence for sugarcane domestication strongly favors the scenario developed by Brandes (1956): *S. officinarum* cultivars were domesticated from *S. robustum,* probably in New Guinea, and *S. barberi* and *S. sinense* cultivars resulted from natural interspecific hybridization between *S. officinarum* and *S. spontaneum.* The scarce molecular data available are compatible with *S. edule* clones being mutants selected from *S. robustum* populations, but to confirm this conclusion more clones must be studied.

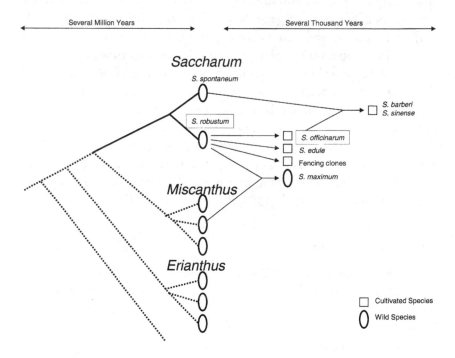

FIGURE 3.3 Scenario compatible with available molecular data for sugarcane evolution and domestication. Ancestors that gave rise to current genera *Saccharum, Erianthus, Miscanthus,* and others diverged in the course of evolution, probably several millions years ago. Only members of the *Saccharum* clade contributed directly to sugarcane cultivars. Allopatric speciation gave rise to two species, *S. spontaneum* west of Sulawesi and *S. robustum* east of Sulawesi. Human-domesticated *S. robustum* in equatorial environment, probably in New Guinea, contributed *S. officinarum* cultivars for sugar, *S. edule* cultivars for vegetables, and possibly other cultivars for others uses (fencing, construction). *S. barberi* and *S. sinense* cultivars resulted from natural hybridization between *S. officinarum* cultivars transported by humans and local *S. spontaneum* populations in subtropical regions. *S. maximum* is at least partially the result of a *Saccharum–Miscanthus* intergeneric hybridization event.

Molecular markers do not favor a contribution from genera other than *Saccharum* for the development of traditional sweet cultivars. In at least one case it has been proven with molecular markers that an *S. maximum* accession was a *Saccharum–Miscanthus* hybrid, but it is not confirmed in every case. These observations are summarized in figure 3.3.

The global picture is now clear. However, it has been established with small samples of materials representing the different germplasm compartments. This permits us to describe the main tendencies in sugarcane evolution but does not allow local exceptions to be revealed. Moreover, independent studies conducted by different researchers were sometimes difficult to correlate because different markers had been used. Therefore a global characterization of the whole *Saccharum* genus, including all traditional cultivars and all wild species, with a common set of molecular descriptors, would be useful. Germplasm present in breeding stations worldwide should be the primary target, but new collections may also be useful, especially for wild species.

Acknowledgments

We acknowledge Dr. Vincent Lebot of the Centre de Coopération Internationale en Recherche Agronomique pour le Développement, Department of Agriculture, Vanuatu, and Dr. Mike Bourke of the Australian National University, Canberra, who were of great help in sharing field experience and giving access to relevant literature; and Mike Butterfield of the South African Sugar Association for critical reading of the manuscript.

References

Alix, K. 1998. *Les Sequences Répétées de la Canne à Sucre: Spécificités Génomique et Apport aux Programmes d'Introgression*. Ph.D. thesis, Ecole National Superieure Agronomique de Rennes, France.

Alix, K., F.C. Baurens, F. Paulet, J.C. Glaszmann, and A. D'Hont. 1998. Isolation and characterization of a satellite DNA family in the *Saccharum* complex. *Genome* 41: 854–864.

Alix, K., F. Paulet, J.C. Glaszmann, and A. D'Hont. 1999. Inter-*Alu* like species-specific sequences in the *Saccharum* complex. *Theoretical and Applied Genetics* 6: 962–968.

Arceneaux, G. 1967. Cultivated sugarcanes of the world and their botanical derivation. *Proceedings of the International Society of Sugar and Sugar Cane Technologists* 12: 844–854.

Baker, C.A. 1874. *Atlas of 220 Weeds of Sugar-Cane Fields in Java*. C.G.G.J. van Steenis (ed.). Greshoff's Rumphius Fund, Amsterdam, The Netherlands.

Barber, C.A. 1920. The origin of the sugar cane. *International Sugar Journal* 22: 249–251.

Barber, C.A. 1922. The classification of Indian canes. *International Sugar Journal* 24: 18–20.

Bayliss-Smith, T. 1996. People–plant interactions in the New Guinea highlands: Agricultural heartland or horticultural backwater? In D.R. Harris (ed.), *The Origins and Spread of Agriculture and Pastoralism in Eurasia*, 499–523. Smithsonian Institution Press, Washington, DC, USA.

Berding, N. and H. Koike. 1980. Germplasm conservation of the *Saccharum* complex: A collection from the Indonesian archipelago. *Hawaiian Planter's Record* 59: 87–176.

Berding, N. and B. T. Roach. 1987. Germplasm collection, maintenance and use. In D. J. Heinz (ed.), *Sugarcane Improvement Through Breeding,* 143–210. Elsevier, New York, NY, USA.

Besse, P., C. L. McIntyre, and N. Berding. 1997. Characterisation of *Erianthus* sect. *Ripidium* and *Saccharum* germplasm (Andropogoneae–Saccharinae) using RFLP markers. *Euphytica* 93: 283–292.

Brandes, E. W. 1956. Origin, dispersal and use in breeding of the Melanesian garden sugarcanes and their derivatives, *Saccharum officinarum* L. *Proceedings of the International Society of Sugar and Sugar Cane Technologists* 9: 709–750.

Burcham, L. T. 1948. Observation on the grass flora of certain Pacific islands. *Contributions from the United States National Herbarium* 30(2). Washington, DC, USA.

Burnquist, W., M. Sorrells, and S. Tanksley. 1992. Characterization of genetic variability in *Saccharum* germplasm by means of RFLP analysis. *Proceedings of the International Society of Sugar and Sugar Cane Technologists* 21: 255–365.

Buzacott, J. H. and C. G. Hughes. 1951. The 1951 cane collecting expedition to New Guinea. *Cane Grower's Quarterly Bulletin* 15: 35–72.

Chen, C. L., L. F. Li, and S. Z. Wu. 1981. Chromosome number distribution of *Saccharum spontaneum* in the southwest region of China. *International Sugar Journal* 83: 264–267.

Coleman, R. E. 1971. Report. *Sugarcane Breeder's Newsletter* 28: 15–17.

Daniels, J. 1967. Sugar canes and related plants of the Fiji islands. *Proceedings of the International Society of Sugar and Sugar Cane Technologists* 12: 1004–1013.

Daniels, J. 1977. 1928 collecting expedition to New Guinea. *Sugarcane Breeder's Newsletter* 40: 15–23.

Daniels, J. and C. A. Daniels. 1976. Buddhism, sugar and sugarcane. *Sugarcane Breeder's Newsletter* 38: 35–60.

Daniels, J. and C. Daniels. 1993. Sugarcane in prehistory. *Archaeologia Oceania* 28: 1–7.

Daniels, J. and B. T. Roach. 1987. Taxonomy and evolution. In D. J. Heinz (ed.), *Sugarcane Improvement Through Breeding,* 7–84. Elsevier, Amsterdam, The Netherlands.

Daniels, J., P. Smith, N. Paton, and C. A. Williams. 1975. The origin of the genus *Saccharum*. *Sugarcane Breeder's Newsletter* 36: 24–39.

Deerr, N. 1949. *The History of Sugar.* Chapman & Hall, London, UK.

D'Hont, A., L. Grivet, P. Feldmann, S. Rao, N. Berding, and J. C. Glaszmann. 1996. Characterisation of the double genome structure of modern sugarcane cultivars (*Saccharum* spp.) by molecular cytogenetics. *Molecular and General Genetics* 250: 405–413.

D'Hont, A., D. Ison, K. Alix, C. Roux, and J. C. Glaszmann. 1998. Determination of basic chromosome numbers in the genus *Saccharum* by physical mapping of ribosomal RNA genes. *Genome* 41: 221–225.

D'Hont, A., Y. H. Lu, P. Feldmann, and J. C. Glaszmann. 1993. Cytoplasmic diversity in sugar cane revealed by heterologous probes. *Sugar Cane* 1: 12–15.

D'Hont, A., F. Paulet, and J. C. Glaszmann. 2002. Oligoclonal interspecific origin of "North Indian" and "Chinese" sugarcanes. *Chromosome Research* 10: 253–262.

D'Hont, A., P. Rao, P. Feldmann, L. Grivet, N. Islam-Faridi, P. Taylor, and J. C. Glaszmann. 1995. Identification and characterisation of intergeneric hybrids, *S. officinarum* × *Erianthus arundinaceus,* with molecular markers and in situ hybridization. *Theoretical and Applied Genetics* 91: 320–326.

Engle, L.M., L.J. Escote, and D.A. Ramirez. 1979. Studies on *Saccharum spontaneum* L., *S. arundinaceum* Retz. and the related species, *Miscanthus floridulus* (Labill.) Warb. in the Philippines. *Proceedings of the International Society of Sugar and Sugar Cane Technologists* 16: 203–210.

Glaszmann, J.C., A. Fautret, J.L. Noyer, P. Feldmann, and C. Lanaud. 1989. Biochemical genetic markers in sugarcane. *Theoretical and Applied Genetics* 78: 537–543.

Glaszmann, J.C., Y.H. Lu, and C. Lanaud. 1990. Variation of nuclear ribosomal DNA in sugarcane. *Journal of Genetics and Breeding* 44: 191–198.

Grassl, C.O. 1946. *Saccharum robustum* and other wild relatives of "Noble" sugar canes. *Journal of the Arnold Arboretum of Harvard University* 27: 234–252.

Hammond, B.W. 1999. *Saccharum spontaneum* (Gramineae) in Panama: The physiology and ecology of invasion. *Journal of Sustainable Forestry* 8: 23–38.

Henty, E.E. 1969. A manual of the grasses of New Guinea. *Botany Bulletin* 1, Lae, New Guinea.

Henty, E.E. 1982. Grasslands and grassland succession in New Guinea. In J.L. Gressitt (ed.), Biogeography and ecology of New Guinea. *Monographiae Biologicae* 42: 459–473.

Krishnamurthi, M. and H. Koike. 1977. Sugarcane collecting expedition: Papua New Guinea, 1977. *Hawaiian Planter's Record* 59: 273–313.

Lennox, C.G. 1939. Sugarcane collection in New Guinea during 1937. *Proceedings of the International Society of Sugar and Sugar Cane Technologists* 6: 171–182.

Lepofsky, D. 2003. The ethnobotany of cultivated plants of the Maohi of the Society Islands. *Economic Botany* 57: 73–92.

Lo, C.C. and S. Sun. 1969. Collecting wild cane in Taiwan. *Proceedings of the International Society of Sugar and Sugar Cane Technologists* 13: 1047–1055.

Lu, Y.H., A. D'Hont, D.I.T. Walker, P.S. Rao, P. Feldmann, and J.C. Glaszmann. 1994. Relationships among ancestral species of sugarcane revealed with RFLP using single copy maize nuclear probes. *Euphytica* 78: 7–18.

Mukherjee, S.K. 1957. Origin and distribution of *Saccharum*. *Botanical Gazette* 19: 55–61.

Nair, N., S. Nair, T. Sreenivasan, and M. Mohan. 1999. Analysis of the genetic diversity and phylogeny in *Saccharum* and related genera using RAPD markers. *Genetic Resources and Crop Evolution* 46: 73–79.

Panje, R.R. 1953. Studies in *Saccharum spontaneum* and allied grasses 3. Recent exploration for *Saccharum spontaneum* and related grasses in India. *Proceedings of the International Society of Sugar and Sugar Cane Technologists* 8: 491–504.

Panje, R.R. and C.N. Babu. 1960. Studies in *Saccharum spontaneum*: distribution and geographical association of chromosome numbers. *Cytologia* 25: 152–172.

Piperidis, G., M.J. Christopher, B.J. Carroll, N. Berding, and A. D'Hont. 2000. Molecular contribution to selection of intergeneric hybrids between sugarcane and the wild species *Erianthus arundinaceus*. *Genome* 43: 1033–1037.

Pohl, R.W. 1983. New records of Mesoamerican grasses. *Iowa State Journal of Research* 58: 191–194.

Poilecot, P. 1999. *Poaceae du Niger*. Boissiera, Geneva, Switzerland.

Price, S. 1965. *Cytology of Saccharum robustum and Related Sympatric Species and Natural Hybrids*. Technical bulletin no. 1337, U.S. Department of Agriculture.

Price, S. and J. Daniels. 1968. Cytology of South Pacific sugarcane and related grasses with special reference to Fiji. *Journal of Heredity* 59: 141–149.

Purseglove, J.W. 1976. The origins and migrations of crops in Tropical Africa. In J.R. Harlan, J.M.J. de Wet, and A.B.L. Stemler (eds.), *Origins of African Plant Domestication*. Mouton, The Hague, The Netherlands.

Roach, B. T. 1972. Chromosome numbers in *Saccharum edule*. *Cytologia* 37: 155–161.

Smith, A. C. 1979. Flora vitiensis nova, *a New Flora of Fiji*. Lawai, Kauai, HI, USA.

Sobral, B. W. S., D. P. Braga, E. S. Lahood, and P. Keim. 1994. Phylogenetic analysis of chloroplast restriction enzyme site mutations in the Saccharinae Griseb. subtribe of the Andropogoneae Dumort. tribe. *Theoretical and Applied Genetics* 87: 843–853.

Sreenivasan, T. V., K. Palanichamy, and M. N. Koppar. 1982. Collection of *Saccharum* germplasm in India. *Plant Genetic Resources Newsletter* 52: 20–24.

Sreenivasan, T. V., K. Palanichamy, and M. N. Koppar. 1985. *Saccharum* germplasm collection in the Sikkim Himalayas of India. *Sugar Cane* 5: 13–15.

Steenis, C. G. G. van. 1950. *Flora Malesiana* serie I. Noordhoff-Kolff N. V., Jakarta, Indonesia.

Tew, T. L., L. H. Purdy, S. Lamadji, and Irawan. 1991. Indonesian sugar cane germplasm collecting expedition: 1984. *Hawaiian Planter's Record* 61: 25–43.

Turner, B. 1992. The colonisation of Anak Krakatau: Interactions between wild sugar cane, *Saccharum spontaneum,* and the antlion, *Myrmeleon frontalis. Journal of Tropical Ecology* 8: 435–449.

Warner, J. N. 1962. Sugar cane: An indigenous Papuan cultigen. *Ethnology* 1: 405–411.

Warner, J. N. and C. O. Grassl. 1958. The 1957 sugar cane expedition to Melanesia. *Hawaiian Planter's Record* 55: 209–236.

Welsh, S. L. 1998. *Flora Societensis: A Summary Revision of the Flowering Plants of the Society Islands.* E.P.S. Inc., Orem, UT, USA.

Westbrooks, R. G. and J. D. Miller. 1993. Investigation of wild sugarcane (*Saccharum spontaneum* L.) escaped from the USDA ARS sugarcane field station at Canal Point, Florida. *Proceedings of the Southern Weed Science Society (*USA*)* 46: 293–295.

Whistler, W. A. 1995. *Wayside Plants of the Islands.* Isle Botanica, Honolulu, HI, USA.

Williams, C. A., J. B. Harborne, and P. Smith. 1974. The taxonomic significance of leaf flavonoids in *Saccharum* and related genera. *Phytochemistry* 13: 1141–1149.

Edward S. Buckler IV
and Natalie M. Stevens

CHAPTER 4

Maize Origins, Domestication, and Selection

Although man does not cause variability and cannot even prevent it, he can select, preserve, and accumulate the variations given to him by the hand of nature almost in any way which he chooses; and thus he can certainly produce a great result.
—Charles Darwin

Wild on a Mexican hillside grows teosinte, its meager ear containing only two entwined rows of small, well-armored kernels. This unassuming grass might easily have been overlooked, were it not for the hand of nature that beckoned with abundant variation, a gift not lost on early agriculturists. Within the last 10,000 years, early Native Americans were able to transform teosinte into a plant whose ear, brimming with row upon row of exposed kernels, feeds the world over. It was a transformation so striking and so complex that some would not believe it possible, leading to years of competing theory and intense debate. But as Darwin himself recognized, when human desires collide with the diversity of nature, the result can be great indeed.

Although controversy still lingers over the origin of maize, the molecular revolution of the last decade has provided compelling evidence in support of teosinte as the progenitor of modern maize. This chapter reviews that evidence in light of several different domestication hypotheses. We also discuss the rich genetic diversity at the source of such a remarkable morphological conversion and examine how human selection has affected this diversity, both at individual loci and for an entire metabolic pathway.

Taxonomy

Maize is a member of the grass family Poaceae (Gramineae), a classification it shares with many other important agricultural crops, including wheat, rice, oats, sorghum, barley, and sugarcane. Based on fossil evidence, it is estimated that these major grass lineages arose from a common ancestor within the last 55–70 million years, near the end of the reign of dinosaurs. Maize is further organized in the genus *Zea,* a group of annual and perennial grasses native to Mexico and Central America. The genus *Zea* includes the wild taxa, known collectively as teosinte (*Zea* ssp.), and domesticated corn, or maize (*Zea mays* L. ssp. *mays*).

For many years, the relationships within genus *Zea* were the subject of much controversy. The central difficulty in the taxonomy of maize and the identification of its closest relatives was the absence of a coblike pistillate inflorescence—or "ear"—in any other known plant. Whereas teosinte produces only 6 to 12 kernels in two interleaved rows protected by a hard outer covering (figure 4.1), modern maize boasts a cob consisting of 20 rows or more, with numerous exposed kernels. In fact, teosinte is so unlike maize in the structure of its ear that 19th-century botanists failed to recognize the close relationship between these plants, placing teosinte in the genus *Euchlaena* rather than in *Zea* with maize (Doebley, 1990b).

Despite these profound physical differences, various morphological, cytological, and genetic studies eventually delineated the relationships within genus *Zea.* H. G. Wilkes (1967) laid the foundation for the current classification scheme in 1967 with the first thorough monograph on teosinte. Wilkes did not attempt a formal hierarchy but instead presented a system of classification using different geographic populations, with separate racial designations based on distinguishing morphological features. In 1980, Hugh Iltis and John Doebley (Doebley and Iltis, 1980; Iltis and Doebley, 1980) produced a system of classification that considered the probable evolutionary relationships between taxa. With the quantitative evaluation of numerous traits and the discovery of many additional populations, Jesus Sanchez (Sanchez G. et al., 1998) provided further characterization of this genus.

Based on the morphological characteristics and geographic delineations established in these systematic treatments, five species of *Zea* are currently recognized:

- *Zea diploperennis* Iltis, Doebley & Guzman, a perennial, diploid teosinte found in very limited regions of the highlands of western Mexico

FIGURE 4.1 The seed spike, or ear, of teosinte (*Zea mays* ssp. *parviglumis*) consists of 2 interleaved rows of 6–12 kernels enclosed in a hard fruitcase (cupule). This female inflorescence, which differs so dramatically from that of maize, has led to much controversy and debate surrounding the origins of maize. (Photo by Hugh Iltis.)

- *Zea perennis* (Hitchcock) Reeves & Mangelsdorf, a perennial tetraploid teosinte, also with a very narrow distribution in the highlands of western Mexico
- *Zea luxurians* (Durieu & Ascherson) Bird, an annual teosinte found in the more equatorial regions of southeastern Guatemala and Honduras
- *Zea nicaraguensis* Iltis & Benz, closely related to *Zea luxurians* and found in mesic environments in Nicaragua (Iltis and Benz, 2000)
- *Zea mays* L., a highly polymorphic, diploid annual species, including both wild teosinte and cultivated maize

This last species, *Zea mays*, is further divided into four subspecies:

- *Z. mays* L. ssp. *huehuetenangensis* (Iltis & Doebley) Doebley, an annual teosinte found in a few highlands of northwestern Guatemala
- *Z. mays* L. ssp. *mexicana* (Schrader) Iltis, an annual teosinte from the highlands of central and northern Mexico
- *Z. mays* L. ssp. *parviglumis* Iltis & Doebley, an annual teosinte, common in the middle and low elevations of southwestern Mexico
- *Z. mays* L. ssp. *mays*, maize or "Indian corn," probably domesticated in the Balsas River Valley of southern Mexico

Origin of Maize

Historical Argument

As scientists labored throughout the mid- to late 1900s to piece together a system of classification for the genus *Zea*, a parallel puzzle surfaced regarding the origin of maize. Despite growing acceptance—reflected in the current taxonomy—of the view that teosinte (*Z. mays*) is the immediate ancestor of maize (*Z. mays* ssp. *mays*), consensus did not come easily. In the struggle to understand the derivation of the enigmatic corn ear, two leading hypotheses emerged.

In the late 1930s, Paul Mangelsdorf and his colleague Robert Reeves proposed a hypothesis known as the tripartite hypothesis (Mangelsdorf, 1974; Mangelsdorf and Reeves, 1938, 1939). This theory stated that maize was domesticated from some unknown wild maize, presumably a plant with structures that resembled the modern maize ear. More specifically, as the name indicates, the hypothesis consisted of three parts: A wild maize prototype from South America, which is now either extinct or undiscovered, was the progenitor of maize; teosinte is the offspring of a cross between maize and *Tripsacum* (another genus of grasses); and sections of *Tripsacum* chromosomes had "contaminated" maize germplasm.

Thus, Mangelsdorf and Reeves invoked a missing ancestor to account for the extreme morphological differences between maize and teosinte while relying on *Tripsacum* to explain their similarities. They pointed to their own successful cross of maize and *Tripsacum* as validation for their hypothesis. Indeed, although the cross entailed significant human intervention, Mangelsdorf and Reeves were able to produce a few, largely sterile maize–*Tripsacum* hybrids. They also analyzed backcross populations of maize–teosinte hybrids and were able to identify four factors (which they interpreted as four *Tripsacum* chromosomal segments) responsible for the morphological differences between maize and teosinte.

For George Beadle, however, the morphological differences between maize and teosinte were not so large as to require an extinct ancestor. In June 1939, less than a year after the publication of the tripartite hypothesis, he responded with his own theory on the origin of maize, an idea he had convinced himself of as a Cornell graduate student under the direction of Rollins Emerson (Doebley, 2001). In his teosinte hypothesis, Beadle (1939) stated that maize is simply a domesticated form of teosinte. He believed that through artificial selection by ancient populations, several small mutations with large effects could have transformed teosinte into maize. Beadle actually used Mangelsdorf and Reeves's own data against them, claiming that their four factors might just as well correspond to four major genes, each of which controlled a single trait that differentiated teosinte from maize. He also challenged their idea that a cross between maize and *Tripsacum,* which took such Herculean efforts on their part, would have occurred in the wild.

On the surface, these dueling hypotheses focused on the origins of a humble ear of corn, but at the core of the controversy was an issue more fundamental and perhaps more far-reaching—a Darwinian debate for the ages. In one corner were evolutionary traditionalists who held that evolution proceeds slowly over time, through the accumulation of many small changes in numerous genes. Thus the dramatic transformation from teosinte to maize was simply not possible in the mere 10,000 years in which humans have been domesticating plants, and a more logical starting point was needed on which selection could act. In the other corner were minds such as Beadle's and Emerson's, where evolution could be more rapid if propelled by changes in a few significant genes. So although teosinte and maize may have looked strikingly different, this difference could be accounted for by only four or five major genes, explaining why the two plants were otherwise genetically similar (so much so that they could be easily crossed to produce fertile offspring).

From its debut in 1938 until the 1960s, the tripartite hypothesis was widely accepted. Through productive collaborations with prominent archaeologists of his day (Mangelsdorf et al., 1964, 1967) and a hemisphere-wide effort targeting maize germplasm conservation (Wellhausen et al., 1952), Mangelsdorf was able to publicize his theory among a wide audience, with his name becoming synonymous with the study of maize evolution. Meanwhile, Beadle temporarily abandoned his teosinte hypothesis for pioneering Nobel work on biochemical genetics and for the presidency of the University of Chicago. During this time his opposing ideas received little attention. Upon his retirement in 1968, however, Beadle rejoined the maize controversy, vigorously pursuing the dispute both in print and in person at several meetings specifically convened to debate the origin of maize. He came armed with additional data that supported his hypothesis (Beadle, 1972, 1977, 1980) and eventually capitalized on the lingering disbelief in the tripartite hypothesis among many maize geneticists. Before Beadle's death in 1989, a host of scientific publications had been issued in support of teosinte as the wild progenitor of maize (see review in Doebley, 1990a).

Modern Argument

The controversy continues. Although the mystery surrounding the origin of maize seemed to be solved, new pieces to the puzzle were added, given time and new technologies. Teosinte and its sister genus *Tripsacum* still take center stage in the modern argument, with one side steadfastly adhering to the teosinte hypothesis while the other revived the idea of a hybridization event. In this section we examine each contemporary hypothesis and its accompanying data in turn, demonstrating that current biological evidence in favor of Beadle's teosinte hypothesis is overwhelming.

Teosinte Hypothesis

The teosinte hypothesis has changed little since Beadle first formalized the idea more than 60 years ago, asserting that teosinte is the wild ancestor of maize. In its modern form, scientists have pinpointed one teosinte in particular, *Zea mays* ssp. *parviglumis,* as the likely progenitor (see figure 4.2 for summary of modern phylogenetics). Because ssp. *parviglumis* is the closest living relative of maize (ssp. *mays*), proponents of this theory reason that maize arose through changes—albeit large changes—to this close ancestor through human selection for specific traits. They point to a wide range of biological data from the 20th century and a wealth of new evidence ushered in with the era of molecular genetics in support of this view.

If maize were simply a domesticated form of teosinte, scientists would need to establish a close relationship between maize and its putative parent. One early indication that maize is strongly allied with *Zea mays* came from

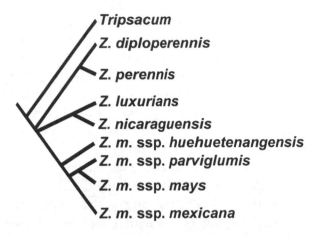

Tripsacum

Z. diploperennis

Z. perennis

Z. luxurians

Z. nicaraguensis

Z. m. ssp. *huehuetenangensis*

Z. m. ssp. *parviglumis*

Z. m. ssp. *mays*

Z. m. ssp. *mexicana*

FIGURE 4.2 The summary phylogeny for the genus *Zea*, based on chromosomal number and morphology (Kato Y., 1976; Kato Y. and Lopez R., 1990), chloroplast (Doebley et al., 1987), ribosomal (Buckler and Holtsford, 1996), isozyme (Doebley et al., 1984), and simple sequence repeat (Matsuoka et al., 2002) data.

studies of chromosome morphology and number. All *Zea* species and sub-species have 10 chromosomes (Kato Y., 1976; Kato Y. and Lopez R., 1990), with the sole exception of *Z. perennis,* which has 20—clearly an example of a complete, duplicated set of chromosomes. On the other hand, most *Tripsacum* species have either 18 or 36 (Mangelsdorf and Reeves, 1938, 1939). Although polyploidy is common in the plant kingdom, either by doubling of a single genome or, more commonly, by combining two or more distinct but related genomes, neither 18 nor 36 chromosomes can easily be derived through normal meiotic associations with the *Zea* genome.

Not only do *Tripsacum* chromosomes differ in number, but they also show marked differences in constitution. Beginning in the 1930s, Barbara McClintock, Paul Mangelsdorf, and collaborators undertook a formal study of chromosome morphology among teosinte plants (Kato Y., 1976; Mangelsdorf, 1974; McClintock et al., 1981). Focusing on chromosomal knobs, or highly repetitive sections of DNA that present as enlarged, deep-staining regions on simple smears, their research revealed that certain grasses such as *Tripsacum* and several *Zea* species had terminal knobs only, whereas others, including three subspecies of *Zea mays,* displayed interstitial knobs. Thus, when coupling basic chromosome numbers with highly conserved chromosomal knob data, maize scientists found early evidence that *Tripsacum* represented a distinct group from *Zea*, with *Z. mays* ssp. *parviglumis, mays,* and *mexicana* forming a natural subgroup within this latter genus.

Chloroplast and ribosomal studies in the late 1980s and 1990s corroborated the story told by earlier chromosomal evidence, showing maize to be only distantly related to *Tripsacum* and more closely aligned within the genus *Zea*. Phylogenies based on the maternally inherited chloroplast clearly place *Z. mays* ssp. *mays* in a group with ssp. *parviglumis* and *mexicana*, along with the fourth subspecies *huehuetenangensis* (Doebley et al., 1987). Phylogenetic studies using nuclear ribosomal internal transcribed spacer (ITS) sequences further delineated these infraspecific *Z. mays* relationships (Buckler and Holtsford, 1996). Ribosomal ITS sequences, which evolve rapidly and are inherited from both parents, indicate that *Zea* species have evolved very recently in comparison to *Zea*'s divergence from *Tripsacum*. In addition, the phylogenetic position of *Z. mays* ssp. *huehuetenangensis* was clearly defined for the first time as being the basal (most diverged) taxon within *Z. mays* (Buckler and Holtsford, 1996).

Thus, the field was narrowing in the quest for maize's wild ancestor. The aforementioned studies had all but eliminated *Tripsacum* as a sister genus that diverged several million years ago. Instead, teosinte fielded the most likely candidates, first as a genus, then within the species *Z. mays,* and finally pared down to just two subspecies, *parviglumis* and *mexicana*. In 1984, isozyme data specifically implicated ssp. *parviglumis* in the origin of maize (Doebley et al., 1984). Simple sequence repeat (SSR) markers—the highest-resolution approach currently available in the arsenal of molecular genetics—later corroborated the isozyme data in naming ssp. *parviglumis* from the Balsas River Valley as the progenitor of maize (Matsuoka et al., 2002). SSR loci, or microsatellite DNA, not only are polymorphic because of the high mutation rate affecting the number of repeat units but also are abundantly distributed throughout broad expanses of eukaryotic DNA. As such, they provide an easily detectable, genome-wide method for determining similarities in evolutionary history between taxa. Comprehensive phylogenetic analyses for maize and teosinte were performed using 99 microsatellite loci from plant samples that encompassed the full geographic range of pre-Columbian maize and Mexican annual teosinte. The study revealed that ssp. *mexicana* is separated from all maize (ssp. *mays*) samples, whereas samples of ssp. *parviglumis* overlap those of maize, documenting the close relationship between ssp. *parviglumis* and maize and supporting the phylogenetic result that the latter subspecies was the sole progenitor of maize (Matsuoka et al., 2002). Furthermore, all maize appears in a single monophyletic lineage that is derived from within ssp. *parviglumis,* thus supporting a single domestication for maize. Using microsatellites that follow a stepwise model and have a known mutation rate, divergence time was estimated at 9188 BP.

Having established *Z. mays* ssp. *parviglumis* as the likely parent of modern maize, and even pinpointing the Balsas River Valley as a candidate for the cradle of maize domestication, research focused on the loci involved in the dramatic transformation from wild grass to cultivated crop. Modern molecular techniques using linkage maps and quantitative trait locus (QTL) analysis have increasingly provided evidence in direct support of another fundamental tenet of the teosinte hypothesis: that a few regions of the maize genome specify the traits that distinguish maize from teosinte. Using basic Mendelian ratios from 50,000 maize and teosinte hybrids, Beadle (1972, 1977, 1980) first recognized that as few as five loci may be involved in important ear and plant morphological changes. More than 20 years later, QTL mapping would validate his idea, identifying five regions of the maize genome with large effects on basic morphology (Doebley et al., 1990; Doebley and Stec, 1991).

Although far from complete, the maize mystery is slowly unraveling through concentrated studies of these important regions. For example, a single major locus, *teosinte glume architecture1* (*tga1*), has been identified that controls the development of the glume, a protective covering on teosinte kernels that is mostly lacking in maize (Dorweiler et al., 1993). Because teosinte's hard glume makes it very difficult to eat, a mutation in this gene leading to a softer glume probably was one of the first targets of selection by Native Americans during domestication. A second locus, *teosinte branched1* (*tb1*), which dictates a difference in plant architecture (long lateral branches terminated by male tassels in teosinte vs. short lateral branches tipped by female ears in maize) has been successfully cloned (Doebley et al., 1995, 1997; Wang et al., 1999). QTLs at genes responsible for three more distinguishing traits (shattering versus solid cobs, single versus paired spikelets, and distichous versus polystichous condition) are the subject of current investigations.

Caution must be exercised in advocating a one-gene, one-trait model. Although a small number of genes, such as *tga1* and *tb1,* clearly have a striking effect on ear and plant morphology and represent major steps in maize evolution, most genes have modest effects. Even Beadle recognized that additional "modifier" genes would be necessary to complete the transition, and perhaps hundreds or even thousands of genes were involved in steps such as increasing the size of the ear, adapting growth to different agricultural environments, and modifying the nutrient content of the maize kernel. However, the essence of the argument remains intact: A small number of single-gene mutations could be sufficient to go from teosinte to a plant that possesses the key morphological features of cultivated maize.

Tripsacum–Z. diploperennis Hypothesis

A modern version of the tripartite hypothesis, formalized in 1995, is Eubanks's *Tripsacum–Z. diploperennis* hypothesis. Still challenging the idea that maize is a domesticated form of teosinte, this theory proposes that maize arose from the progeny of a cross between *Z. diploperennis* and *T. dactyloides* (Eubanks, 1995, 1997, 2001). At the heart of this proposal are two putative hybrids, dubbed Tripsacorn and Sundance, that originated from these two grasses (figure 4.3). Unlike the parents, the rudimentary ear of these hybrids has exposed kernels attached to a central rachis, or cob. If such hybrids once

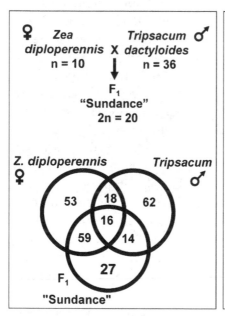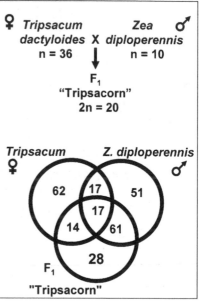

FIGURE 4.3 Sundance (*left*) and Tripsacorn (*right*) are the putative hybrids from a cross between *Z. diploperennis* and *T. dactyloides*. RFLP molecular analysis for these hybrids calls into dispute the successful hybridization of these plants because 23% of polymorphisms in the F_1 generation were not found in either parent. Overlapping regions of the Venn diagrams correspond to the number of shared bands between parent and putative offspring, whereas the numbers that appear in a single circle represent unique RFLP bands (data from Eubanks, 1997).

occurred naturally, then—at least according to proponents of the *Tripsacum–Z. diploperennis* hypothesis—the evolutionary puzzle of the origin of maize and its unparalleled architecture is solved.

However, there are several fundamental problems with the *Tripsacum–Z. diploperennis* theory. First, although producing a *Tripsacum–Z. diploperennis* hybrid may very well be possible, the documentation provided by Eubanks (1995, 1997) in support of these hybrids does not demonstrate that these two grasses were successfully hybridized. The chromosome number of both Tripsacorn and Sundance is $2n = 20$. If *Tripsacum* ($2n = 36$ or 72) had indeed been one of the parents, then these hybrids would be expected to have 28 or 46 chromosomes, as evidenced by previous crosses between maize and *Tripsacum*. For example, successful experimental crosses between *T. dactyloides* and *Z. mays* ssp. *mays* by Mangelsdorf and Reeves (1939) produced hybrids with $2n = 28$. Many other *Zea* and *Tripsacum* crosses were made by de Wet (de Wet and Harlan, 1974; de Wet et al., 1972), and a single generation conversion to $2n = 20$ was never seen. Although the creation of a *Z. diploperennis* doubled haploid—in which all 10 *Zea* chromosomes are spontaneously doubled and all 36 *Tripsacum* chromosomes are immediately eliminated from the embryo—might be invoked to explain such a hybrid, the $2n = 20$ condition is more likely to be the result of a contaminated cross. Indeed, $2n = 20$ is also the chromosome number of maize and thus the number one would expect in a maize–*Z. diploperennis* hybrid.

A second concern regarding the validity of the *Tripsacum–Z. diploperennis* hypothesis centers on the analysis of RFLP data for the putative hybrids (Eubanks, 1997). Because these molecular markers are inherited directly from the parents, restriction fragments present in a true hybrid must be traced back to at least one parent. Of the polymorphisms identified in Tripsacorn and Sundance, there was indeed some sharing between putative parent and offspring. It is interesting to note that the hybrids shared four times as many bands with *Z. diploperennis* as with *Tripsacum*, indicating a much closer relationship with teosinte than with *Tripsacum*. Perhaps more telling, however, is that 23% of the molecular markers surveyed were not found in either parent (figure 4.3). How does one account for these novel bands?

Proponents of the *Tripsacum–Z. diploperennis* hypothesis would argue that these restriction fragments are a consequence of the hybridization event itself: interactions between the combined genomes causing novel patterns of gene sequence. However, producing such novel gene sequences would entail either a point mutation at 2% of DNA sites in one generation, or about 120 mutations per gene;[1] or a large insertion every 17,800 base pairs in one generation, or 168,000 total insertions across the genome.[2] Such genome activity is extremely unlikely and almost certainly lethal. Roughly 120 point mutations per gene in one generation is more than 3 million times the normal rate of mutation (6×10^{-9} substitutions per site per year from Gaut et al., 1996). And although the combination of two novel genomes may activate a few transposons here or there, it is doubtful that a genome could survive a rearrangement on the order of 168,000 large insertions because it would most certainly interfere with vital gene function. It seems far more plausible, as suggested earlier, that these novel bands are the product of a contaminated cross.

Even if these experimental hybrids are indeed true hybrids, they do not in themselves constitute proof that maize arose from the progeny of a cross between *Z. diploperennis* and *T. dactyloides*. Problems also exist with an argument often cited in support of the *Tripsacum–Z. diploperennis* hypothesis that attempts to tie together maize and *Tripsacum* evolution. The argument is based on shared ancestral polymorphisms between samples of teosinte (*Z. mays*), *Tripsacum*, and maize (*Z. mays* ssp. *mays*). A recent RFLP study by Eubanks (2001) found that maize and *Tripsacum* share 92 unique polymorphisms (figure 4.4). From these data, it was inferred that "polymorphisms uniquely shared between *Tripsacum* and maize were likely derived from a

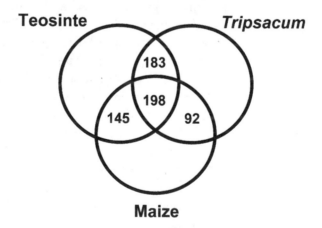

FIGURE 4.4 Shared ancestral polymorphisms between samples of teosinte (*Z. mays*), *Tripsacum*, and maize (*Z. mays* ssp. *mays*) as reported by Eubanks (2001). RFLP data revealed 92 polymorphisms unique to maize and *Tripsacum* and 198 shared by all three samples. The unique sharing of bands between maize and *Tripsacum* results from poor sampling of teosinte and the impossibility of sampling extinct alleles.

Tripsacum ancestor" (Eubanks, 2001:507). However, this can be true only if *all* alleles—both extant and extinct—are sampled from the three taxa, obviously an impossible feat. On the contrary, rather than providing proof of a *Tripsacum* origin, these shared polymorphisms are simply what one would expect to see between two grasses that share 93.5% of sites by vertical descent; indeed, 45% of RFLP bands should be shared between any *Zea* and *Tripsacum* pair because these grasses diverged from a common ancestor several million years ago.

Furthermore, there is also some question as to whether these 92 polymorphisms are uniquely shared between *Tripsacum* and maize. The teosinte sample used for the study is not reflective of the extremely high diversity inherent in the *Zea* genome (a closer look at this diversity follows later in the chapter). It included only one *Z. mays* ssp. *parviglumis* individual, thus underrepresenting a group that is not only one of the most diverse grasses but also is the one group most likely to possess alleles in common with maize (Doebley et al., 1984; Matsuoka et al., 2002). If the ssp. *parviglumis* sample had been larger and the teosinte alleles already extinct could also be considered, it is certain that many of the 92 bands would no longer be uniquely shared between *Tripsacum* and maize. Additionally, the *Tripsacum* sample can be called into question because it included *T. andersonii*, a natural, sterile *Zea–Tripsacum* hybrid with 64 chromosomes (Dewet et al., 1983). Thus, the *Tripsacum* sample already captured some *Zea* alleles, leading to inflated band sharing with both the maize and teosinte samples and calling into dispute the extent of *Tripsacum's* unique contribution to the maize genome.

Finally, time itself tells a story inconsistent with the *Tripsacum–Z. diploperennis* hypothesis. Regardless of the progenitor involved, the domestication of maize cannot be older than the significant human migrations to the New World, which occurred roughly 15,000 years ago (Dillehay, 1989). By using the 18 currently sequenced genes in both maize and *Tripsacum* (Tenaillon et al., 2001; Whitt et al., 2002), we found that, on average, the genes diverged by 6.5% at noncoding and silent sites. If a mutation rate of 6×10^{-9} substitutions per site per year (Gaut et al., 1996) is assumed, this suggests that maize and *Tripsacum* alleles diverged around 5.2 million years ago, long before Native Americans could have combed the Mexican hillsides in search of food. In contrast, ssp. *parviglumis* and maize have an average divergence time of 9188 BP (Matsuoka et al., 2002). This date is consistent with the date of 6250 BP for the oldest known maize fossil (Piperno and Flannery, 2001).

Thus, from improbable hybrids to incongruous timelines, it appears that a *Tripsacum* key will not unlock the mystery of the origin of maize. However, we would be remiss not to acknowledge its potential contribution to the development of the maize genome. Because horizontal transfer of mitochondrial genes has been demonstrated between distantly related plants (Bergthorsson et al., 2003), there is a chance that some *Tripsacum* alleles could have introgressed into maize, but the contribution, if any, probably was very small. No phylogenetic, cytological, or molecular evidence exists in support of the *Tripsacum–Z. diploperennis* hypothesis, but the horizontal transfer of perhaps a handful of genes cannot formally be ruled out. If such a genome "jump" did occur, the genes involved probably conferred disease resistance rather than drove domestication because pathogens can provide intense selection pressure over billions of plants, making defense genes ideal candidates for transfer.

The Final Verdict

In short, the teosinte hypothesis best fits the evidence. For most maize geneticists and evolutionists (Bennetzen et al., 2001) familiar with the issues and data surrounding the origin of maize, there is little doubt that maize is a domesticated derivative of the wild Mexican grass teosinte (*Z. mays* ssp. *parviglumis*). However, questions persist in regard to the precise morphogenetic steps needed to complete the extreme transition from wild teosinte to cultivated maize. Just how did early Native American farmers achieve what is arguably the most remarkable breeding accomplishment of all time?

Domestication

The evolution of maize and the development of Native American societies were intimately connected; indeed, maize has been credited as the grain that civilized the New World. These early farming communities used corn not only for food but also for art and religious inspiration. Maize probably was domesticated over a period of a few thousand years in south central Mexico, the principal habitat of its immediate ancestor, *Z. mays* ssp. *parviglumis*. Archaeological remains of the earliest maize cob, found at Guila Naquitz Cave in the Oaxaca Valley of Mexico, date back roughly 6250 years (Piperno and Flannery, 2001). There is also much microfossil evidence suggesting dispersal to Central and South America by 7000–5000 BP (Piperno and Pearsall, 1998). Therefore maize probably was domesticated between 12,000 and 7500 years ago, as the first steps of domestication necessarily preceded this evidence, and its initiation cannot be older than the significant human migrations to the New World in roughly 15,000 BP (Dillehay, 1989).

Although the extraordinary morphological and genetic diversity among the maize landraces led some researchers to propose multiple, independent origins for maize (Kato Y., 1984), recent phylogenetic analyses based on comprehensive samples of maize and teosinte indicate a single domestication event. As noted earlier, a microsatellite-based phylogeny for a sample of 264 maize and teosinte plants showed all maize in a single monophyletic lineage that is derived from within ssp. *parviglumis* (Matsuoka et al., 2002). After this domestication, maize spread from Mexico over the Americas along two major paths (Matsuoka et al., 2002).

Domesticated maize was the result of repeated interaction with humans, with early farmers selecting and planting seed from plants with beneficial traits while eliminating seed from plants with less desirable features. As a result, alleles at genes controlling favored traits increased in frequency within the population, whereas less favored alleles decreased. Thus with each succeeding generation these ancient agriculturists produced a plant more like modern maize and less like the wild grass of their ancestors.

This human selection process probably was both conscious and unconscious (Rindos, 1984). Native Americans may have combed the Mexican hillsides in search of teosinte plants with promising mutations, deliberately choosing the plants that provided more of and easier access to the sustenance they needed. For example, teosinte kernels are surrounded by a hard protective covering, or glume. Because this glume makes them very difficult to eat, plants with a softer glume were conceivably targeted during domestication. However, loss of shattering (a natural mechanism for seed dispersal) was more likely to be an inadvertent consequence of the harvesting process because early farmers could only plant the seeds that arrived home with them, still attached to the central rachis, or eventual maize cob.

Over time, these ancient agriculturists were able to select, consciously or not, the combination of major and many minor gene mutations that now distinguish maize from its wild ancestor. As it turns out, many of the same genes involved in this transformation might also be involved in that of other grasses, including wheat, rice, and sorghum (Paterson et al., 1995). Despite the independent domestication of these cereal complexes, it now appears that the earliest plant selectors desired the same sets of traits, as evidenced by selection at a common set of loci. QTLs for seed size, seed dispersal (shattering), and photoperiod have been mapped in maize, rice, and sorghum. These QTLs correspond to homologous regions between taxa more often than would be expected by chance and provide further evidence that domestication of these grasses was the result of mutations in a small number of genes with large effects (Buckler et al., 2001).

Diversity

The ability of Native Americans to transform a wild grass into the world's largest production grain crop is not only the product of skillful breeding but also a tribute to the tremendous diversity of the teosinte genome. Years before his time, these ancient farmers first practiced what Darwin later preached: that selection must be combined with natural variation in order for evolution to take place. As it turns out, teosinte is extremely diverse, with modern molecular studies measuring nucleotide diversity at silent sites in *Z. mays* ssp. *parviglumis* at roughly 2–3% (Eyre-Walker et al., 1998; Goloubinoff et al., 1993; Hilton and Gaut, 1998; White and Doebley, 1999; Whitt et al., 2002). Maize retained much of the diversity of its wild ancestor, with any two maize varieties differing from one another in 1.4% of their DNA (silent sites) (Tenaillon et al., 2001). For the sake of comparison, this level of nucleotide diversity found in maize is 2–5 times higher than that of other domesticated grass crops and is 14 times higher than that of humans; indeed, the divergence between two maize lines is roughly equivalent to the difference between humans and chimpanzees (Chen and Li, 2001).

This begs the question as to why *Z. mays* ssp. *parviglumis* has such high levels of diversity. Population genetics theory shows that levels of molecular diversity are the product of high mutation rates coupled with large effective population size. New alleles appear in a population by the natural process of mutation, and the random loss of these alleles (genetic drift) affects small populations more severely than large ones, as alleles are drawn from a smaller parental gene pool. *Z. mays* ssp. *parviglumis* conforms to both these criteria: A high rate of mutation has been documented in grasses (Gaut et al., 1996), and population size for this wild grass historically has been quite large. Scientific literature documents such high diversity in several other species that also enjoy large population size, including *Drosophila simulans* (the fruit fly), with measures as high as 3.5% (Begun and Whitley, 2000).

In contrast, humans, whose founding populations in Africa were quite small in comparison, have only 0.1% diversity (Cargill et al., 1999).

Like most other grasses, maize maintained a substantial proportion of the variation of its wild progenitor, with only a 30% drop in diversity at the average locus (Buckler et al., 2001). This is probably because humans—both ancient and modern—depend on domesticated grains as a basis for subsistence, so large quantities of plants are needed before they are useful. If 10 people derive 10% of their calories from maize, it is estimated that roughly 250,000–350,000 plants would have to be grown annually (Buckler et al., 2001; Hillman and Davies, 1990).

Such abundant variation in the maize genome presents an intriguing paradox in light of the dramatic morphological differences between it and its closest living relative. On one hand, the extreme phenotypic and molecular variation found in maize is consistent with a large historical population size, as discussed in the preceding paragraph. On the other hand, maize is so unlike teosinte in ear morphology and plant architecture as to suggest strong selection during domestication, a decidedly diversity-limiting process. In other words, the initial steps of most domestication events probably included a population bottleneck.

Coalescent theory has been used to study the likelihood of such a domestication bottleneck in maize. Based on sequence diversity at the neutral *Adh1* locus in maize (*Z. mays* ssp. *mays*), its progenitor (*Z. mays* ssp. *parviglumis*), and a more distant relative (*Zea luxurians*), current diversity in maize can indeed be explained by a founding population with a modest number of diverse teosinte individuals (Eyre-Walker et al., 1998). However, the exact size of this founding population depends on the duration of the domestication event (the more founding individuals, the longer the bottleneck), something archaeological evidence has yet to elucidate with any certainty. Despite the virtual necessity of a population bottleneck to initiate maize domestication, its effects probably were limited by high rates of outcrossing and the impressive diversity among the founding teosinte population.

Targets of Selection

Individual Loci

Although the maize genome as a whole is extremely diverse, individual targets of selection can be identified because domestication should strongly reduce sequence diversity at genes controlling traits of human interest. As previously discussed, *tb1* is responsible for some of the major distinguishing morphological differences between teosinte and modern maize. Because this locus represents a key step in maize domestication, its nucleotide polymorphism should be lower than that of neutral sites. Indeed, within the promoter region of *tb1,* maize possesses only 3% of the diversity found in teosinte, or 61-fold lower diversity in the domesticated crop than in the closest wild relative (Wang et al., 1999). The timing and sequence of such character selection by early farmers is now being revealed by the fusion of molecular biology and archaeological research. Surveys of *tb1* in ancient DNA suggest that selection at this locus occurred before 4400 BP (Jaenicke et al., 2003).

A recent large survey of 1772 maize loci suggests that roughly 3–5% of these genes have undergone selection since domestication (Vigouroux et al., 2002). Coalescent simulations were used to compare the genetic diversity (or divergence) at a locus with what one would expect under a neutral model that incorporates the domestication bottleneck. This approach to screening large numbers of loci for the signature of selection appears to offer a powerful method for identifying new candidate genes of agronomic importance.

Starch Pathway

Whereas changes in plant shape and ear morphology were the initial focus of Beadle and his successors, many additional traits have been the target of human selection over the last few thousand years. Some of these traits of particular significance were yield, ear size (which increased from 2 cm to 30 cm), and grain quality. Starch is the key product of maize, accounting for 73% of the kernel's total weight. Therefore the genes involved in starch synthesis are among the most important for grain production, critical to both the yield and the quality of the grain.

A simplified pathway of starch production in maize is outlined in figure 4.5. Amylopectin makes up roughly three-quarters of the total product, with amylose comprising the remainder. Amylopectin is primarily responsible for granule swelling and eventual thickening of pastes upon addition of heat, and amylose typically is thought to affect the gelling of starch, all chemical and structural properties important in food processing. For example, starch pasting modifies the ability of foods to hold fat and protein molecules that enhance flavor and texture,

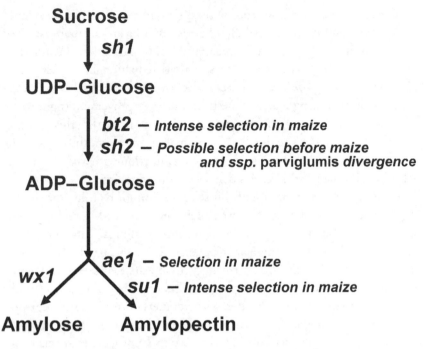

FIGURE 4.5 A simplified pathway of starch production in maize, indicating the relative position of the 6 sampled genes in the pathway: *amylose extender1 (ae1)*, *brittle2 (bt2)*, *shrunken1 (sh1)*, *shrunken2 (sh2)*, *sugary1 (su1)*, and *waxy1 (wx1)*. The genes *bt2, sh1*, and *sh2*, located upstream in the pathway, aid in the formation of glucose, whereas the enzymes coded by *ae1, su1*, and *wx1* produce the final products of starch metabolism: amylose and amylopectin. The signature of selection at each locus is also noted, as revealed by low nucleotide diversity. ADP = adenosine diphosphate; UDP = uridine diphosphate.

certainly an aspect of maize that Native American breeders might have included in the domestication and improvement process.

Although plant genetics and biochemistry have thus far identified more than 20 genes involved in starch production, Whitt et al. (2002) focused on six key genes known to play major roles in starch production: *amylose extender1 (ae1), brittle2 (bt2), shrunken1 (sh1), shrunken2 (sh2), sugary1 (su1)*, and *waxy1 (wx1)*. For each locus, diversity estimates (π) were performed by sequencing 6–13 kb from 30 diverse maize lines, along with 1–2 kb from *Z. mays* ssp. *parviglumis* and 2–4 kb from *Tripsacum dactyloides* for comparison. The Hudson–Kreitman–Aquade (HKA) test (Hudson et al., 1987), a test that compares rates of divergence between species to levels of polymorphism within species, was then used to formally test for selection.

The results were striking: Four of these six starch loci exhibited evidence of selection (Whitt et al., 2002), whereas random loci in maize showed almost no proof of selection. Three maize loci in particular, *su1, bt2,* and *ae1,* revealed a dramatic three- to sevenfold reduction in diversity over *Z. mays* ssp. *parviglumis,* which is consistent with artificial selection in the starch pathway during maize domestication and improvement (figure 4.5). The significant HKA results for both *bt2* and *su1* indicate that this selection probably occurred before the dispersal of maize germplasm throughout the world, whereas with *ae1* the HKA test (in conjunction with a second test of selection, Tajima's test) suggests that selection is ongoing.

Although the exact nature of this selection cannot be fully understood until a wide range of teosinte starch alleles are examined in maize genetic backgrounds, our results provide an intriguing glimpse into the preferences of early Native American breeders. Given the particular roles of *ae1, bt2,* and *su1* in the starch pathway, it appears that selection favored increased yield and different amylopectin qualities. Because starch (unlike protein) is often lacking in hunter–gatherer diets of tropical and subtropical societies, it is reasonable to presume that early cultivators of maize focused on improving starch yield. Starch pasting properties are also logical targets of selection in maize because the ratio of amylose to amylopectin and the chemical structure of amylopectin (specifically the length of branched glucose chains) affect everything from porridge to tortilla texture.

A timeline indicating when these early breeders selected for starch production and other advantageous traits is being constructed with help from archaeology. Ancient DNA analysis from maize samples unearthed in Mexico and the southwestern United States has revealed that *su1* alleles known to occur in modern maize probably were under selection between 1800 and 900 years ago (Jaenicke et al., 2003). Future studies that integrate important archeological questions, such as when and how ancient peoples used maize, with molecular evidence of selection will make it possible to trace the genetic consequences of domestication over time.

The enduring legacy of ancient maize agriculturalists is far more than the germplasm for a softer tortilla, however. As evidenced by our research, the reduction of diversity in starch loci is dramatic and should motivate a paradigm shift in maize breeding. Although tremendous variation at most loci has allowed maize to respond to centuries of artificial selection and industrial farming practices, limited diversity in the starch pathway and perhaps other pathways of critical importance may prevent current breeding practices from reaching their full potential. The ability of plant breeders and scientists to improve current maize lines and develop new products to meet future needs depends on useful variation within the maize germplasm. Perhaps the most efficient way to introduce this potentially useful diversity into maize is to introgress or transform the abundant allelic variation present in teosinte for selected genomic regions or specific genes. By using this raw genetic material from maize's wild relatives, the next generation can continue what the early Mexican natives so deftly began: the most impressive feat of genetic modification and morphological evolution ever accomplished in any plant or animal domesticate.

Acknowledgments

We would like to thank Sherry Whitt and Larissa Wilson for their research contributions to this chapter; both played important roles in establishing the signature of selection in the maize starch pathway. This work was supported by the U.S. Department of Agriculture-Agricultural Research Service and the National Science Foundation (DBI-0321467).

Notes

1. Necessary point mutation frequency was obtained by dividing the frequency of novel bands in the putative hybrids (0.23) by the length of nucleotides in RFLP cut sites (6 + 6).

2. Insertion number was obtained by dividing the average band size (4096 for restriction enzymes with 6 bp recognition sites) by the frequency of novel bands in the putative hybrids (0.23).

References

Beadle, G.W. 1939. Teosinte and the origin of maize. *Journal of Heredity* 30: 245–247.

Beadle, G.W. 1972. The mystery of maize. *Field Museum of Natural History Bulletin* 43: 2–11.

Beadle, G.W. 1977. The origin of *Zea mays*. In C.A. Reed (ed.), *Origins of Agriculture*, 615–635. Mouton Press, The Hague, The Netherlands.

Beadle, G.W. 1980. The ancestry of corn. *Scientific American* 242: 112–119.

Begun, D.J. and P. Whitley. 2000. Reduced X-linked nucleotide polymorphism in *Drosophila simulans*. *Proceedings of the National Academy of Sciences (USA)* 97: 5960–5965.

Bennetzen, J., E. Buckler, V. Chandler, J. Doebley, J. Dorweiler, B. Gaut, M. Freeling, S. Hake, E. Kellogg, R.S. Poethig, V. Walbot, and S. Wessler. 2001. Genetic evidence and the origin of maize. *Latin American Antiquity* 12: 84–86.

Bergthorsson, U., K. Adams, B. Thomason, and J. Palmer. 2003. Widespread horizontal transfer of mitochondrial genes in flowering plants. *Nature* 424: 197–201.

Buckler, E. S. IV and T. P. Holtsford. 1996. *Zea* systematics: Ribosomal ITS evidence. *Molecular Biology and Evolution* 13: 612–622.

Buckler, E. S. IV, J. M. Thornsberry, and S. Kresovich. 2001. Molecular diversity, structure and domestication of grasses. *Genetical Research* 77: 213–218.

Cargill, M., D. Altshuler, J. Ireland, P. Sklar, K. Ardlie, N. Patil, N. Shaw, C. R. Lane, E. P. Lim, N. Kalyanaraman, J. Nemesh, L. Ziaugra, L. Friedland, A. Rolfe, J. Warrington, R. Lipshutz, G. Q. Daley, and E. S. Lander. 1999. Characterization of single-nucleotide polymorphisms in coding regions of human genes. *Nature Genetics* 22: 231–238.

Chen, F. C. and W. H. Li. 2001. Genomic divergences between humans and other hominoids and the effective population size of the common ancestor of humans and chimpanzees. *American Journal of Human Genetics* 68: 444–456.

de Wet, J. M. J., L. M. Engle, C. A. Grant, and S. T. Tanaka. 1972. Cytology of maize–*Tripsacum* introgression. *American Journal of Botany* 59: 1026–1029.

de Wet, J. M. J., G. B. Fletcher, K. W. Hilu, and J. R. Harlan. 1983. Origin of *Tripsacum*–Andersonii (Gramineae). *American Journal of Botany* 70: 706–711.

de Wet, J. M. J. and J. R. Harlan. 1974. *Tripsacum*–maize interaction: A novel cytogenetic system. *Genetics* 78: 493–502.

Dillehay, T. D. 1989. *Monte Verde, a Late Pleistocene Settlement in Chile.* Smithsonian Institution Press, Washington, DC, USA.

Doebley, J. 1990a. Molecular evidence and the evolution of maize. *Economic Botany* 44: 6–27.

Doebley, J. 1990b. Molecular systematics of *Zea* (Gramineae). *Maydica* 35: 143–150.

Doebley, J. 2001. George Beadle's other hypothesis: One-gene, one-trait. *Genetics* 158: 487–493.

Doebley, J. F., M. M. Goodman, and C. W. Stuber. 1984. Isoenzymatic variation in *Zea* (Gramineae). *Systematic Botany* 9: 203–218.

Doebley, J. F. and H. H. Iltis. 1980. Taxonomy of *Zea* (Gramineae). I. A subgeneric classification with key to taxa. *American Journal of Botany* 67: 982–993.

Doebley, J., W. Renfroe, and A. Blanton. 1987. Restriction site variation in the *Zea* chloroplast genome. *Genetics* 117: 139–147.

Doebley, J. F. and A. Stec. 1991. Genetic analysis of the morphological differences between maize and teosinte. *Genetics* 129: 285–295.

Doebley, J., A. Stec, and C. Gustus. 1995. *Teosinte branched1* and the origin of maize: Evidence for epistasis and the evolution of dominance. *Genetics* 141: 333–346.

Doebley, J., A. Stec, and L. Hubbard. 1997. The evolution of apical dominance in maize. *Nature* 386: 485–488.

Doebley, J., A. Stec, J. Wendel, and M. Edwards. 1990. Genetic and morphological analysis of a maize–teosinte F_2 population: Implications for the origin of maize. *Proceedings of the National Academy of Sciences (USA)* 87: 9888–9892.

Dorweiler, J., A. Stec, J. Kermicle, and J. Doebley. 1993. *Teosinte glume architecture 1:* A genetic locus controlling a key step in maize evolution. *Science* 262: 233–235.

Eubanks, M. 1995. A cross between two maize relatives: *Tripsacum dactyloides* and *Zea diploperennis*. *Economic Botany* 49: 172–182.

Eubanks, M. W. 1997. Molecular analysis of crosses between *Tripsacum dactyloides* and *Zea diploperennis* (Poaceae). *Theoretical and Applied Genetics* 94: 707–712.

Eubanks, M. W. 2001. The mysterious origin of maize. *Economic Botany* 55: 492–514.

Eyre-Walker, A., R. L. Gaut, H. Hilton, D. L. Feldman, and B. S. Gaut. 1998. Investigation of the bottleneck leading to the domestication of maize. *Proceedings of the National Academy of Sciences (USA)* 95: 4441–4446.

Gaut, B. S., B. R. Morton, B. C. McCaig, and M. T. Clegg. 1996. Substitution rate comparisons between grasses and palms: Synonymous rate differences at the nuclear gene *Adh* parallel rate differences at the plastid gene *rbcL. Proceedings of the National Academy of Sciences (USA)* 93: 10274–10279.

Goloubinoff, P., S. Pääbo, and A. C. Wilson. 1993. Evolution of maize inferred from sequence diversity of an *Adh2* gene segment from archaeological specimens. *Proceedings of the National Academy of Sciences (USA)* 90: 1997–2001.

Hillman, G. and M. S. Davies. 1990. Domestication rates in wild-type wheats and barley under primitive cultivation. *Biological Journal of the Linnean Society* 39: 39–78.

Hilton, H. and B. S. Gaut. 1998. Speciation and domestication in maize and its wild relatives: Evidence from the *globulin-1* gene. *Genetics* 150: 863–872.

Hudson, R. R., M. Kreitman, and M. Aquade. 1987. A test of neutral molecular evolution based on nucleotide data. *Genetics* 116: 153–159.

Iltis, H. and B. Benz. 2000. *Zea nicaraguensis* (Poaceae), a new teosinte from Pacific coastal Nicaragua. *Novon* 10: 382–390.

Iltis, H. H. and J. F. Doebley. 1980. Taxonomy of *Zea* (Gramineae). II. Subspecific categories in the *Zea mays* complex and generic synopsis. *American Journal of Botany* 67: 994–1004.

Jaenicke, V., E. S. Buckler, B. D. Smith, M. T. P. Gilbert, A. Cooper, J. Doebley, and S. Pääbo. 2003. Early allelic selection in maize as revealed by ancient DNA. *Science* 302: 1206–1208.

Kato Y., T. A. 1976. Cytological studies of maize (*Zea mays* L.) and teosinte (*Zea mexicana* Schrader Kuntze) in relation to their origin and evolution. *Massachusetts Agricultural Experiment Station Bulletin* 635: 1–185.

Kato Y., T. A. 1984. Chromosome morphology and the origin of maize and its races. *Evolutionary Biology* 17: 219–253.

Kato Y., T. A. and A. Lopez R. 1990. Chromosome knobs of the perennial teosintes. *Maydica* 35: 125–141.

Mangelsdorf, P. C. 1974. *Corn: Its Origin, Evolution, and Improvement.* Harvard University Press, Cambridge, MA, USA.

Mangelsdorf, P. C., R. S. MacNeish, and W. C. Galinat. 1964. Domestication of corn. *Science* 143: 538–545.

Mangelsdorf, P. C., R. S. MacNeish, and W. C. Galinat. 1967. Prehistoric wild and cultivated maize. In D. S. Byers (ed.), *The Prehistory of the Tehuacan Valley: Environment and Subsistence,* Vol. 1, 178–200. University of Texas Press, Austin, TX, USA.

Mangelsdorf, P. C. and R. G. Reeves. 1938. The origin of maize. *Proceedings of the National Academy of Sciences (USA)* 24: 303–312.

Mangelsdorf, P. C. and R. G. Reeves. 1939. The origin of Indian corn and its relatives. *Texas Agricultural Experiment Station Bulletin* 574: 1–315.

Matsuoka, Y., Y. Vigouroux, M. M. Goodman, J. Sanchez G., E. S. Buckler, and J. F. Doebley. 2002. A single domestication for maize shown by multilocus microsatellite genotyping. *Proceedings of the National Academy of Sciences (USA)* 99: 6080–6084.

McClintock, B., T. A. Kato Y., and A. Blumenschein. 1981. *Chromosome Constitution of Races of Maize.* Colegio de Postgraduados, Chapingo, Mexico.

Paterson, A. H., Y. R. Lin, Z. Li, K. F. Schertz, J. F. Doebley, S. R. M. Pinson, S. C. Liu, J. W. Stansel, and J. E. Irvine. 1995. Convergent domestication of cereal crops by independent mutations at corresponding genetic loci. *Science* 269: 1714–1718.

Piperno, D. R. and K. V. Flannery. 2001. The earliest archaeological maize (*Zea mays* L.) from highland Mexico: New accelerator mass spectrometry dates and their implications. *Proceedings of the National Academy of Sciences (USA)* 98: 2101–2103.

Piperno, D. R. and D. M. Pearsall. 1998. *The Origins of Agriculture in the Lowland Neotropics*. Academic Press, San Diego, CA, USA.

Rindos, D. 1984. *The Origins of Agriculture: An Evolutionary Perspective*. Academic Press, San Diego, CA, USA.

Sanchez, G. J., T. A. Kato Y., M. Aguilar S., J. M. Hernandez C., A. Lopez R., and J. A. Ruiz C. 1998. *Distribucion y Caracterizacion del Teocintle*. Instituto Nacional de Investigaciones Forestales, Agricolas y Pecuarias, Guadalajara, Mexico.

Tenaillon, M. I., M. C. Sawkins, A. D. Long, R. L. Gaut, J. F. Doebley, and B. S. Gaut. 2001. Patterns of DNA sequence polymorphism along chromosome 1 of maize (*Zea mays* ssp. *mays* L.). *Proceedings of the National Academy of Sciences (USA)* 98: 9161–9166.

Vigouroux, Y., M. McMullen, C. T. Hittinger, K. Houchins, L. Schulz, S. Kresovich, Y. Matsuoka, and J. Doebley. 2002. Identifying genes of agronomic importance in maize by screening microsatellites for evidence of selection during domestication. *Proceedings of the National Academy of Sciences (USA)* 99: 9650–9655.

Wang, R. L., A. Stec, J. Hey, L. Lukens, and J. Doebley. 1999. The limits of selection during maize domestication. *Nature* 398: 236–239.

Wellhausen, E. J., L. M. Roberts, E. Hernandez X., and P. C. Mangelsdorf. 1952. *Races of Maize in Mexico: Their Origin, Characteristics and Distribution*. Bussey Institute of Harvard University, Cambridge, MA, USA.

White, S. E. and J. F. Doebley. 1999. The molecular evolution of *terminal ear1*, a regulatory gene in the genus *Zea*. *Genetics* 153: 1455–1462.

Whitt, S. R., L. M. Wilson, M. I. Tenaillon, B. S. Gaut, and E. S. Buckler. 2002. Genetic diversity and selection in the maize starch pathway. *Proceedings of the National Academy of Sciences (USA)* 99: 12959–12962.

Wilkes, H. G. 1967. *Teosinte: The Closest Relative of Maize*. Bussey Institute of Harvard University, Cambridge, MA, USA.

Mary W. Eubanks

CHAPTER 5

Contributions of *Tripsacum* to Maize Diversity

Although more maize (*Zea mays* L.) is grown around the globe than any other crop today, scientists are still discovering how a wild grass with a small, few-seeded, shattering spike was transformed into the large maize ear with hundreds of kernels, a phenomenon unparalleled in the botanical kingdom. Under domestication maize lost its ability for self-propagation and became dependent on humans for survival. Therefore, the story of its biological evolution is tightly intertwined with cultural evolution. The maize genome, which is a diploidized allopolyploid (Gaut and Doebley, 1997) that contains many duplicate genes (Rhoades, 1951; Helentjaris et al., 1988) and large-scale chromosomal rearrangements (McClintock, 1984; Wilson et al., 1999), is as puzzling and complex as the morphogenesis of the ear is mysterious. The archaeological record thus far has been silent on this piece of the origin puzzle because the earliest remains found to date have all the basic characteristics of domesticated maize (Galinat, 1985; Eubanks, 2001a, 2001b, 2001c). This has been corroborated by analysis of ancient DNA, which revealed that early maize had the same alleles as modern maize (Jaenicke-Després et al., 2003). Elucidation of the sources and development of maize diversity with more than 300 landraces in Latin America depends on how well we can reconstruct its origin and trace the biocultural pathways of its radiation.

Origin of Maize

Evidence from molecular and crossing studies has resolved the contentious debate about the ancestry of maize by confirming that it originated from teosinte (*Zea* spp.), a wild grass endemic to Central America (Doebley, 1990; Eubanks, 1995). Two more questions regarding its origin recently have been resolved by new molecular evidence. A study of microsatellite data (Matsuoka et al., 2002) supports the hypothesis that the cradle of maize probably was in the highlands of southern Mesoamerica (MacNeish and Eubanks, 2000), and ancient DNA studies (Jaenicke-Després et al., 2003) support the hypothesis that teosinte was rapidly transformed into maize around 9000–7000 years ago (MacNeish and Eubanks, 2000). Although there is scientific consensus that maize traces its descent to teosinte, and many (Bennetzen et al., 2001) concur that it traces directly to *Z. mays* ssp. *parviglumis* Iltis & Doebley, others contend that a different teosinte species was involved in the origin of maize. Possible alternative species are *Z. mays mexicana* (Schrader) Iltis (Beadle, 1980; Kato Y., 1984; Galinat, 1988), *Z. luxurians* (Asch. & Dur.) Bird (Bird, 1979), or *Z. diploperennis* Iltis, Doebley & Guzmán (Wilkes, 1979). How then was the small, shattering teosinte spike transformed into the maize ear with many exposed kernels on a firm cob? The teosinte hypothesis states that accumulation of intrinsic mutations for a few key genes within annual teosinte (*Z. m.* ssp. *parviglumis*) resulted in the evolution of maize (Doebley, 1992). The recombination hypothesis proposes that maize arose from human selection of novel phenotypes among intergenomic recombinants between teosinte and gamagrass (*Tripsacum* spp.) (Eubanks, 1995; MacNeish and Eubanks, 2000). Understanding the evolutionary mechanisms that led to maize speciation and the role of humans in its transformation and dispersal will have important implications for identifying genetic resources for crop improvement (de Wet, 1979; Ladizinsky, 1989; Berthaud et al., 1996; Taylor, 2001; Eubanks, 2002a, 2002b). It is also relevant in assessing concerns about the flow of transgenes from genetically engineered corn into Mexican landraces and wild relatives (Ortiz-García and Ezcurra, 2003).

Zea **Taxonomy**

Maize, along with its wild relatives *Zea* spp. and *Tripsacum* spp., are in the Tripsacinae (Clayton, 1973, 1981), formerly Maydeae, an American subtribe of the Andropogoneae: warm season, tropical C4 grasses. The Tripsacinae are wind-pollinated and monoecious (i.e., they have separate male and female flowers on the same plant). In maize and wild *Zea* species the pollen is produced in tassels at the apices of the stalks, and the female flowers are in the leaf axils, whereas in *Tripsacum* species the staminate and pistillate flowers are borne on a single spike, with the male flowers above the female flowers. The geographic range of extant species of wild *Zea*, or teosinte, is west of the Sierra Madre Oriental Mountains in Mexico, Guatemala, and Honduras (see Eubanks, 2001c, figure 2.5 for a distribution map). These include *Z. mays* ssp. *huehuetenangensis* (Iltis & Doebley) Doebley, *Z. luxurians* (Asch. & Dur.) Bird, *Z. nicaraguensis* Iltis & Benz, *Z. perennis* (Hitch.) Mangelsdorf & Reeves, *Z. diploperennis* Iltis, Doebley & Guzmán, *Z. mays* ssp. *mexicana* (Schrader) Iltis, and *Z. mays* ssp. *parviglumis* Iltis & Doebley. With the exception of the perennial *Z. diploperennis* and *Z. perennis,* the teosintes are annuals. Apart from *Z. perennis,* a 40-chromosome tetraploid, the teosintes are diploid ($2n = 20$). They are naturally cross-fertile with each other and maize, and there is much introgression among *Zea* species (Wilkes, 1967). Some Mexican farmers still plant teosinte along the margins of their maize fields every few years because they believe it crosses with their maize and improves the hardiness of their crop (Wilkes, 1967).

Tripsacum Taxonomy

Tripsacum, commonly called gamagrass, is the sister genus of *Zea* and has a much broader geographic distribution, ranging from North America to South America. Gamagrass is adapted to a wide variety of habitats ranging from mountains to lowlands, temperate to tropical regions, and dry prairies to wetlands (see Eubanks, 2001c, figure 2.6, for a distribution map). There are at least 12 species of this rhizomatous perennial, which is a polyploid ($x = 18$) with 36–108 chromosomes (Randolph, 1970; de Wet et al., 1976, 1981; Brink and de Wet, 1983). *Tripsacum* is divided into two taxonomic sections: section *Tripsacum,* in which the paired staminate flowers are sessile, and section *Fasciculata,* in which, as in *Zea,* one of the staminate flowers of the pair is pedicellate and the other sessile (Hitchcock, 1906; Brink and de Wet, 1983). Section *Tripsacum* includes *T. andersonii* Gray ($2n = 64$), *T. australe* Cutler & Anderson ($2n = 36$), *T. bravum* Gray ($2n = 36, 72$), *T. cundinamarce* de Wet & Timothy ($2n = 36$), *T. dactyloides* (L.) L. ($2n = 36, 72$), *T. floridanum* Porter & Vasey ($2n = 36$), *T. latifolium* Hitchcock ($2n = 36$), *T. peruvianum* ($2n = 72, 90, 108$), and *T. zopilotense* ($2n = 36, 72$). Section *Fasciculata* includes *T. lanceolatum* Rupr. & Fournier ($2n = 72$), *T. laxum* Nash ($2n = 36$), *T. maizar* Hernandez & Randolph

(2n = 36, 72), *T. fasciculatum* Trin. & Ascherson (2n = 36), and *T. pilosum* Scribn. & Merrill (2n = 72).

Tripsacum Ethnobotany

Tripsacum is cultivated as fodder for guinea pigs and used to mark property boundaries in Mesoamerica and South America (de Wet et al., 1983). It is also an important forage plant grazed by bison and other large ungulates. There are still places in the United States where farmers maintain gamagrass hay meadows that were originally fenced off and preserved by pioneer farmers who first settled the land (Cadenhead, 1975; Eaheart, 1992). *Tripsacum* kernels have three times the protein of maize and are higher in linoleic acid and the amino acids glutamine, alanine, methionine, and leucine (Bargman et al., 1988). Florida hunters carry gamagrass kernels for trail food (Galinat and Craighead, 1964). Caches of gamagrass seeds found in Ozark bluff dweller sites and a dry cave in northeastern Mexico suggest that humans may have used gamagrass for food in prehistory (Gilmore, 1931; Mangelsdorf et al., 1967). This nutritious grain is currently under development for high-quality flour and cooking oil.

Zea–Tripsacum Crossability

The F_1 progeny of crosses between maize and gamagrass have a high degree of female sterility and are male sterile (Mangelsdorf, 1974). The typical chromosome constitution of maize–*Tripsacum* hybrids includes the full gametic complement from both parents, that is, 10 chromosomes from maize and 36 chromosomes from *Tripsacum* (Mangelsdorf and Reeves, 1939). These plants appear more like *Tripsacum* than maize and are perennial. In rare cases, hybrids contain a total of 20 chromosomes (James, 1979). Such plants are more maizelike and are annual. No crosses between *Tripsacum* and teosinte had succeeded until Eubanks (1995) crossed eastern gamagrass (*T. dactyloides*) with the diploid perennial teosinte (*Z. diploperennis*) that was discovered on the verge of extinction in the mountains of Jalisco, Mexico, in the late 1970s (Iltis et al., 1979). The 2n = 20 teosinte–gamagrass hybrids are fully fertile. They are also cross-fertile with maize and thus provide a genetic bridge to move *Tripsacum* genes into maize (Eubanks, 2002a, 2002b). This is a significant breakthrough for corn improvement because *Tripsacum* is a rich resource of beneficial traits, such as pest and disease resistance, drought tolerance, adaptation to acid soils and waterlogged soils, and salt tolerance (de Wet, 1979), that can be introduced into maize by recurrent backcross selection using conventional breeding methods that avoid the high costs, regulations, and other issues associated with transgenic corn (Eubanks, 2002a, 2002b).

Experimental Crosses

Crosses between diploid perennial teosinte (*Z. diploperennis*, 2n = 20) and eastern gamagrass (*T. dactyloides*, 2n = 36, 72) produced phenotypes closely resembling ancient maize remains (MacNeish and Eubanks, 2000; Eubanks, 2001a, 2001b, 2001c). Because there have been no finds of teosinte with key mutations involved in its transformation into maize (Iltis, 2000), this is the first empirical demonstration of how teosinte could have been transformed into maize via *Tripsacum* introgression. The cross can be made with either genus as the pollen recipient, and some segregating phenotypes exhibit certain "missing links" (Galinat, 1985) in the transition to maize: the rachis segments are partially fused and do not break apart easily; there are two kernels per rachis segment instead of a single kernel, as in teosinte; and the kernels are slightly exposed at the tips (Eubanks, 2001b, figure 10). These F_1 recombinants simulate reconstructed prototypes of "wild maize" because they demonstrate an intermediate form in the transition from teosinte to maize not found in the archaeological record. Phenotypes resembling the oldest archaeological specimens from the valleys of Tehuacán and Oaxaca in southern Mexico (Eubanks, 2001b, figures 11, 13, and 14) were observed in a population of segregating F_2 plants in which the pollen recipient was gamagrass (*T. dactyloides*). This suggests that gamagrass may confer a maternal inheritance effect on gene expression that converts the basal glume into the cupule, changes the hard outer glumes into soft, papery chaff, exposing the kernels and making them easy to shell, and converts the distichously arranged spike into a multirowed ear.

Comparative Genomics

Although there have been numerous molecular studies of maize, teosinte, and *Tripsacum,* sampling has varied significantly from one study to the next, and there has been incongruence between data sets. For example, Matsuoka et al. (2002) sampled 193 accessions of maize but only one plant from each accession. Their sampling of teosinte included one plant from each of 33 accessions of *Z. m.* ssp. *mexicana,* 34 of *Z. m.* ssp. *parviglumis,* and four plants of *Z. m.* ssp. *huehuetenangensis.* No other teosinte or *Tripsacum* species were included in the microsatellite analysis. Noteworthy for the question of *Tripsacum* introgression into *Zea* is that most studies have focused almost exclusively on *Zea* (see Eubanks, 2001c, for a review). Therefore, in order to collect requisite data for contrasting analyses of the teosinte and recombination hypotheses, a preliminary comparative genomics study was conducted to examine allelic diversity in four ancient indigenous maize races, six teosinte (*Zea*) species (the seventh *Zea* species, *Z. nicaraguensis,* was discovered after this work was completed), and seven gamagrass (*Tripsacum*) species. Four extant popcorns that resemble maize identified in the archaeological record (Eubanks, 1999) were fingerprinted: Chapalote and Nal Tel from Mexico and Pollo and Pira from South America. Six teosinte species and seven *Tripsacum* species were genotyped. The DNA from 10–13 plants of one accession of the following *Zea* species was sampled: *Z. m.* ssp. *parviglumis, Z. m.* ssp. *mexicana, Z. m.* ssp. *huehuetenangensis, Z. luxurians, Z. diploperennis,* and *Z. perennis.* One clonal colony of each of the following *Tripsacum* species was sampled: *T. dactyloides* and *T. lanceolatum* from North America, *T. maizar* from Guatemala, and *T. dactyloides meridionale, T. andersonii, T. peruvianum,* and *T. cundinmarce* from South America. One accession of *Manisuris selloana* (Hack.) Kuntz, another grass in the American Andropogoneae, served as the outgroup for cladistic analysis. See table 5.1 for taxa provenance and accession information. Because sampling was restricted to a single accession of each taxon, the *Zea* species were selected from regions where they grow in greatest isolation from maize. These taxa therefore are expected to have the least number of introgressed alleles from other *Zea* species. Theoretically, they are the purest, most representative populations of the extant species. The *Tripsacum* species were selected to represent a wide geographic range. The four popcorns were selected because among extant land races their genomes probably most closely approximate the ancient maize gene pool.

The DNA fingerprinting method restriction fragment length polymorphism (RFLP) genotyping was chosen because of its high degree of accuracy and diagnostic power in maize (Helentjaris et al., 1986). This DNA fingerprinting technique is routinely used for genetic identity analysis of closely related species, to estimate genetic distance, to determine paternity, and to complement conventional pedigree records in commercial hybrid production (Melchinger et al., 1991; Smith and Smith, 1992; Messmer et al., 1993). Bulked total genomic DNA harvested from plants of each species grown in a

Table 5.1 Taxa Included in the Comparative Genomic Analysis

Taxon	Region	Source	Accession
Z. mays ssp. *mays*			
Nal Tel	Mexico	M. M. Goodman	"Yuc7, 72–73"
Chapalote	Mexico	M. M. Goodman	"Sin2, 70–75"
Pira	Colombia	USDA	P.I. 44512
Pollo	Colombia	M. M. Goodman	71–72
Z. m. ssp. *parviglumis*	Mexico	USDA	P.I. 384061
Z. m. ssp. *mexicana*	Mexico D. F.	USDA	P.I. 566683
Z. huehuetenangensis	Guatemala	USDA	P.I. 441934
Z. luxurians	Guatemala	USDA	P.I. 306615
Z. diploperennis	Mexico	H. H. Iltis	Iltis no. 1250
Z. perennis	Mexico	USDA	Ames 21875
T. dactyloides	Kansas	USDA	MIA 34680
T. d. meridionale	Colombia	USDA	MIA 34597
T. andersonii	Venezuela	USDA	MIA 34435
T. maizar	Guatemala	USDA	MIA 34744
T. lanceolatum	Arizona	USDA	MIA 34713
T. peruvianum	Peru	USDA	MIA 34503
T. cunidnamarce	Colombia	USDA	MIA 34631

greenhouse was digested using the restriction enzymes *Eco*RI, *Eco*RV, *Hind*III, and *Bam*HI. These restriction enzymes are six-base cutters that produced 1–10 bands across all of the taxa surveyed. The Southern blots were probed with 140 publicly available molecular markers mapped to the 10 linkage groups of maize (Gardiner et al., 1993) and six mitochondrial loci. Figure 5.1 illustrates the order and approximate locus of each nuclear probe on its respective *Zea* linkage group. Some unmapped nuclear probes and the mitochondrial markers are not indicated in the figure. Each locus represents a gene based on clone identification because the molecular markers were mapped by recombination analyses based on proof of the identity of a clone (Neuffer et al., 1997). Each polymorphic band is therefore equivalent to an allele. Such broad genomic coverage is not practical using DNA sequencing.

The operating assumption of the DNA fingerprinting test is that if maize is directly descended from the teosinte *Z. m.* ssp. *parviglumis,* then maize and teosinte are expected to exclusively share a large proportion of the same alleles not present in other species. On the other hand, if

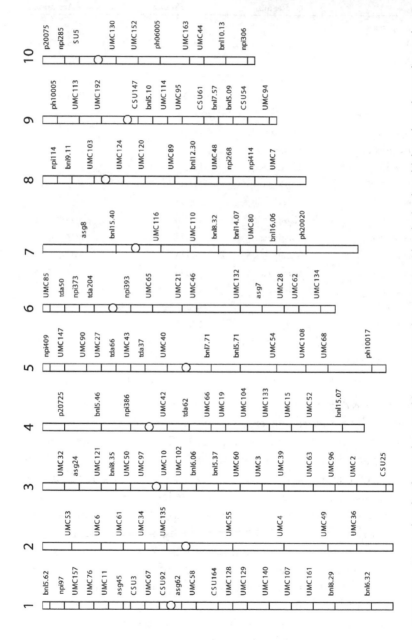

FIGURE 5.1 Mapped molecular markers used in RFLP genotyping. Unmapped and mitochondrial markers that make up the 140-marker genome coverage not shown.

maize and *Tripsacum* share alleles not present in teosinte, then those maize alleles could have been derived from one or more *Tripsacum* progenitors. Though not conclusive, the presence of alleles specific to both teosinte and *Tripsacum* in maize would support a maize hybrid origin.

Phylogenetic Analysis

A matrix of the nuclear and mitochondrial data was constructed in which each character state (i.e., band) was coded as present (1) or absent (0) and input into a NEXUS file for maximum parsimony analysis. A heuristic search was performed with default search parameters in PAUP version 4.0b10 (Swofford, 2002). The tree was rooted with *Manisuris* as the outgroup. Bootstrap support was determined using 1000 replicates. Three most parsimonious trees (MPTs) were recovered with 2876 steps, 858 informative characters, and a consistency index (C.I.) of 0.497. One of three MPTs (figure 5.2) shows a maize clade that resolves the two Mexican and two South American maize races in accordance with their respective geographic areas, and that clade is sister to the Mexican annual teosintes, *Z. m.* ssp. *mexicana* and *Z. m.* ssp. *parviglumis*. *Z. huehuetenangensis* from highland Guatemala is sister to this clade, and *Z. luxurians* from southern Guatemala and the two perennial teosintes form a clade sister to it. The two perennial teosintes are grouped together. These divisions are reasonably congruent with restriction site variation in the *Zea* chloroplast genome (Doebley et al., 1987) and Buckler and Holtsford's (1996) phylogeny based on nuclear ribosomal internal transcribed spacer sequence data. The shallow interior nodes support a rapid radiation of maize and the Mexican annual teosintes, as indicated by recent ancient DNA evidence (Jaenicke-Després et al., 2003). The long terminal branches indicate that this radiation was followed by much differentiation of the subspecies, perhaps through human selection. Each taxon (terminal) has many changes not found in the sister taxa. There is 100% bootstrap support for a monophyletic maize clade, 81% support for the South American group within it, and strong support (92%) for a *Zea* clade that includes maize, *Z. parviglumis, Z. mexicana,* and *Z. huehuetenangensis*. However, resolution among these three *Zea* taxa is not robust. Bootstrap support for a separate *Tripsacum* clade is 71%. Within the clade there is 100% bootstrap support for a subclade containing diploid (2x) and tetraploid (4x) *T. dactyloides* from North America and 89% support for the clade containing the three *Tripsacum* species from

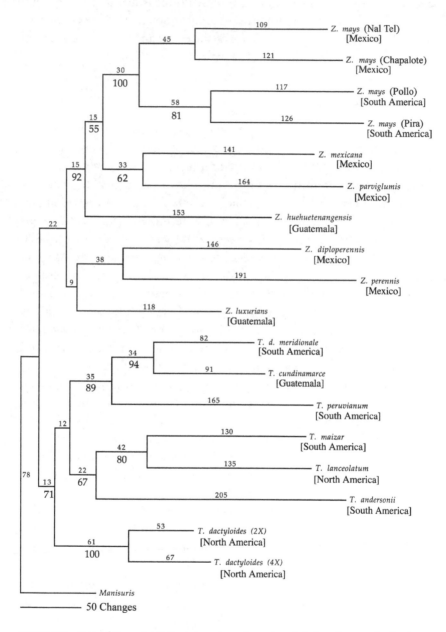

FIGURE 5.2 One of 3 most parsimonious trees based on total nuclear and mitochondrial polymorphisms from RFLP genotyping. Branch lengths are denoted above and bootstrap values below the branches.

northwestern South America. The grouping of the two species from section *Fasciculatum, T. lanceolatum* from Arizona and *T. maizar* from South America (80% bootstrap), appears to support the subsectional divisions within *Tripsacum* (Hitchcock, 1906; Brink and de Wet, 1983; Li et al., 2000). The low C.I. of 0.497 signals a fair degree of homoplasy in the data set. This may result from lineage sorting, hybridization, lack of enough informative characters, or incomplete population or taxon sampling. Broader taxon and character sampling is needed to sort out the history of these taxa. These data are not sufficient to confirm or reject the teosinte hypothesis or the recombination hypothesis.

Genetic Similarity

Another way to examine the data set that may shed more light on the evolutionary history of maize is to look at the genetic relatedness between sampled taxa. Genetic similarity was calculated as the percentage of bands shared between two taxa. The proportion of shared alleles was calculated by multiplying the number of polymorphic bands shared between taxon X and taxon Y by 2 and dividing that quantity by the sum of the total number of alleles in X and Y (Avise, 1994, see p. 95). The results for the total RFLP data set including the mitochondrial polymorphisms are summarized in table 5.2 and those for the mitochondrial alleles alone are in table 5.3. Looking at the total DNA evidence, the wild *Zea* most like maize appears to be *Z. m.* ssp. *mexicana,* which shares 41% of its alleles with Pollo and 40% with Pira, maize races from South America. Because these South American races are geographically isolated from Mexican teosintes, this finding appears to support the hypothesis that *Z. m.* ssp. *mexicana* is the ancestor of maize (Beadle, 1939; Galinat, 1977). However, it should be noted that at 40% the frequency of alleles shared between *Z. m.* ssp. *parviglumis* and the ancient indigenous Mexican race Chapalote is nearly as high and not significantly different. This could be interpreted as lending equal support to the hypothesis that teosinte from the Rio Balsas region of Mexico is the progenitor (Doebley, 1990), or it may indicate that there was more than one origin of maize (Kato Y., 1984; Galinat, 1992). Alternatively, because of cross-fertility between maize and teosinte, which grows in and at the margins of maize fields throughout Mexico today, modern introgressive hybridization cannot be ruled out as contributing to the genetic similarities.

Table 5.2 Genetic Similarity Matrix of Shared Alleles in Combined RFLP and Mitochondrial Data Set

Total Alleles	Zd	Zmex	Zlux	Zhue	Zpar	Zper	Td	Tdmer	Tand	Tcun	Tmaiz	Tper	Tlan	Msel	Nal Tel	Chap	Pollo	Pira
Z. diploperennis	**227**	0.33	0.34	0.27	0.27	0.36	0.23	0.19	0.22	0.21	0.21	0.19	0.14	0.11	0.26	0.29	0.28	0.27
Z. mexicana	80	**255**	0.31	0.31	0.23	0.37	0.21	0.19	0.23	0.17	0.2	0.21	0.17	0.12	0.37	0.38	0.41	0.4
Z. luxurians	68	67	**177**	0.3	0.33	0.33	0.14	0.21	0.27	0.19	0.18	0.2	0.12	0.11	0.24	0.25	0.25	0.26
Z. huehuetenangensis	58	79	57	**210**	0.35	0.31	0.19	0.21	0.23	0.16	0.17	0.2	0.15	0.12	0.3	0.33	0.29	0.31
Z. parviglumis	67	113	73	82	**266**	0.38	0.18	0.24	0.26	0.19	0.22	0.24	0.17	0.13	0.34	0.4	0.38	0.35
Z. perennis	86	94	71	72	99	**257**	0.18	0.2	0.31	0.16	0.24	0.24	0.18	0.15	0.28	0.31	0.31	0.31
T. dactyloides	46	45	25	37	39	39	**175**	0.25	0.25	0.23	0.27	0.25	0.16	0.13	0.16	0.15	0.12	0.14
T. d. meridionale	39	41	37	41	54	43	45	**183**	0.28	0.55	0.25	0.38	0.2	0.17	0.17	0.15	0.15	0.17
T. andersonii	54	59	58	53	67	79	54	61	**256**	0.29	0.38	0.31	0.32	0.17	0.16	0.21	0.2	0.22
T. cundinamarce	45	39	36	33	43	37	43	104	66	**194**	0.25	0.46	0.19	0.16	0.17	0.15	0.15	0.16
T. maizar	47	48	37	37	55	59	54	51	93	54	**232**	0.36	0.41	0.1	0.16	0.18	0.16	0.18
T. peruvianum	46	53	43	45	63	60	53	83	79	102	86	**252**	0.34	0.19	0.17	0.19	0.17	0.19
T. lanceolatum	30	40	53	31	42	42	32	40	75	38	92	80	**216**	0.19	0.12	0.18	0.13	0.15
Manisuris selloana	14	19	14	17	21	24	16	21	28	21	15	30	27	**70**	0.12	0.12	0.12	0.11
Nal Tel	61	91	51	67	86	69	34	36	39	37	38	43	27	19	**244**	0.54	0.52	0.45
Chapalote	74	101	56	80	109	82	33	35	55	36	46	49	45	20	142	**279**	0.54	0.49
Pollo	71	109	56	71	102	82	27	35	54	35	41	45	33	20	135	151	**276**	0.51
Pira	68	106	59	76	94	82	32	38	57	38	46	50	36	18	117	136	140	**274**

The bold numbers on the diagonal are the total number of alleles in a particular taxon. Numbers below the diagonal are the total number of alleles shared between taxa. Numbers above the diagonal indicate number of alleles shared between taxa. Numbers above the diagonal indicate frequency of alleles shared between taxa.

Zd = Z. diploperennis, Zmex = Z. mexicana, Zlux = Z. luxurians, Zhue = Z. huehuetenangensis, Zpar = Z. parviglumis, Zper = Z. perennis, Td = T. dactyloides, Tdmer = T. d. meridionale, Tand = T. andersonii, Tcun = T. cundinamarce, Tmaiz = T. maizar, Tper = T. peruvianum, Tlan = T. lanceolatum, Msel = M. selloana, Chap = Chapalote.

Table 5.3 Genetic Similarity Matrix for Shared Mitochondrial Alleles

Mitochondrial Alleles	Zd	Zmex	Zlux	Zhue	Zpar	Zper	Td	Tdmer	Tand	Tcun	Tmaiz	Tper	Tlan	Msel	Nal Tel	Chap	Pollo	Pira
Z. diploperennis	**18**	0.432	0.541	0.368	0.452	0.27	0.15	0.19	0.368	0.36	0.457	0.302	0.222	0.09	0.34	0.38	0.294	0.286
Z. mexicana	8	**19**	0.526	0.821	0.833	0.62	0.21	0.25	0.256	0.14	0.389	0.392	0.324	0.25	0.67	0.7	0.51	0.555
Z. luxurians	10	10	**19**	0.513	0.441	0.59	0.21	0.318	0.359	0.21	0.34	0.343	0.22	0.17	0.43	0.402	0.229	0.33
Z. huehuetenangensis	7	16	10	**20**	0.809	0.614	0.21	0.3	0.3	0.2	0.38	0.476	0.368	0.32	0.7	0.732	0.44	0.601
Z. parviglumis	9	17	9	17	**22**	0.628	0.26	0.296	0.333	0.203	0.363	0.455	0.404	0.288	0.62	0.651	0.428	0.466
Z. perennis	6	14	10	14	15	**26**	0.23	0.218	0.307	0.187	0.381	0.46	0.324	0.175	0.49	0.472	0.443	0.428
T. dactyloides	2	3	3	3	4	4	**9**	0.36	0.28	0.319	0.46	0.357	0.3	0	0.13	0.13	0.08	0.23
T. d. meridionale	3	4	5	5	5	4	4	**13**	0.497	0.702	0.541	0.473	0.461	0	0.17	0.182	0	0.135
T. andersonii	7	5	7	6	7	8	4	8	**20**	0.571	0.543	0.383	0.476	0.2	0.09	0.098	0.112	0.054
T. cundinamarce	5	2	3	3	3	3	3	8	8	**10**	0.615	0.608	0.448	0.143	0	0	0	0
T. maizar	8	7	6	7	8	10	6	8	10	8	**17**	0.622	0.692	0.109	0.2	0.213	0.183	0.235
T. peruvianum	6	8	7	10	11	8	5	8	8	9	12	**22**	0.707	0.192	0.36	0.372	0.214	0.207
T. lanceolatum	4	6	4	7	8	7	4	7	9	6	9	14	**18**	0.143	0.2	0.206	0.117	0.171
Manisuris selloana	1	3	2	4	3	2	0	0	2	1	1	2	1	**5**	0.19	0.196	0.225	0.217
Nal Tel	7	14	9	15	14	12	2	3	2	0	4	8	4	2	**23**	0.955	0.632	0.711
Chapalote	7	14	8	15	14	11	2	3	2	0	4	8	4	2	21	**21**	0.601	0.691
Pollo	5	9	4	8	8	9	1	0	0	0	3	4	2	2	12	11	**16**	0.727
Pira	5	10	6	11	9	9	3	2	1	0	4	4	3	2	14	13	12	**17**

The bold numbers on the diagonal are the total number of alleles in a particular taxon. Numbers below the diagonal indicate frequency of alleles shared between taxa. Numbers above the diagonal indicate number of alleles shared between taxa.

Zd = Z. diploperennis, Zmex = Z. mexicana, Zlux = Z. luxurians, Zhue = Z. huehuetenangensis, Zpar = Z. parviglumis, Zper = Z. perennis, Td = T. dactyloides, Tdmer = T. d. meridionale, Tand = T. andersonii, Tcun = T. cundinamarce, Tmaiz = T. maizar, Tper = T. peruvianum, Tlan = T. lanceolatum, Msel = M. selloana, Chap = Chapalote.

Maize domestication genes and the number of alleles maize shares with teosinte and *Tripsacum* at these genetic loci are summarized in table 5.4. The domestication genes include *tb1* for teosinte branched (Burnham and Yagyu, 1961; Doebley et al., 1997), *tr1* for two-ranked (Langham, 1940; Rodgers, 1950), *te1/pd1* for terminal ear and paired female spikelets (Langham, 1940; Rodgers, 1950; Matthews et al., 1974), *tga1* for teosinte glume architecture (Dorweiler et al., 1993; Galinat, 1970), *su1* for sugary (Jaenicke-Després et al., 2003), *tu1* for tunicate (Mangelsdorf and Galinat, 1964), *ri1/ph1* for rind and pith abscission (Galinat, 1975, 1978), *pbf1* for prolamine box binding factor (Jaenicke-Després et al., 2003), multiple effects (Mangelsdorf, 1947; Doebley and Stec, 1991), and four-ranked (Mangelsdorf, 1947). Of the 104 alleles in maize at these loci, 24 were not found in any of the wild relatives. Seven of those 24 polymorphisms, which appear to be specific to maize (indicated in table 5.4 with an asterisk: UMC140-M2, UMC61-MI, UMC50-M2, BNL5.37-M3, BNL6.06-MI, UMC40-M5, AND UMC52-M3), are formed as new recombinant alleles in teosinte–*Tripsacum* hybrids. Recombinant alleles refer to new bands intermediate in size between bands found in the parents. Because there is no loss or gain in the number of bands inherited from the parents, the new alleles are not created by point mutations in the DNA sequence homologous to the RFLP probe or in the restriction cut sites. A possible explanation for the formation of the new recombinant alleles is unequal crossing over in repetitive DNA that accommodates differences in parental chromosome architecture and facilitates proper pairing during cell division. Repeated recovery of the same recombinant alleles in crosses between different *Tripsacum* and *Z. diploperennis* individuals from different populations indicates that the mechanism for this genomic reorganization is highly precise and suggests that it could be the source of the primordial genes of maize domestication. Table 5.5 summarizes the allele frequency distributions for the domestication genes. With 29 domestication alleles shared between *Z. m.* ssp. *mexicana* and *Z. m.* ssp. *parviglumis* and maize, the Mexican annual teosintes stand out as most similar to maize. *T. peruvianum* shares the second highest number of domestication alleles with maize at 23. It appears that *Tripsacum* introgression may be pronounced in genomic regions carrying domestication genes. This finding seems to corroborate other indicators for a South American connection and role for *Tripsacum* maternal inheritance in the early evolution of maize.

Table 5.4 Maize Alleles at Loci for Domestication Genes Shared with Other *Zea* and *Tripsacum* Taxa

Alleles	Taxa with Shared Maize Polymorphism
UMC107 (*tb1*)	
M1	*Z. m.* ssp. *parviglumis, Z. huehuetenangensis*
UMC140 (*tb1*)	
M1	*T. andersonii, T. maizar*
M2	*
M3	*Z. m.* ssp. *mexicana, T. cundinamarce, T. peruvianum*
M4	*Z. m.* ssp. *mexicana, Z. m.* ssp. *parviglumis, Z. huehuetenangensis, Z. perennis, T. andersonii*
UMC6 (*tr1*)	
M1	*Z. m.* ssp. *parviglumis, Z. luxurians*
M2	*Z. huehuetenangensis, Z. diploperennis, T. dactyloides, T. d. meridionale, T. cundinamarce, T. maizar, T. perunvianum*
M3	*Z. m.* ssp. *parviglumis, Z. luxurians, T. maizar, T. peruvianum*
UMC34 (*tr1*)	
M1	Allele did not appear in any other taxa included in study
M2	Allele did not appear in any other taxa included in study
M3	*Z. m.* ssp. *parviglumis, Z. m.* ssp. *mexicana*
UMC53 (*tr1*)	
M1	*Z. m.* ssp. *mexicana, Z. m.* ssp. *parviglumis, Z. huehuetenangensis, Z. luxurians, Z. perennis, T. dactyloides, T. andersonii*
M2	Allele did not appear in any other taxa included in study
UMC61 (*tr1*)	
M1	*
M2	*Z. diploperennis, T. maizar*
M3	*Z. m.* ssp. *parviglumis, Z. perennis, T. d. meridionale, T. cundinamarce, T. peruvianum*
M4	*Z. m.* ssp. *mexicana, Z. huehuetenangensis, Z. luxurians, T. andersonii, T. maizar, T. peruvianum*
UMC50 (*te1*)	
M1	*Z. luxurians, Z. diploperennis, T. peruvianum*
M2	*
M3	*T. dactyloides, T. peruvianum*
M4	*Z. m.* ssp. *mexicana, Z. perennis, T. maizar*
M5	*Z. m.* ssp. *parviglumis, Z. huehuetenangensis, T. maizar*
M6	*Z. m.* ssp. *parviglumis, Z. m.* ssp. *mexicana, Z. huehuetenangensis, Z. luxurians, Z. diploperennis, Z. perennis, T. dactyloides, T. d. meridionale, T. andersonii, T. cundinamarce, T. maizar, T. peruvianum, T. lanceolatum*
BNL5.37 (*te1*)	
M1	Allele did not appear in any other taxa included in study

(continued)

Table 5.4 (continued)

Alleles	Taxa with Shared Maize Polymorphism
M2	Allele did not appear in any other taxa included in study
M3	*
UMC102 (*te1*)	
M1	*Z. m.* ssp. *parviglumis, T. andersonii*
M2	*Z. huehuetenangensis*
M3	Allele did not appear in any other taxa included in study
M4	*Z. m.* ssp. *mexicana, Z. diploperennis*
BNL6.06 (*te1*)	
M1	*
M2	*Z. m.* ssp. *parviglumis, Z. m.* ssp. *mexicana, Z. huehuetenangensis, Z. diploperennis, Z. perennis, T. d. meridionale, T. andersonii, T. maizar, T. peruvianum*
UMC63 (*te1/pd1*)	
M1	*Z. diploperennis, Z. perennis*
M2	*Z. m.* ssp. *mexicana*
M3	*Z. m.* ssp. *parviglumis, Z. m.* ssp. *mexicana, Z. diploperennis, T. dactyloides, T. d. meridionale, T. andersonii, T. cundinamarce, T. maizar*
M4	*Z. perennis, T. cundinamarce, T. peruvianum*
M5	Allele did not appear in any other taxa included in study
UMC42 (*tga1, su1*)	
M1	*Z. m.* ssp. *mexicana, T. peruvianum*
M2	*Z. m.* ssp. *mexicana, Z. m.* ssp. *parviglumis, T. peruvianum*
M3	*Z. m.* ssp. *mexicana*
BNL5.46 (*tga1*)	
M1	*Z. m.* ssp. *parviglumis, Z. m.* ssp. *mexicana, Z. luxurians*
M2	*Z. m.* ssp. *mexicana, T. peruvianum, T. lanceolatum*
M3	*Z. huehuetenangensis, T. andersonii, T. maizar, T. lanceolatum*
M4	*Z. m.* ssp. *parviglumis, Z. huehuetenangensis, Z. perennis*
M5	*T. d. meridionale*
M6	*Z. m.* ssp. *parviglumis, Z.* m. ssp. *mexicana, Z. luxurians, Z. diploperennis, Z. perennis, T. dactyloides, T. d. meridionale, T. cundinamarce, T. maizar, T. perunvianum*
UMC66 (*su1/tu1*)	
M1	*Z. huehuetenangensis*
M2	*Z. m.* ssp. *parviglumis, Z. perennis, T. andersonii, T. lanceolatum*
Tda62 (*su1*)	
M1	Allele did not appear in any other taxa included in study
M3	*Z. diploperennis, Z. m.* ssp. *mexicana, Z. m.* ssp. *parviglumis, Z. huehuetenangensis, Z. luxurians, Z. perennis*

(continued)

Table 5.4 (continued)

Alleles	Taxa with Shared Maize Polymorphism
M2	*Z. diploperennis, T. dactyloides, T. andersonii, T. cundinamarce*
M4	*Z. diploperennis, Z. m.* ssp. *mexicana, Z. m.* ssp. *parviglumis, Z. huehuetenangensis, Z. luxurians, Z. perennis, T. dactyloides, T. d. meridionale, T. andersonii, T. cundinamarce, T. maizar, T. peruvianum*
ph20725 (*ri1/ph1*)	
M1	*T. dactyloides*
M2	Allele did not appear in any other taxa included in study
M3	*Z. perennis, T. maizar, T. peruvianum, T. lanceolatum*
M4	*Z. m.* ssp. *mexicana, Z. huehuetenangensis, Z. luxurians, Z. diploperennis, T. dactyloides*
UMC55 (*pbf1*)	
M1	*Z. m.* ssp. *mexicana, Z. m.* ssp. *parviglumis, Z. huehuetenangensis, Z. perennis, T. dactyloides, T. d. meridionale, T. andersonii*
M2	*Z. diploperennis, Z. perennis*
UMC15 (multiple effects)	
M1	*Z. m.* ssp. *mexicana, T. andersonii, T. peruvianum*
M2	*Z. luxurians, T. dactyloides, T. maizar*
M3	*Z. huehuetenangensis, T. d. meridionale, T. andersonii, T. cundinamarce*
M4	*Z. m.* ssp. *parviglumis*
UMC40 (multiple effects)	
M1	*T. dactyloides*
M2	*T. d. meridionale*
M3	*T. dactyloides*
M4	*Z. huehuetenangensis, Z. diploperennis*
M5	*
M6	*T. dactyloides*
UMC52 (multiple effects)	
M1	Allele did not appear in any other taxa included instudy
M2	*Z. m.* ssp. *parviglumis, Z. huehuetenangensis*
M3	*
M4	Allele did not appear in any other taxa included in study
M5	*Z. m.* ssp. *mexicana, Z. diploperennis*
npi409 (multiple effects)	
M1	*Z. m.* ssp. *mexicana, Z. diploperennis, Z. perennis*
M2	*T. dactyloides, T. d. meridionale, T. andersonii, T. cundinamarce, T. lanceolatum*
M2	*T. dactyloides*
M3	*T. maizar*
M4	Allele did not appear in any other taxa included in study

(*continued*)

Table 5.4 (continued)

Alleles	Taxa with Shared Maize Polymorphism
Tda66 (multiple effects)	
M1	Allele did not appear in any other taxa included in study
M5	Allele did not appear in any other taxa included in study
M6	*T. peruvianum, T. lanceolatum*
M7	*Z. m.* ssp. *parviglumis, Z. m.* ssp. *mexicana, Z. huehuetenangensis, Z. luxurians, Z. diploperennis, Z. perennis, T. dactyloides*
UMC27 (multiple effects)	
M1	*Z. m.* ssp. *parviglumis, T. d. meridionale, T. peruvianum*
M2	Allele did not appear in any other taxa included in study
M3	*T. cundinamarce, T. peruvinaum*
M4	*Z. m.* ssp. *mexicana, Z. luxurians, T. d. meridionale, T. cundinamarce*
M5	*Z. m.* ssp. *parviglumis, Z. huehuetenangensis, T. peruvianum*
M6	*Z. perennis*
UMC90 (multiple effects)	
M1	*T. dactyloides, T. d. meridionale, T. maizar*
M2	*T. peruvianum*
M3	*T. lanceolatum*
M4	*Z. m.* ssp. *parviglumis, Z. m.* ssp. *mexicana, Z. diploperennis, Z. perennis*
M5	*Z. m.* ssp. *parviglumis, Z. m.* ssp. *mexicana, Z. diploperennis, Z. perennis*
M6	*Z. m.* ssp. *mexicana, Z. huehuetenangensis, Z. luxurians, Z. diploperennis, T. andersonii*
M7	*Z. m.* ssp. *parviglumis*
UMC114 (4-ranked)	
M1	*Z. diploperennis*
M2	*T. lanceolatum*
M3	*T. d. meridionale, T. cundinamarce, T. maizar, T. peruvianum*
M4	*Z. m.* ssp. *mexicana*
M5	*Z. m.* ssp. *parviglumis*
UMC95 (4-ranked)	
M1	Allele did not appear in any other taxa included in study
M2	Allele did not appear in any other taxa included in study
M3	*Z. luxurians, T. d. meridionale*
M4	*Z. m.* ssp. *parviglumis, T. andersonii, T. lanceolatum*

* RFLP band was recovered in cross between perennial teosinte and eastern gama grass.

Table 5.5 Distribution of Alleles Among Key Genes in Maize Domestication

Taxa	Total Alleles	*tbl*	*trl*	*tel*	*tgal*	*sul/tul*	*ril/phl*	*pbfl*	Multiple Effects	4-Ranked
Maize	104	5	12	20	9	6	4	2	37	9
Z. m. ssp. *parviglumis*	29	2	5	5	4	2	0	1	8	2
Z. m. ssp. *mexicana*	29	2	3	6	6	1	1	1	8	1
Z. *huehuetenangensis*	21	2	3	4	2	2	1	1	6	0
Z. *luxurians*	14	0	4	2	2	1	1	0	4	0
Z. *diploperennis*	20	0	2	6	1	2	1	1	6	1
Z. *perennis*	20	1	2	5	2	2	1	2	5	0
T. *dactyloides*	18	0	2	3	1	1	2	1	8	0
T. *d. meridionale*	20	0	2	5	2	2	0	1	6	2
T. *andersonii*	18	2	2	4	1	3	0	1	4	1
T. *cundinamarce*	14	1	2	3	1	2	0	0	4	1
T. *maizar*	18	1	4	5	2	1	1	0	3	1
T. *peruvianum*	23	1	4	5	4	1	1	0	6	1
T. *lanceolatun*	9	0	0	0	2	1	1	0	3	2

An intriguing finding is that for every *Tripsacum* species, there are cases in which the frequency of shared alleles is closer to a *Zea* species than to other *Tripsacum* species. A striking example of this phenomenon is *T. peruvianum*, which shares 48% of its mitochondrial alleles with *Z. huehuetenangensis*. This suggests that these taxa have hybridized at some time in the past, or they have a common maternal ancestor. The recombination hypothesis (MacNeish and Eubanks, 2000), which proposes that maize is derived from hybridization between teosinte and *Tripsacum* with maternal descent through *Tripsacum,* can be examined by looking at the mitochondrial DNA frequencies for maize. With 73% of its mitochondrial alleles shared with Chapalote and 70% shared with Nal Tel, the maternally inherited DNA of *Z. huehuetenangensis* is more like maize than either *Z. m.* ssp. *mexicana* or *Z. m.* ssp. *parviglumis*. In addition to lending support to the recombination hypothesis, it raises the possibility of movement of *Tripsacum* from South America into highland Guatemala, where it hybridized with teosinte and produced the initial recombinant genetic diversity that provided the foundation for maize domestication. Alternatively, because *Tripsacum* species from Guatemala were not included in the DNA fingerprinting assays, natural introgression between

FIGURE 5.3   *Z. diploperennis × T. laxum* F$_1$ hybrid.

Tripsacum and teosinte endemic to Guatemala, where large stands of these taxa grow sympatrically (Kempton and Popenoe, 1937; Galinat, 1976), could have provided the initial influx of recombinant alleles leading to domesticated maize. Feasibility of the latter scenario is underscored by experimental hybrids derived from crossing *Z. diploperennis* with *T. laxum* from Guatemala (figures 5.3 and 5.4).

FIGURE 5.4   Inflorescence of *Z. diploperennis* × *T. laxum* F$_1$ hybrid.

Chromosomal Structure of Maize and Wild Relatives

Stebbins (1950) proposed that introgressive hybridization is as potent an evolutionary mechanism as divergence through mutation, recombination, and natural selection. An important difference is that the genes enter the genome of one species through transfer from another species across an isolating reproductive barrier. Another difference is that groups of genes, rather than single genes, are added to the genetic complement of an organism. Mangelsdorf (1947) clearly showed a critical difference between maize and teosinte with respect to genomic organization. Certain segments of the maize chromosomes are not homeologous with teosinte chromosomes. Subsequent studies revealed regions of the maize genome are homeologous with segments of *Tripsacum* chromosomes (Maguire, 1960a, 1960b, 1961, 1962, 1963a, 1963b; Galinat, 1973; Newell and de Wet, 1973; Blakey, 1993). The chromosome architecture and genetic profiles of diploid perennial teosinte have many similarities to *Tripsacum* that they do not share with the other *Zea* species. Through highly precise translocations and chromosome fusions (Eubanks, 2001c), viable recombinant progeny with the same chromosome number as *Zea* (2n = 20) are consistently recovered. The chromosomal fusions and rearrangements that produce viable teosinte–*Tripsacum* hybrids apparently provide a mechanism for switching from a tetrasomic to disomic condition (Wilson et al., 1999), signaling that progenitor maize is a cryptic as well as segmental polyploid (Gaut and Doebley, 1997). Such reduction in chromosome number has been documented in experimental crosses among widely divergent taxa such as *Lolium, Hordeum, Hypochoeris, Festuca, Solanum, Lycopersicon,* and *Brassica* (see Eubanks, 2001c). It has also been documented in natural hybrids between diploids and polyploids of *Antennaria* (Bayer and Stebbins, 1987), *Dactylorhiza* (Lord and Richards, 1977), and various ferns. Divergent cross-hybrids often are used as a genetic bridge to break sterility barriers in order to transfer beneficial agronomic traits in crop improvement breeding programs (Tsitsin, 1960; Singh, 1993).

Conclusions

A broad overview of the RFLP genotyping results revealed that maize shared 456 alleles with *Tripsacum* or teosinte (Eubanks, 2001b). Of those, 20.2% (92) were shared only between maize and *Tripsacum,* 36.4% (166) were specifically shared between maize and teosinte, and 43.4% (198) were shared by all three taxa. Because more than one-fifth of the alleles in maize are only in maize and *Tripsacum,* it can be inferred that those maize alleles may have derived from one or more *Tripsacum* ancestors. Likewise, because more than one-third of the maize alleles are shared only with teosinte, it can be inferred that those alleles probably were inherited from one or more teosinte ancestors. Alleles shared by all three taxa could have been inherited either from

Tripsacum or teosinte species or a common ancestor of both taxa. This comparative genomic investigation suggests that the maize genome may be a complex chimera of genes from teosinte and gamagrass. Thus domesticated maize could have arisen from recombination between one or more ancient populations of teosinte and gamagrass. DNA fingerprinting and experimental crosses between teosinte and *Tripsacum* offer preliminary evidence for a reticulate evolutionary history of maize involving intergeneric hybridization. They also point to the intriguing possibility, originally suggested by Mangelsdorf and Cameron (1942), that the highlands of Guatemala were a pivotal crossroads in maize domestication. Natural hybrids between wild grasses growing in the foothills of southeast Turkey and western Iran produced the recombinant raw material for the domestication of wheat (Diamond, 1997). How closely this possible scenario for maize domestication resembles the origin of the cereal grain that gave rise to the birth of Western civilization may ultimately prove to be no coincidence. The results indicate that broader population and taxon sampling in DNA fingerprinting assays, further experimental crosses, explorations to search for natural hybrids between *Zea* and *Tripsacum* in the field, and expanded archaeological reconnaissance and excavations are needed to further test the teosinte and recombination hypotheses.

Acknowledgments

A special thank you to Jennifer Arrington, who assisted with the cladistic analysis. I thank Major M. Goodman for providing seed of the maize landraces; Hugh H. Iltis and the U.S. Department of Agriculture (USDA) Germplasm Research Initiative Network, Ames, Iowa, for seed of the teosinte species; Ray Schnell and W. C. Wasik of the USDA-ARS South Atlantic Subtropical Horticultural Research Station, Miami, Florida, and Indiana University at Bloomington for the *Tripsacum* clonal material; and Walton C. Galinat for the *Manisuris* material. I am grateful for the helpful comments of anonymous reviewers and for National Science Foundation grant numbers 9660146 and 9801386, which provided funding for this research.

References

Avise, J. C. 1994. *Molecular Markers, Natural History and Evolution*. Chapman & Hall, New York, NY, USA.

Bargman, T., G. Hanners, R. Becker, R. Saunders, and J. Rupnow. 1988. Compositional and nutritional evaluation of eastern gamagrass (*Tripsacum dactyloides* L.), a perennial relative of maize (*Zea mays* L.). *Nebraska Agricultural Research Division, Lincoln* 8687: 1–21.

Bayer, R. J. and G. L. Stebbins. 1987. Chromosome numbers, patterns of distribution and apomixis in *Antennaria* (Asteraceae: Inuleae). *Systematic Botany* 12: 305–319.

Beadle, G. W. 1939. Teosinte and the origin of maize. *Journal of Heredity* 30: 245–247.

Beadle, G. W. 1980. The ancestry of corn. *Scientific American* 242: 112–119.

Bennetzen, J., E. Bucker, V. Chandler, J. Doebley, J. Dorweiler, B. Gaut, M. Freeling, S. Hake, E. Kellogg, R. S. Poethig, V. Walbot, and S. Wessler. 2001. Genetic evidence and the origin of maize. *Latin American Antiquity* 12: 84–86.

Berthaud, J., Y. Savidan, and O. LeBlanc. 1996. *Tripsacum:* Diversity and conservation. In S. Taba (ed.), *Maize Genetic Resources,* 74–85. CIMMYT, Mexico City, Mexico.

Bird, R. M. 1979. The evolution of maize: A new model for the early stages. *Maize Genetics Newsletter* 53: 53–54.

Blakey, C. A. 1993. *A Molecular Map in* Tripsacum dactyloides, *Eastern Gamagrass.* Unpublished Ph.D. dissertation, University of Missouri, Columbia, MO, USA.

Brink, D. and J. M. J. de Wet. 1983. Supraspecific groups in *Tripsacum* (Gramineae). *Systematic Botany* 8: 243–249.

Buckler, E. S. IV and T. P. Holtsford. 1996. *Zea* systematics: Ribosomal ITS evidence. *Molecular Biology and Evolution* 13: 612–622.

Burnham, C. R. and P. Yagyu. 1961. Linkage relations of teosinte branched. *Maize Genetics Cooperation Newsletter* 35: 87.

Cadenhead, J. F. III. 1975. *Influence of Burning and Mowing on an Eastern Gamagrass* (Tripsacum dactyloides) *Community in South Central Texas.* Unpublished thesis, Texas A&M University, College Station, TX, USA.

Clayton, W. D. 1973. The awnless genera of Andropogoneae. *Kew Bulletin* 28: 49–58.

Clayton, W. D. 1981. Notes on the tribe Andropogoneae (Gramineae). *Kew Bulletin* 35: 813–818.

de Wet, J. M. J. 1979. *Tripsacum* introgression and agronomic fitness in maize (*Zea mays* L.). In *Proceedings of the Conference on Broadening the Genetic Basis of Crops,* 203–210. Pudoc, Wageningen, The Netherlands.

de Wet, J. M. J., G. B. Fletcher, K. W. Hilu, and J. R. Harlan. 1983. Origin of *Tripsacum andersonii* (Gramineae). *American Journal of Botany* 70: 706–711.

de Wet, J. M. J., J. R. Gray, and J. R. Harlan. 1976. Systematics of *Tripsacum* (Gramineae). *Phytologia* 33: 203–227.

de Wet, J. M. J., D. H. Timothy, K. W. Hilu, and G. B. Fletcher. 1981. Systematics of South American *Tripsacum* (Gramineae). *American Journal of Botany* 68: 269–276.

Diamond, J. 1997. Location, location, location: The first farmers. *Science* 278: 1243–1244.

Doebley, J. F. 1990. Molecular evidence for the evolution of maize. *Economic Botany* 44(suppl.): 6–27.

Doebley, J. F. 1992. Mapping the genes that made maize. *Trends in Genetics* 8: 302–307.

Doebley, J. F., W. Renfroe, and A. Blanton. 1987. Restriction site variation in the *Zea* chloroplast genome. *Genetics* 117: 139–147.

Doebley, J. F. and A. Stec. 1991. Genetic analysis of the morphological differences between maize and teosinte. *Genetics* 129: 285–295.

Doebley, J. F., A. Stec, and L. Hubbard. 1997. The evolution of apical dominance in maize. *Nature* 386: 485–488.

Dorweiler, J., A. Stec, J. Kermicle, and J. Doebley. 1993. Teosinte glume architecture 1: A genetic locus controlling a key step in maize evolution. *Science* 262: 233–235.

Eaheart, D. 1992. Mystery grass turns into business. *Rangelands* 14: 103–104.

Eubanks, M. W. 1995. A cross between two maize relatives: *Tripsacum dactyloides* and *Zea diploperennis* (Poaceae). *Economic Botany* 49: 172–182.

Eubanks, M. W. 1999. *Corn in Clay: Maize Paleoethnobotany in Pre-Columbian Art*. University of Florida Press, Gainesville, FL, USA.

Eubanks, M. W. 2001a. An interdisciplinary perspective on the origin of maize. *Latin American Antiquity* 12: 91–98.

Eubanks, M. W. 2001b. The mysterious origin of maize. *Economic Botany* 55: 492–514.

Eubanks, M. W. 2001c. The origin of maize: Evidence for *Tripsacum* ancestry. *Plant Breeding Reviews* 20: 15–66.

Eubanks, M. W. 2002a. Investigation of novel genetic resource for rootworm resistance in corn. In *2002 Proceedings of the NSF Design, Service and Manufacturing Grantees & Research Conference*, 2544–2550. College of Engineering, Iowa State University, Ames, IA, USA.

Eubanks, M. W. 2002b. Tapping ancestral genes in plant breeding. In J. R. Stepp, F. S. Wyndham, and R. K. Zarger (eds.), *Ethnobiology and Biocultural Diversity*, 225–238. University of Georgia Press, Athens, GA, USA.

Galinat, W. C. 1970. A comparison between the chromosome 4 syndrome of *Zea* and the Q segment of *Triticum* (wheat). *Maize Genetics Cooperation Newsletter* 44: 107–108.

Galinat, W. C. 1973. Intergenomic mapping of maize, teosinte and *Tripsacum*. *Evolution* 7: 644–655.

Galinat, W. C. 1975. Abscission layer development in the rachis of *Zea:* Its nature, inheritance and linkage. *Maize Genetics Cooperation Newsletter* 49: 100–102.

Galinat, W. C. 1976. Preserve Guatemala teosinte, a relict link in corn's evolution. *Science* 180: 323.

Galinat, W. C. 1977. The origin of corn. In G. F. Sprague (ed.), *Corn and Corn Improvement*, 1–47. American Society of Agronomy, Madison, WI, USA.

Galinat, W. C. 1978. The inheritance of some traits essential to maize and teosinte. In D. B. Walden (ed.), *Maize Breeding and Genetics*, 93–111. J. Wiley, New York, NY, USA.

Galinat, W. C. 1985. The missing links between teosinte and maize: A review. *Maydica* 30: 137–160.

Galinat, W. C. 1988. The origin of corn. In G. F. Sprague (ed.), *Corn and Corn Improvement*, 1–31. American Society of Agronomy, Madison, WI, USA.

Galinat, W. C. 1992. Evolution of corn. *Advances in Agronomy* 47: 203–231.

Galinat, W. C. and F. C. Craighead. 1964. Some observations on the dissemination of *Tripsacum*. *Rhodora* 66: 371–374.

Gardiner, J., E. H. Coe, S. Melia-Hancock, D. A. Hoisington, and S. Chao. 1993. Development of a core RFLP map in maize using an immortalized F_2 population. *Genetics* 134: 917–930.

Gaut, B. S. and J. F. Doebley. 1997. DNA evidence for the segmental allotetraploid origin of maize. *Proceedings of the National Academy of Sciences (USA)* 94: 6809–6814.

Gilmore, M. R. 1931. Vegetal remains of the Ozark bluff-dweller culture. *Michigan Academy of Sciences Arts and Letters* 14: 83–102.

Helentjaris, T., M. Slocum, S. Wright, A. Schaefer, and J. Nienhuis. 1986. Construction of genetic linkage maps in maize and tomato using restriction fragment length polymorphisms. *Theoretical and Applied Genetics* 72: 761–769.

Helentjaris, T., D. F. Weber, and S. Wright. 1988. Identification of the genomic locations of duplicate nucleotide sequences in maize by analysis of restriction fragment length polymorphisms. *Genetics* 118: 353–363.

Hitchcock, A. S. 1906. Notes on North American grasses. VI. Synopsis of *Tripsacum*. *Botanical Gazette* 41: 92–298.

Iltis, H. H. 2000. Homeotic sexual translocations and the origin of maize (*Zea mays,* Poaceae): A new look at an old problem. *Economic Botany* 54: 7–42.

Iltis, H. H., J. F. Doebley, R. Guzmán, and B. Pazy. 1979. *Zea diploperennis* (Gramineae): A new teosinte from Mexico. *Science* 203: 186–188.

Jaenicke-Després, V., E. S. Buckler, B. D. Smith, M. T. P. Gilbert, A. Cooper, J. Doebley, and S. Pääbo. 2003. Early allelic selection in maize as revealed by ancient DNA. *Science* 303: 1206–1208.

James, J. 1979. New maize × *Tripsacum* hybrids for maize improvement. *Euphytica* 28: 239–247.

Kato Y., T. A. 1984. Chromosome morphology and the origin of maize and its races. *Evolutionary Biology* 17: 219–253.

Kempton, J. H. and W. Popenoe. 1937. Teosinte in highland Guatemala: Report of an expedition to Guatemala, El Salvador, and Chiapas, Mexico. *Carnegie Institution of Washington, Contributions to American Archaeology No. 23.* 483: 199–218.

Ladizinsky, G. 1989. Ecological and genetic considerations in collecting and using wild relatives. In A. D. H. Brown, O. H. Frankel, D. R. Marshall, and J. T. Williams (eds.), *The Use of Plant Genetic Resources,* 297–305. Cambridge University Press, Cambridge, UK.

Langham, D. G. 1940. The inheritance of intergeneric differences in *Zea–Euchlaena* hybrids. *Genetics* 25: 88–107.

Li, Y. G., C. L. Dewald, and V. A. Sokolov. 2000. Sectional delineation of sexual *Tripsacum dactyloides–T. maizar* allotriploids. *Annals of Botany* 85: 845–850.

Lord, R. M. and A. J. Richards. 1977. A hybrid swarm between the diploid *Dactylorhiza uchsii* (Druce) Soó and the tetraploid *D. purpurella* (T. & T. A. Steph.) Soó in Durham. *Watsonia* 11: 205–211.

MacNeish, R. S. and M. W. Eubanks. 2000. Comparative analysis of the Río Balsas and Tehuacán models for the origin of maize. *Latin American Antiquity* 11: 3–20.

Maguire, M. P. 1960a. A study of homology between a terminal portion of *Zea* chromosome 2 and a segment derived from *Tripsacum*. *Genetics* 45: 195–209.

Maguire, M. P. 1960b. A study of pachytene chromosome pairing in a corn–*Tripsacum* hybrid derivative. *Genetics* 45: 651–664.

Maguire, M. P. 1961. Divergence in *Tripsacum* and *Zea* chromosomes. *Evolution* 15: 394–400.

Maguire, M. P. 1962. Common loci in corn and *Tripsacum*. *Journal of Heredity* 53: 87–88.

Maguire, M. P. 1963a. Chromatid interchange in allodiploid maize–*Tripsacum* hybrids. *Canadian Journal of Genetics and Cytology* 5: 414–420.

Maguire, M. P. 1963b. High transmission frequency of a *Tripsacum* chromosome in corn. *Genetics* 48: 1185–1194.

Mangelsdorf, P. C. 1947. The origin and evolution of maize. *Advances in Genetics* 1: 161–207.

Mangelsdorf, P. C. 1974. *Corn, Its Origin, Evolution and Improvement.* Harvard University Press, Cambridge, MA, USA.

Mangelsdorf, P. C. and J. W. Cameron. 1942. Western Guatemala a secondary center of origin of cultivated maize varieties. *Harvard University Botanical Museum Leaflets* 10: 217–252.

Mangelsdorf, P. C. and W. C. Galinat. 1964. The tunicate locus in maize dissected and reconstituted. *Proceedings of the National Academy of Sciences (USA)* 51: 147–150.

Mangelsdorf, P. C., R. S. MacNeish, and W. C. Galinat. 1967. Prehistoric maize, teosinte and *Tripsacum* from Tamaulipas, Mexico. *Harvard University Botanical Museum Leaflets* 22: 33–63.

Mangelsdorf, P. C. and R. G. Reeves. 1939. The origin of Indian corn and its relatives. *Texas Agricultural Experiment Station Bulletin* 574: 1–315.

Matsuoka, Y., Y. Vigouroux, M. M. Goodman, J. Sanchez G., E. Buckler, and J. Doebley. 2002. A single domestication for maize shown by multilocus microsatellite genotyping. *Proceedings of the National Academy of Sciences (USA)* 99: 6080–6084.

Matthews, D. L., C. O. Grogan, and C. E. Manchester. 1974. Terminal ear mutant of maize (*Zea mays* L.). *Journal of Agricultural Sciences* (Cambridge) 82: 433–435.

McClintock, B. 1984. The significance of responses of the genome to challenge. *Science* 226: 792–801.

Melchinger, A. E., M. M. Messmer, M. Lee, W. L. Woodman, and K. R. Lamkey. 1991. Diversity and relationships among U.S. maize inbreds revealed by restriction fragment length polymorphisms. *Crop Science* 31: 669–678.

Messmer, M. M., A. E. Melchinger, R. Herrmann, and J. Boppenmaier. 1993. Relationships among early European maize inbreds: II. Comparison of pedigree and RFLP data. *Crop Science* 33: 944–950.

Neuffer, M. G., E. H. Coe, and S. R. Wessler. 1997. *Mutants of Maize*. Cold Spring Harbor Laboratory Press, Cold Spring Harbor, NY, USA.

Newell, C. A. and J. M. J. de Wet. 1973. A cytological survey of *Zea–Tripsacum* hybrids. *Canadian Journal of Genetics and Cytology* 15: 763–778.

Ortiz-García, S. and E. Ezcurra. 2003. Transgenic maize in Mexico: Risks and reality. In *Botany 2003 Abstracts* 4 (www.2003.botanyconference.org).

Randolph, L. F. 1970. Variation among *Tripsacum* populations of Mexico and Guatemala. *Brittonia* 22: 305–337.

Rhoades, M. M. 1951. Duplicate genes in maize. *American Naturalist* 85: 105–110.

Rodgers, J. S. 1950. The inheritance of inflorescence characters in maize–teosinte hybrids. *Genetics* 35: 541–558.

Singh, R. J. 1993. *Plant Cytogenetics*. CRC Press, Boca Raton, FL, USA.

Smith, O. S. and J. S. C. Smith. 1992. Measurement of genetic diversity among maize hybrids: A comparison of isozymic, RFLP, pedigree, and heterosis data. *Maydica* 37: 53–60.

Stebbins, G. L. 1950. *Variation and Evolution in Plants*. Columbia University Press, New York, NY, USA.

Swofford, D. L. 2002. *PAUP*: Phylogenetic Analysis Using Parsimony (and Other Methods)*, Version 4.0b10. Sinauer Associates, Sunderland, MA, USA.

Taylor, D. A. 2001. Ancient teachings, modern lessons. *Environmental Health Perspectives* 109: A207–215.

Tsitsin, N. V. 1960. *Wide Hybridization in Plants*. Translated from Russian. Akademiya Nauk SSSR, Moscow, Russia.

Wilkes, H. G. 1967. *Teosinte: The Closest Relative of Maize*. Harvard University, Bussey Institution, Cambridge, MA, USA.

Wilkes, H. G. 1979. Mexico and Central America as a center for the origin of maize. *Crop Improvement (India)* 6: 1–18.

Wilson, W. A., S. E. Harrington, W. L. Woodman, M. Lee, M. E. Sorrells, and S. R. McCouch. 1999. Inferences on the genome structure of progenitor maize through comparative analysis of rice, maize and the domesticated panicoids. *Genetics* 153: 453–473.

PART 2

SYSTEMATICS AND THE ORIGIN OF CROPS

Phylogenetic and Biogeographic Relationships

Roberto Papa, Laura Nanni, Delphine Sicard,
Domenico Rau, and Giovanna Attene

CHAPTER 6

Evolution of Genetic Diversity
in *Phaseolus vulgaris* L.

Among domesticated plant species, the common bean (*Phaseolus vulgaris* L.) is the most important protein source for direct human consumption (Singh, 2001; Broughton et al., 2003). It is a diploid ($2n = 2x = 22$), annual species and is predominantly self-pollinating, with the occasional occurrence of cross-pollination by pollinators such as the bumblebee, *Bombus* spp. (Free, 1966). Many studies have been aimed at determining the origins, domestication, and evolution of the genetic diversity of *P. vulgaris*. Since seed storage proteins first became important in bean research, the advent of molecular techniques has had a major impact on our understanding of the *P. vulgaris* evolutionary history (Gepts, 1988b). The presence of geographically isolated gene pools in *P. vulgaris* that originated from at least two independent domestication events and the overlapping distribution with other domesticated and wild species that have different mating systems and are at various degrees of reproductive isolation make *P. vulgaris* and the genus *Phaseolus* a unique model for studies of plant evolution. Therefore, in addition to a brief illustration of the major aspects of the evolutionary history of *P. vulgaris* (for further details, see Gepts, 1996, 1988a; Debouck, 1999; Singh, 2001; Broughton et al., 2003; Snoeck et al., 2003), we focus here on recent studies highlighting the roles of the various evolutionary forces in shaping the genetic diversity of *P. vulgaris*. These include the potential role of introgressive hybridization between *P. vulgaris* and *P. coccineus* in Mesoamerica, the effects of gene flow and selection between wild and domesticated bean populations, the evolution of disease resistance, and the effects of the introduction of the bean into the Old World.

The genus belongs to the tribe *Phaseolae* (subfamily *Papilionoideae*, family *Leguminosae*), which includes two other genera with domesticated species: *Glycine* (soybean) and *Vigna* (cowpea). Verdecourt (1970) redefined *Phaseolus* as a large, diverse genus of at least 50 species, as was later confirmed by further studies (Maréchal et al., 1978; Lackey, 1981, 1983). *Phaseolus* is strictly of the New World, and it grows naturally in the warm tropical and subtropical regions from Mexico (Sousa and Delgado-Salinas, 1993) to Argentina (Delgado-Salinas, 1985; Debouck et al., 1987).

Phaseolus includes five domesticated species: *P. vulgaris* (common bean), *P. lunatus* (lima bean), *P. acutifolius* A. Gray (tepary bean), *P. coccineus* ssp. *coccineus* (runner bean), and *P. coccineus* L. ssp. *polyanthus* Greenman = *P. polyanthus* (= *P. coccineus* ssp. *darwinianus*) (year-long bean). Each of these has a distinct geographic distribution, life history, and reproductive system (Maréchal et al., 1978; Delgado-Salinas, 1985). The phylogenetic relationships between these *Phaseolus* species have been investigated using a number of morphological (Maréchal et al., 1978; Debouck, 1991), biochemical (Sullivan and Freytag, 1986; Jaaska, 1996; Pueyo and Delgado-Salinas, 1997), and molecular (Delgado-Salinas et al., 1993; Schmit et al., 1993; Llaca et al., 1994; Hamann et al., 1995; Vekemans et al., 1998) tools. In particular, a recent phylogenetic analysis of *Phaseolus* and its close relatives combined molecular (internal transcribed spacer [ITS]/5.8S DNA sequences) and nonmolecular data (vegetative, floral, and fruit morphological characters and chromosome numbers) (Delgado-Salinas et al., 1999) and confirmed that *Phaseolus* is monophyletic. This is consistent with several studies of both wild and domesticated species of *Phaseolus* that have used a wide range of tools, including seed proteins, isozymes, and nuclear, chloroplast, and mitochondrial DNA (Debouck, 1999). Delgado-Salinas et al. (1999) also revealed that there may be anywhere from two to nine subclades within *Phaseolus*, with the cultivated species falling into two distinct lineages. In one, the domesticated species *P. vulgaris*, *P. coccineus*, *P. polyanthus*, and *P. acutifolius* are found together with two wild species, *P. albescens* and *P. costaricensis*. Another clade contains *P. lunatus* and wild species of both Andean and Mesoamerican distributions (Fofana et al., 1999; Maquet and Baudoin, 1996; Delgado-Salinas et al., 1999).

FIGURE 6.1   Distributions of the wild populations of *P. vulgaris* and *P. coccineus.*

The intraspecific organization of genetic variation in *P. vulgaris* has been investigated in detail. The presence of two distinct gene pools was suggested by analyses of seed morphology (Evans, 1973, 1980), of hybrid nonviability in crosses between *P. vulgaris* from Mesoamerica and South America, and of outbreeding depression (see Singh, 2001, for review). The analyses of variations in seed storage proteins (e.g., phaseolin) also supported the presence of distinct Mesoamerican and Andean gene pools, with the presence of parallel geographic patterns in both the domesticated and the wild beans indicating the occurrence of independent domestication in Mesoamerica and South America (Gepts et al., 1986; Gepts and Bliss, 1988; Koenig and Gepts,

1989; Koenig et al., 1990; Singh et al., 1991). A different type of phaseolin (type I) has been observed in wild accessions from north Peru and Ecuador, and sequence analyses of the locus coding for these proteins revealed that type I phaseolin is the ancestral form from which the other phaseolins evolved. This indicated that the populations from north Peru and Ecuador were the closest descendants of the ancestor of the common bean (Kami et al., 1995). Overall, these studies indicated three different wild gene pools (Mesoamerican, Andean, and Ancestral) (figure 6.1), with evidence of domestication events only in the Mesoamerican and Andean gene pools. Both the independent domestication and the origins of wild *P. vulgaris* have been confirmed by various studies based on other molecular markers (Khairallah et al., 1992; Becerra and Gepts, 1994; Caicedo et al., 2000; Papa and Gepts, 2003).

The Andean and Mesoamerican gene pools have different structures and levels of genetic diversity in both the wild and domesticated populations, where the occurrence of different races has also been described (Singh, 2001). Indeed, there is a higher genetic diversity in the Mesoamerican than the Andean gene pool for both wild and domesticated populations (Koenig and Gepts, 1989; Beebe et al., 2000, 2001; Papa and Gepts, 2003; McClean et al., 2004). Additionally, a higher interpopulation component of genetic variance has been indicated for the Mesoamerican wild populations (using amplified fragment length polymorphism [AFLP]; Papa and Gepts, 2003), in comparison with the Andean wild populations (using random amplified polymorphic DNA [RAPD]; Cattan-Toupance et al., 1998). A much higher level of genetic differentiation has also been observed between the domesticated races from Mesoamerica (using RAPD; Beebe et al., 2000) than between those from South America (using AFLP; Beebe et al., 2001). However, further direct comparisons may be needed because of the use of different types of molecular markers.

Interspecific Hybridization

In contrast to South America, in Mesoamerica *P. vulgaris* often is sympatric with other species that are partially sexually compatible. For this reason, one possible explanation for the differences in the levels of genetic diversity between the gene pools is the occurrence of introgressive hybridization between *P. vulgaris* and the other *Phaseolus* species. Indeed, in Mesoamerica the distribution of *P. vulgaris* overlaps with that of *P. coccineus* and *P. polyanthus*. Molecular studies have shown that *P. polyanthus*, which was formerly included in *P. coccineus*, is intermediate in its morphological features between these other two species (Hernandez-Xolocotzi et al., 1959), and a hybrid origin has indeed been suggested (Piñero and Eguiarte, 1988; Kloz, 1971; Llaca et al., 1994). At the molecular level, *P. polyanthus* is closer to *P. coccineus* by nuclear DNA comparison (Piñero and Eguiarte, 1988; Delgado-Salinas et al., 1999) but more similar to *P. vulgaris* by chloroplast DNA comparison (Llaca et al., 1994). Thus *P. polyanthus* probably originated from a cross that involved *P. vulgaris* as the maternal parent, with successive backcrosses to *P. coccineus* as the paternal donor (Schmit et al., 1993; Llaca et al., 1994). This interpretation is consistent with studies showing that in artificial crosses between *P. coccineus* and *P. vulgaris,* fertile F_1 progeny can be produced, particularly when *P. vulgaris* is the maternal parent (Singh, 2001; Broughton et al., 2003). This suggests that introgression between *P. coccineus* and *P. vulgaris* occurred in the evolutionary history of both species in Mesoamerica.

Using nuclear and chloroplast microsatellites (simple sequence repeats; SSRs), there is evidence of introgression in sympatric populations of *P. coccineus* and *P. vulgaris* from Morelos, Mexico (Sicard and Papa, unpublished data), which suggests that gene flow might still be important in shaping the structure of the genetic diversity of these two species in Mesoamerica. Through an analysis that used the same SSR loci of wild and domesticated germplasm accessions of these two species and included the Andean gene pool of *P. vulgaris,* the level of introgression was seen to be highly locus specific. Thus loci that displayed higher similarities between *P. vulgaris* and *P. coccineus* from Mesoamerica also showed a stronger differentiation between Andean and Mesoamerican *P. vulgaris*. Because only microsatellites designed from genic regions were used, it was not possible to discriminate between the effects of selection and gene flow in driving this introgression. Nevertheless, these results may have strong implications for our understanding of the structure and level of genetic diversity in the common bean. In particular, they suggest that introgression from *P. coccineus* probably was one of the causes of both the higher genetic diversity present in Mesoamerica (as compared with the Andes) and the partial reproductive isolation between the gene pools. However, other possible explanations, such as homoplasy and convergent evolution, remain to be investigated.

Gene Flow and Selection Between Wild and Domesticated *P. vulgaris*

For beans, as for many other species (Harlan and de Wet, 1971), the wild and domesticated forms belong to the same biological species and are completely cross-fertile (Koinange et al., 1996). The domestication process has led to a reduction in genetic diversity within each of the bean gene pools (Sonnante et al., 1994), as has been seen for other species (e.g., *Zea mays*: Doebley et al., 1990; Ladizinsky, 1998). This effect, called a domestication bottleneck, is a function of the small samples of individuals that founded the domesticated populations. In addition to this founder effect, which has generally affected the whole genome diversity, selection for specific traits probably has also contributed to reductions in genetic diversity at target loci and in the surrounding genomic regions. This results from the combined

FIGURE 6.2 Close-range sympatry between wild and domesticated common bean (*P. vulgaris* L.) in Teopisca, Chiapas, Mexico. Wild and the domesticated common beans have a similar climbing growth habit and phenology. Pods of wild and domesticated beans. (Photo courtesy of Papa and Gepts.)

effects of selection and recombination (e.g., hitchhiking; Maynard Smith and Haigh, 1974; Kaplan et al., 1989). Thus the effects of domestication at neutral loci that are linked to those selected during domestication are likely to be strictly related to the breeding system of a given species (allogamous versus autogamous), along with other factors affecting the amount of recombination (e.g., population size). For instance, in the allogamous plant species *Zea mays,* the role of hitchhiking appears to have affected restricted genomic regions around selected sites (Wang et al., 1999, 2001; Tenaillon et al., 2001; Clark et al., 2003). A higher level of linkage disequilibrium probably would be expected in autogamous species, such as the common bean. The traits that distinguish the domesticated from the wild form are collectively called the domestication syndrome (Hammer, 1984), and they are shared by most domesticated crop species. These key traits include the lack of seed dispersal and dormancy, a compact plant architecture, a higher yield, a synchronicity, and an early flowering. The majority of these domestication traits have simple Mendelian determinism with, in most cases, complete or semidominance of the wild allele. Indeed, with few exceptions, domesticated alleles are associated with a lack of gene function (Gepts, 2002; Gepts and Papa, 2002).

Wild and domesticated forms often are found in sympatry throughout the distribution of the common bean (figure 6.2), from North Mexico to Argentina. Several examples of introgression have been documented, along with the occurrence of weedy populations that colonize highly disturbed areas, such as abandoned fields (Freyre et al., 1996; Beebe et al., 1997). Even if the autogamous breeding system is a limiting factor, the observed level of outcrossing (2–3%) (Ibarra-Pérez et al., 1997; Ferreira et al., 2000) suggests that, as found in other highly selfing species (Ellstrand et al., 1999), gene flow is likely to limit the independent evolution of wild and domesticated populations. A significant level of gene flow between wild and domesticated *P. vulgaris* has recently been observed in Puebla, Mexico, using inter–simple sequence repeats (ISSRs) (González et al., 2005), and in Michoacán and Guanajuato, Mexico, using phenotypic markers and ISSRs (Payró de la Cruz et al., in press).

The introgression between the wild and the domesticated common bean (*P. vulgaris* L.) in Mesoamerica has also been studied using genetically mapped AFLP markers (Papa and Gepts, 2003; Papa et al., in press). AFLPs have been positioned on a molecular linkage map (Freyre et al., 1998) where several genes and quantitative trait loci have been located, including those responsible for the genetic control of the domestication syndrome (Koinage et al., 1996). Diversity for the same markers was thus analyzed in two samples of wild and domesticated populations from Mexico. Gene flow occurred principally in close-range sympatry, that is, when two populations grew in close proximity (figure 6.2). Through both phenetic and admixture population analyses, introgression was found to be about three to four times higher from domesticated to wild populations than in the reverse direction (Papa and Gepts, 2003). Mapping of AFLP markers has also shown that differentiation between wild and domesticated populations is highest near the genes for domestication and is lower farther from these genes. Concurrently, the genetic bottleneck induced by domestication was strongest around these genes. Therefore selection may be a major evolutionary factor in the maintenance of the identities of wild and domesticated populations in sympatric situations. Furthermore, domesticated alleles appear to have displaced wild alleles in sympatric wild populations, thus leading to a reduction in genetic diversity in such populations (Papa et al., in press).

Evolution of Disease Resistance

The common bean is one of the few plant species for which population genetics and molecular genetics have both been used to study the evolution of resistance and the defense against parasites at both the ecological and molecular levels (de Meaux and Mitchell-Olds, 2003; Seo et al., 2004).

At the phenotypic level, genetic variation for resistance against parasites has been reported between and within *Phaseolus vulgaris* gene pools. The two cultivated common bean gene pools are differentiated by their resistance to the fungi responsible for anthracnose, *Colletotrichum lindemuthianum* (Sicard et al., 1997a, 1997b); for rust, *Uromyces appendiculatus* (Steadman et al., 1995); and for angular leaf spot, *Phaeoisariopsis griseola* (Guzman et al., 1995). In each of these interactions, the plants of one cultivated gene pool were more resistant to the fungus coming from the other gene pool than to the fungus isolated from the same gene pool. Similar results were obtained in natural populations where different sets of resistance genes against *C. lindemuthianum* were found in the three gene pools (Geffroy et al., 1999). Natural populations of the three gene pools maintained resistance genes that were overcome by local fungi but remained useful against possible invaders (Geffroy et al., 1999). Within centers of diversity, natural populations of *P. vulgaris* were differentiated for resistance to the fungus

C. lindemuthianum in both Mexico and Argentina (Cattan-Toupance et al., 1998; Sicard, unpublished data). In Mexico, natural populations of *P. vulgaris* were maladapted to the fungus *C. lindemuthianum* and had a greater resistance to allopatric strains than to local strains (Sicard, unpublished data).

The effects of parasite selection pressure on the molecular diversity of *P. vulgaris* have been studied by comparing the diversity between phenotypic resistance, neutral markers, and molecular markers located on both resistance candidate and defense-related genes. For resistance genes, restriction fragment length polymorphism (RFLP) markers located in a nucleotide-binding site (NBS) and AFLPs located on a leucine-rich repeat (LRR) domain of two families of resistance genes have been developed (Neema et al., 2001; de Meaux and Neema, 2003). For defense-related genes, three microsatellites located in genes encoding pathogenesis-related protein and located in different linkage groups have been used (Yu et al., 2000; Sicard and Papa, unpublished data). Population structures (i.e., the population differentiation) at the gene pool level and on the regional scale were conserved for all three: the phenotypic resistance markers, the resistance or defense gene-tagged markers, and the neutral markers. This suggests that the history of the common bean and its lifecycle (autogamous, low seed migration) influences molecular polymorphism at both neutral and defense or resistance loci (Neema et al., 2001; de Meaux et al., 2003; de Meaux and Neema, 2003). The levels of population differentiation and the levels of within-population diversity differed between the neutral and resistance gene-tagged markers. Plants of the Mesoamerican and Andean centers of diversity were shown to be more differentiated for RAPD markers than for NBS-tagged RFLP markers, which suggests a homogenizing effect of selection on the NBS region of two resistance gene candidate families, as was also found from DNA sequence data (Neema et al., 2001; Ferrier-Cana et al., 2003). In Mexico, a comparison of neutral markers and markers tagged on the LRR domain of one resistance gene family revealed that the average level of diversity within populations was higher for resistance gene candidate–tagged markers than for RAPD markers, suggesting diversifying selection or higher mutation rates in the LRR region of these resistance loci. This is consistent with the hypothesis that the LRR domains of resistance proteins form a versatile binding domain that is involved in parasite recognition (de Meaux and Neema, 2003).

Altogether, these data show that population history, population dynamics, and parasite selection pressure are all shaping the phenotypic and molecular polymorphism at resistance genes.

Introduction into the Old World

After Columbus's voyage in 1492, intense biological exchanges occurred between the Old World and the New World. Several crops were introduced, mainly into the Iberian Peninsula, from which they spread into the rest of Europe and around the world (Simmonds, 1976). The common bean probably arrived in Spain and Portugal from Central America in 1506 (Ortwin-Sauer, 1966). In 1528, Pizarro explored Peru, and the introduction of accessions from the Andes probably started after 1532 (Brucher and Brucher, 1976). The first description of the common bean in a European herbal was by Fuchs (1543) in Germany, around which time it also started its expansion into the Mediterranean area. Birri and Coco (2000) report on the contents of a manuscript published by Pierio Valeriano Bolsanio in 1550 (Biblioteca Vaticana Codice Latino 5215 C 8–9) that described his travels in 1532, from Rome to Belluno (northeast Italy); a bag of beans was received from the pope, Giuliano de Medici (Pope Clemente VII, 1523–1534), with the specific objective of its introduction as a crop plant. As Gepts (2002) notes, the bronze portals of the cathedral of Pisa, which have been dated to 1595, include realistic representations of the common bean. This all suggests that *P. vulgaris* was well known in Italy by the end of the 16th century. *P. vulgaris* probably arrived in Turkey and Iran at the beginning of the 1600s. In the 17th and 18th centuries, the Arabs introduced the common bean into East Africa, and in 1669 it was being cultivated on a large scale in the Netherlands (Van der Groen, 1669). Overall, this demonstrates that the pathways of dissemination of beans into Europe were very complex, with several introductions from the New World combined with direct exchanges between European and other Mediterranean countries.

In recent years, molecular markers have contributed to our understanding of the origins and dissemination pathways of *P. vulgaris* from its areas of domestication into Europe. The phaseolins have been used to characterize a European collection of *P. vulgaris* that was mainly from Portugal, Spain, France, and the Netherlands. This revealed that the European common bean arose from the introduction of domesticated beans from both of the American gene pools, with a higher frequency of Andean phaseolin types (76%; T, C, and H types) than of the Mesoamerican types (24%; S and

FIGURE 6.3 Distribution of phaseolin types across Europe (%). *White background:* Andean phaseolin types (T, C, and H). *Black background:* Mesoamerican phaseolin types (S and B). The sample sizes are given in parentheses after the country names. For the Iberian Peninsula, the data were obtained as weighted means of the results of the experiments of Gepts and Bliss (1988), Lioi (1989), Ocampo et al. (2002), and Rodiño et al. (2003). The data for France and the Netherlands are from Gepts and Bliss (1988). The data for Germany, Italy, Greece, Cyprus, Turkey, and the former Soviet Union are from Lioi (1989). When pooled samples were used, the calculations did not take into account the possible redundancy between different collections.

B types) (Gepts and Bliss, 1988). This was confirmed by Lioi (1989) in an analysis of a large collection of accessions that were mainly from Italy, Greece, and Cyprus (66% Andean types) and by Masi and Spagnoletti (unpublished data), who analyzed 544 accessions collected throughout Europe (76% Andean types). Despite a large variance in sample sizes and sampling strategies within and between these studies, at the single-country level along the Mediterranean Arch (from the Iberian Peninsula to Turkey, throughout France, Italy, Greece, and Cyprus) a prevalence of the Andean phaseolin type has always been observed, with a minimum of 54% for Greece (Gepts and Bliss, 1988; Lioi, 1989; Rodiño et al., 2001, 2003; Ocampo et al., 2002) (figure 6.3). The lack of information for the countries of Central Europe should be noted. When regions within a country are considered, this prevalence of the Andean gene pool is also confirmed for studies in Galicia, Spain (Escribano et al., 1998), Abruzzo in central Italy (Piergiovanni et al., 2000a), Basilicata in southern Italy (Limongelli et al., 1996; Piergiovanni et al., 2000b), and the Marche region in central Italy (using ISSRs and nuclear and chloroplast SSRs; Sicard et al., in press). Thus, the overall frequencies of the Mesoamerican and Andean gene pools appear to be very similar on the continental, country, and regional scales, suggesting large seed exchanges between the European countries.

Differences in the frequencies of each Andean phaseolin type have also been discussed. Gepts and Bliss (1988) showed that in the Iberian Peninsula, phaseolin C was the most common. The prevalence of the C type within Portuguese and Spanish landraces was also observed by Rodiño et al. (2001) and Ocampo et al. (2002). In contrast, Escribano et al. (1998) analyzed landraces from Galicia, Spain, and observed that type T was the most common. This was also seen with a collection of 388 accessions from the Iberian Peninsula (Rodiño et al., 2003). Overall, five phaseolins have been observed in the Iberian Peninsula, including type H (15%) and type B (1%). This may suggest a higher diversity for phaseolin types in this area than in the rest of Europe, although this greater phaseolin variability in the Iberian Peninsula may just be related to the greater number of samples analyzed or differences in the sampling strategies between the studies (figure 6.3).

On a smaller geographic scale, a study conducted in the Abruzzo region of central Italy showed a prevalence of type C (Piergiovanni et al., 2000a), as has also been seen in the Basilicata region in southern Italy (Limongelli et al., 1996; Piergiovanni et al., 2000a). Interestingly, the Hellenic Peninsula has the highest frequency of phaseolin S (46%), a strictly Mesoamerican type; the frequency of phaseolin S, when compared with that of the rest of Europe, is also high (38%) in Cyprus and Turkey (figure 6.3; Lioi, 1989). Therefore, the overall data indicate that in the eastern Mediterranean area there is a high frequency of type S. Finally it should be noted that in France and the Netherlands, type T appears at a very high frequency (Gepts and Bliss, 1988), as in Germany and in the former Soviet Union (Lioi, 1989). It has also been suggested that as well

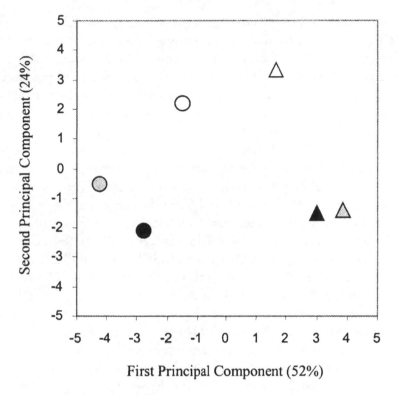

First Principal Component (52%)

FIGURE 6.4   Relationships between the wild American (*black*), domesticated American (*gray*), and domesticated Iberian (*white*) germplasm of the Mesoamerican (*triangles*) and Andean (*circles*) gene pools. The graph summarizes the differences in isozyme allele frequencies at the eight loci that are common among the studies of Koenig and Gepts (1989), Singh et al. (1991), and Santalla et al. (2002) (Diap-1, Diap-2, Me, Mdh-1, Mdh-2, Prx, Rbcs, and Skdh) and was obtained using JMP 3.1.5 software (SAS Institute, Inc., 1995). For the wild Mesoamerican gene pool, the weighted averages of the Mexican and Central American frequencies (Koenig and Gepts, 1989, table 3) were calculated, but for the wild Andean, only the frequencies for Argentina were considered.

as migration and selection, the phaseolin geographic distribution may be affected by the differential distribution of phaseolin patterns among consumption categories (e.g., dry beans vs. green pod cultivars) (Brown et al., 1982; Gepts and Bliss, 1988). Several studies have shown the occurrence in Europe of markers pertaining to both Andean and Mesoamerican gene pools within the same bean landrace (Piergiovanni et al., 2000a, 2000b;

Rodiño et al., 2001), and molecular evidence of hybridization between gene pools has been obtained by analyzing germplasm from the Marche region in central Italy (Sicard et al., in press). Recently, using isozymes, introgression between the Mesoamerican and the Andean gene pools was observed in the Iberian Peninsula, and two groups of intermediate or putative recombinants (25% of the accessions) between the two gene pools were found (Santalla et al., 2002).

It has been suggested that crop expansion from America to Europe resulted in a reduction in the diversity of the European common bean because of strong founder effects, adaptation to a new environment, and consumer preferences (Gepts, 1999). Isozyme loci have been used to characterize domesticated common beans both from the Americas (Singh et al., 1991) and from the Iberian Peninsula (Santalla et al., 2002). Recalculation of the diversity values using the eight isozyme loci in common between these two studies reveals that the Iberian Peninsula diversity ($H_T = 0.25$) is about 30% lower than that of the Americas ($H_T = 0.37$). The difference in diversity (H) of the two gene pools was larger in the Americas (Mesoamerican = 0.23; Andean = 0.16) than in the Iberian Peninsula (Mesoamerican origin = 0.20; Andean origin = 0.21), which results in a much stronger genetic difference between the two gene pools in the Americas ($G_{ST} = 0.47$) than in the Iberian Peninsula ($G_{ST} = 0.18$). This has also been shown using principal component analysis (PCA) of the allelic frequencies (figure 6.4), where wild germplasm was also used as the reference (Koenig and Gepts, 1989). Of note, within gene pools, domesticated American germplasm is closer to the wild germplasm than to the domesticated germplasm from the Iberian Peninsula (figure 6.4). This lower differentiation in Europe can be explained by the combined actions of greater gene flow between different gene pools caused by the lack of geographic barriers and convergent evolution.

Overall, the data suggest that the structure of genetic diversity of common bean in Europe has been highly influenced by hybridization between the two gene pools together with homogeneous selection for adaptation to the European environments. For example, this is likely to have been the case for photoperiod insensitivity. In addition, the bottleneck effect of the introduction of the common bean into Europe might not have been as strong as was previously suspected (Gepts, 1999), and it appears that hybridization between the two gene pools of *P. vulgaris* has had a significant impact on the maintenance of the overall level of genotypic diversity. Second, heterogeneous selection for different uses and local adaptation to a wide range of environments and agronomic practices in Europe might also have counteracted the effects of drift and homogeneous selection for adaptation to European environmental conditions. Third, the founding populations might have been highly representative of the diversity present in the American gene pools. This could be because there were several different introductions from the Americas or because the attractiveness of various types of seed color and shape probably has favored the capture of different alleles and genotypes. Extensive studies on the genetic diversity of the European bean populations are still needed to test these hypotheses.

These data suggest that the expansion of *P. vulgaris* into Europe and introgression between different gene pools (probably because of the lack of geographic barriers) have had a significant impact on the shaping of the genetic diversity of this species. However, because evidence of germplasm exchange between Mesoamerica and the Andes has been documented (Gepts, 1988a), a strict relationship between the gene pools and the areas from which the common bean was introduced into Europe cannot be assumed; similarly, hybrids between gene pools could also have originated in the Americas and the progeny later introduced into the Old World. To obtain a comprehensive picture of the origins, levels, and structures of the common bean diversity in Europe, representative samples from different European and Mediterranean countries should be compared with an appropriate large sample from the Americas using different types of molecular markers.

Conclusions

We have shown how the advent of molecular techniques has greatly improved our ability to understand the complex evolutionary history of the common bean and how various evolutionary forces have contributed to the structure of its genetic diversity in the New World and, more recently, in the Old World. New molecular tools have been developed recently for the bean, and others are likely to become available in the near future (Broughton et al., 2003), which will expand our capacity for investigation. For instance, along with nuclear markers, the development in the bean of SSRs and sequence-tagged sites (STSs) specific for chloroplast DNA (Sicard et al., in press) and mitochondrial DNA (Arrieta-Montiel et al., 2001) could be of particular interest in tracking the migration pathways. Indeed, migration would be better studied using molecular markers that differ in their inheritance patterns (uniparental vs. biparental; Provan et al., 2001). Moreover, we have shown how a combination of molecular maps and gene-tagging markers and neutral markers can distinguish the evolutionary role played by selection from that caused by drift and migration.

The relative roles of evolutionary forces should be resolved if it is possible to compare the information from the gene-tagging and neutral markers. As was first pointed out by Cavalli-Sforza (1966), whereas migration and drift affect loci similarly across the entire genome, selection affects only specific loci because of recombination. Today, readily available sequence information and genetic and physical maps open new perspectives for the possibility of tracking the signatures of evolutionary forces along the genome, even if several methodological problems remain to be resolved. The use of molecular markers tagging specific gene family domains, such as those that are AFLP derived and that have been developed to study wild bean populations (Neema et al., 2001), would also be particularly interesting, and they could also be developed for other gene families (van Tienderen et al., 2002). Similarly, SSRs and single nucleotide polymorphisms located in genic regions (Yu et al., 2000; Gaitán-Solís et al., 2002; McClean et al., 2002; Blair et al., 2003; Guerra-Sanz, 2004) and STSs linked to genes of interest (Murray et al., 2002; McClean et al., 2002; Erdmann et al., 2002) would be of particular interest when used in combination with putative neutral markers such as SSRs developed from genomic libraries (Gaitán-Solís et al., 2002). The development of gene-tagging markers for *Phaseolus* will also increase with the growing expressed sequence tag (EST) sequencing efforts (Broughton et al., 2003). These opportunities should be enhanced by the location of molecular markers and sequence data within genetic (Kelly et al., 2003; Broughton et al., 2003) and physical (Vanhouten and MacKenzie, 1999; Kami and Gepts, 2000; Melotto et al., 2004) maps.

As long as we are able to interpret the increasing amounts of data that are being generated, the development of genomics studies should allow not just the development of new research tools but also an improved understanding of the genome organization and structure and its evolution.

Acknowledgments

This work was partially supported by the Italian Government, Project MURST-Cofin2002, "Beans in Europe," # 2002075423.

References

Arrieta-Montiel, M., A. Lyznik, M. Woloszynska, H. Janska, J. Tohme, and S. Mackenzie. 2001. Tracing evolutionary and developmental implications of mitochondrial stoichiometric shifting in the common bean. *Genetics* 158(6): 851–64.

Becerra, V. L., and P. Gepts. 1994. RFLP diversity of common bean (*Phaseolus vulgaris*) in its centres of origin. *Genome* 37: 256–263.

Beebe, S., J. Rengifo, E. Gaitan, M. C. Duque, and J. Tohme. 2001. Diversity and origin of Andean landraces of common bean. *Crop Science* 41: 854–862.

Beebe, S., P. W. Skroch, J. Tohme, M. C. Duque, F. Pedraza, and J. Nienhuis. 2000. Structure of genetic diversity between common bean land races of Middle American origin based on correspondence analysis of RAPD. *Crop Science* 40: 264–273.

Beebe, S., O. Toro, A. V. González, M. I. Chacon, and D. G. Debouck. 1997. Wild-weed–crop complexes of common bean (*Phaseolus vulgaris* L, Fabaceae) in the Andes of Peru and Colombia, and their implications for conservation and breeding. *Genetic Resources and Crop Evolution* 44: 73–91.

Birri, F. and C. Coco. 2000. *Cade a Fagiolo. Dal mondo antico alla nostra tavola. Storia e miti della carne dei poveri.* Marsilio, Venice, IT.

Blair, M. W., F. Pedraza, H. F. Buendia, E. Gaitán-Solís, S. E. Beebe, P. Gepts, and J. Thome. 2003. Development of a genome-wide anchored microsatellite map of common bean (*Phaseolus vulgaris* L.). *Theoretical and Applied Genetics* 107: 1362–1374.

Broughton, W. J., G. Hernandez, M. Blair, S. Beebe, P. Geptsand, and J. Vanderleyden. 2003. Beans (*Phaseolus* spp.): Model food legumes. *Plant and Soil* 252: 55–128.

Brown, J. W. S., F. A. Bliss, and T. C. Hall. 1982. Genetic variation in the subunits of globulin-1 storage protein of French bean. *Theoretical and Applied Genetics* 59: 83–88.

Brucher, B. and H. Brucher. 1976. The South American wild bean (*Phaseolus aborigenus* Burk.) as an ancestor of the common bean. *Economic Botany* 30: 257–272.

Caicedo, A. L., E. Gaitan, M. C. Duque, O. Toro Chica, D. G. Debouck, and J. Tohme. 2000. AFLP fingerprinting of *Phaseolus lunatus* L. and related wild species from South America. *Crop Science* 39: 1497–1507.

Cattan-Toupance, I., Y. Michalakis, and C. Neema. 1998. Genetic structure of wild bean populations in their South-Andean centre of diversity. *Theoretical and Applied Genetics* 96: 844–851.

Cavalli-Sforza, L. L. 1966. Population structure and human evolution. *Proceedings of the Royal Society, London* Ser. B 164: 362–379.

Clark, R. M., E. Linton, J. Messing, and J. F. Doebley. 2003. Pattern of diversity in the genomic region near the maize domestication gene *tb1*. *Proceedings of the National Academy of Sciences (USA)* 101: 700–707.

Debouck, D. G. 1991. Systematics and morphology. In A. van Schoonhoven and O. Voysest (eds.), *Common Beans, Research for Crop Improvement,* 55–118. CAB International, Oxon, UK.

Debouck, D. G. 1999. Biodiversity, ecology and genetic resources of *Phaseolus* beans: Seven answered and unanswered questions. In *Wild Legumes*. National Institute of Agrobiological Resources Tsukuba, Ibaraki, Japan.

Debouck, D. G., J. H. Linan Jara, S. A. Campana, and R. J. H. de la Cruz. 1987. Observations on the domestication of *Phaseolus lunatus* L. *FAO/IBPGR Plant Genetic Resources Newsletters* 70: 26–32.

Delgado-Salinas, A. 1985. Systematics of the genus *Phaseolus* (Leguminosae) in North and Central America. Ph.D. dissertation, University of Texas, Austin, TX, USA.

Delgado-Salinas, A., A. Bruneau, and J. J. Doyle. 1993. Chloroplast DNA phylogenetic studies in New World Phaseolinae (Leguminosae: Papilionoideae: Phaseoleae). *Systematic Botany* 18: 1, 6–17.

Delgado-Salinas, A., T. Turley, A. Richman, and M. Lavin. 1999. Phylogenetic analysis of the cultivated and wild species of *Phaseolus* (Fabaceae). *Systematic Botany* 24: 438–460.

de Meaux, J., I. Cattan-Toupance, C. Lavigne, T. Langin, and C. Neema. 2003. Polymorphism of a complex resistance gene candidate family in wild populations of common bean (*Phaseolus vulgaris*) in Argentina: Comparison with phenotypic resistance polymorphism. *Journal of Molecular Evolution* 12: 263–273.

de Meaux, J. and T. Mitchell-Olds. 2003. Evolution of plant resistance at the molecular level: Ecological context of species interactions. *Heredity* 91: 345–352.

de Meaux, J. and C. Neema. 2003. Spatial patterns of the putative recognition domain of resistance gene candidates in wild bean populations. *Journal of Molecular Evolution* 57: S90–S102.

Doebley, J., A. Stec, J. Wendel, and M. Edwards. 1990. Genetic and morphological analysis of a maize–teosinte F_2 population: Implications for the origin of maize. *Proceedings of the National Academy of Sciences (USA)* 87: 9888–9892.

Ellstrand, N., H. Prentice, and J. Hancock. 1999. Gene flow and introgression from domesticated plants into their wild relatives. *Annual Review of Ecology and Systematics* 30: 539–563.

Erdmann, P. M., R. K. Lee, M. J. Bassett, and P. E. McClean. 2002. A molecular marker tightly linked to P, a gene required for flower and seedcoat color in common bean (*Phaseolus vulgaris* L.), contains the *Ty3*–gypsy retrotransposon *Tpv3g*. *Genome* 45: 728–736.

Escribano, M. R., M. Santalla, P. A. Casquero, and A. M. De Ron. 1998. Patterns of genetic diversity in landraces of common bean (*Phaseolus vulgaris* L.) from Galicia. *Plant Breeding* 117: 49–56.

Evans, A. M. 1973. Plant architecture and physiological efficiency in the field bean. In D. Wall (ed.), *Potentials of Field Bean and Other Food Legumes in Latin America,* 279–284. CIAT, Cali, Colombia.

Evans, A. M. 1980. Structure, variation, evolution and classification in *Phaseolus*. In R. J. Summerfield and A. H. Bunting (eds.), *Advances in Legume Science,* 337–347. Royal Botanic Gardens, Kew, UK.

Ferreira, J. J., E. Alvarez, M. A. Fueyo, A. Roca, and R. Giraldez. 2000. Determination of the outcrossing rate of *Phaseolus vulgaris* L. using seed protein markers. *Euphytica* 113: 259–263.

Ferrier-Cana, E., V. Geffroy, C. Macadre, F. Creusot, P. Imbert-Bollore, M. Sevignac, and T. Langin. 2003. Characterization of expressed NBS–LRR resistance gene candidates from common bean. *Theoretical and Applied Genetics* 106: 251–261.

Fofana, B., J. P. Baudoin, X. Vekemans, D. G. Debouck, and P. du Jardin. 1999. Molecular evidence for an Andean origin and a secondary gene pool for the lima bean (*Phaseolus lunatus* L.) using chloroplast DNA. *Theoretical and Applied Genetics* 98: 202–212.

Free, J. B. 1966. The pollination of the beans *Phaseolus multiflorus* and *Phaseolus vulgaris* by honeybees. *Journal of Apicultural Research* 5: 87–91.

Freyre, R., R. Ríos, L. Guzmán, D. Debouck, and P. Gepts. 1996. Ecogeographic distribution of *Phaseolus* spp. (Fabaceae) in Bolivia. *Economic Botany* 50: 195–215.

Freyre, R., P. Skeoch, V. Geffroy, A.-F. Adam-Blondon, A. Shirmohamadali, W. Johnson, V. Llaca, R. Nodari, P. Pereira, S.-M. Tsai, J. Tohme, M. Dron, J. Nienhuis, C. Vallejos, and P. Gepts. 1998. Towards an integrated linkage map of common bean. 4. Development of a core map and alignment of RFLP maps. *Theoretical and Applied Genetics* 97: 847–856.

Fuchs, L. 1543. *Von Welschen Bonen. Cap.* CCLXIX *in Neu Kreuternuch.* Michal Isengrin, Basel. Reprinted in 1964. K. Kölb, Munich, Germany.

Gaitán-Solís, E., M. C. Duque, K. J. Edwards, and J. Thome. 2002. Microsatellite repeats in common bean (*Phaseolus vulgaris*): Isolation, characterization, and cross-species amplification in *Phaseolus* ssp. *Crop Science* 97: 847–856.

Geffroy, V., D. Sicard, J. C. F. De Oliveira, M. Sevignac, S. Cohen, P. Gepts, C. Neema, T. Langin, and M. Dron. 1999. Identification of an ancestral resistance gene cluster involved in the coevolution process between *P. vulgaris* and its fungal pathogen *Colletotrichum lindemuthianum*. *Molecular Plant–Microbe Interactions* 12: 774–784.

Gepts, P. 1988a. A Middle American and Andean gene pool. In P. Gepts (ed.), *Genetic Resources of* Phaseolus *Beans,* 375–390. Kluwer Academic Publishers, Dordrecht, The Netherlands.

Gepts, P. 1988b. Phaseolin as an evolutionary marker. In P. Gepts (ed.), *Genetic Resources of* Phaseolus *Beans,* 215–241. Kluwer Academic Publishers, Dordrecht, The Netherlands.

Gepts, P. 1996. Origin and evolution of cultivated *Phaseolus* species. In B. Pickersgill and J. M. Lock (eds.), *Advances in Legume Systematics 8: Legumes of Economic Importance,* 65–74. Royal Botanic Gardens, Kew, UK.

Gepts, P. 1999. Development of an integrated genetic linkage map in common bean (*Phaseolus vulgaris* L.) and its use. In S. Singh (ed.), *Bean Breeding for the 21st Century,* 53–91, 389–400. Kluwer, Dordrecht, The Netherlands.

Gepts, P. 2002. A comparison between crop domestication, classical plant breeding, and genetic engineering. *Crop Science* 42: 1780–1790.

Gepts, P. and F. A. Bliss. 1988. Dissemination pathways of common bean (*Phaseolus vulgaris,* Fabaceae) deduced from phaseolin electrophoretic variability. II. Europe and Africa. *Economic Botany* 42: 86–104.

Gepts, P., T. C. Osborne, K. Rashka, and F. A. Bliss. 1986. Electrophoretic analysis of phaseolin protein variability in wild forms and landraces of the common bean, *Phaseolus vulgaris* L.: Evidence for two centers of domestications. *Economic Botany* 40: 451–468.

Gepts, P. and R. Papa. 2002. Evolution during domestication. In *Encyclopedia of Life Science.* Macmillan, Nature Publishing Group, London.

González, A., A. Wong, A. Delgado-Salinas, R. Papa, and P. Gepts. 2005. Assessment of inter simple sequence repeat markers to differentiate sympatric wild and domesticated populations of common bean (*Phaseolus vulgaris* L.). *Crop Science* 45: 606–615.

Guerra-Sanz, J. M. 2004. New SSR markers of *Phaseolus vulgaris* from sequence databases. *Plant Breeding* 123: 87–89.

Guzman, P., R. L. Gilbertson, R. Nodori, W. C. Johnson, S. R. Temple, D. Mandala, A. B. C. Mkandawire, and P. Gepts. 1995. Characterization of variability in the fungus *Phaeoisariopsis griseola* suggests coevolution with the common bean (*Phaseolus vulgaris* L.). *Phytopathology* 85: 600–607.

Hamann, A., D. Zink, and W. Nagl. 1995. Microsatellite fingerprinting in the genus *Phaseolus*. *Genome* 38: 507–515.

Hammer, K. 1984. Das Domestikations syndrome. *Kulturpflanze* 32: 11–34.

Harlan, J. R. and J. M. J. de Wet. 1971. Towards a rational classification of cultivated plants. *Taxon* 20: 509–517.

Hernandez-Xolocotzi, E., S. Miranda Colin, and C. Prwyer. 1959. El origen de *Phaseolus coccineus* L. *darwinianus* Hernandez X. & Miranda C. subspecies nova. *Revista de la Sociedad Mexicana de Historia Natural* 20: 99–121.

Ibarra-Pérez, F., B. Ehadaie, and G. Waines. 1997. Estimation of outcrossing rate in common bean. *Crop Science* 37: 60–65.

Jaaska, V. 1996. Isoenzyme diversity and phylogenetic affinities among the *Phaseolus* beans (Fabaceae). *Plant Systematics and Evolution* 200: 3–4, 233–252.

Kami, J., V. Becerra-Velasquez, D. G. Debouck, and P. Gepts. 1995. Identification of presumed ancestral DNA sequences of phaseolin *Phaseolus vulgaris. Proceedings of the National Academy of Sciences (USA)* 92: 1101–1104.

Kami, J. and P. Gepts. 2000. Development of a BAC library in common bean genotype BAT93. *Annual Report Bean Improvement Cooperative* 43: 208–209.

Kaplan, N., R. Hudson, and C. Langley. 1989. The "hitchhiking" effect revisited. *Genetics* 123: 887–899.

Kelly, J. D., P. Gepts, P. N. Miklas, and D. P. Coyne. 2003. Tagging and mapping of genes and QTL and molecular marker–assisted selection for traits of economic importance in bean and cowpea. *Field Crops Research* 82: 135–154.

Khairallah, M. M., B. B. Sears, and M. W. Adams. 1992. Mitochondrial restriction fragment length polymorphisms in wild *Phaseolus vulgaris* L.: Insights on the domestication of the common bean. *Theoretical and Applied Genetics* 84: 915–922.

Kloz, J. 1971. Serology of the Leguminosae. In J. P. Harborne, D. Boulter, and B. L. Turner (eds.), *Chemotaxonomy of the Leguminosae,* 309–365. Academic Press, London, UK.

Koenig, R. and P. Gepts. 1989. Allozyme diversity in wild *Phaseolus vulgaris:* Further evidence for two major centers of genetic diversity. *Theoretical and Applied Genetics* 78: 809–817.

Koenig, R., S. P. Singh, and P. Gepts. 1990. Novel phaseolin types in wild and cultivated common bean (*P. vulgaris,* Fabaceae). *Economic Botany* 44: 50–60.

Koinange, E. M. K., S. P. Singh, and P. Gepts. 1996. Genetic control of the domestication syndrome in common bean. *Crop Science* 36: 1037–1045.

Lackey, J. A. 1981. Phaseolae. In R. M. Polhill and P. H. Raven (eds.), *Advances in Legume Systematics,* Part 2, 301–327. Royal Botanic Gardens, Kew, UK.

Lackey, J. A. 1983. A review of generic concepts in American Phaseolinae (Fabaceae, Faboideae). *Iselya* 2: 21–64.

Ladizinsky, G. 1998. *Plant Evolution and Origin of Crop Species.* Prentice Hall, Englewood Cliffs, NJ, USA.

Lioi, L. 1989. Geographical variation of phaseolin patterns in an old world collection of *Phaseolus vulgaris. Seed Science & Technology* 17: 317–324.

Llaca, V., S. A. Delgado, and P. Gepts. 1994. Chloroplast DNA as an evolutionary marker in the *Phaseolus vulgaris* complex. *Theoretical and Applied Genetics* 88: 646–652.

Maquet, A. and J. P. Baudoin. 1996. New lights on phyletic relations in the genus *Phaseolus. Annual Report Bean Improvement Cooperative* 39: 203–204.

Maréchal, R., J. Mascherpa, and F. Stanier. 1978. Etude taxonomique d'un groupe complexe d'espècies des genres *Phaseolus* et *Vigna* (Papilonaceae) sur la base de donées morphologiques, traitées par l'analyse informatique. *Boissiera* 28: 1–273.

Maynard Smith, J. and J. Haigh. 1974. The hitch-hiking effect of a favourable gene. *Genetical Research* 23: 23–35.

McClean, P. E., R. K. Lee, and P. N. Miklas. 2004. Sequence diversity analysis of dihydroflavonol 4-reductase intron 1 in common bean. *Genome* 47: 266–280.

McClean, P. E., R. Lee, C. Otto, P. Gepts, and M. J. Bassett. 2002. STS and RAPD mapping of genes controlling seed coat color and patterning in common bean. *Journal of Heredity* 93: 148–152.

Melotto, M., M. F. Coelho, A. Pedrosa-Harand, J. D. Kelly, and L. E. A. Camargo. 2004. The anthracnose resistance locus Co-4 of common bean is located on chromosome 3 and contains putative disease resistance–related genes. *Theoretical and Applied Genetics* 109: 690–699.

Murray, J., J. Larsen, T. E. Michaels, A. Schaafsma, C. E. Vallejos, and K. P. Pauls. 2002. Identification of putative genes in bean (*Phaseolus vulgaris*) genomic (Bng) RFLP clones and their conversion to STSs. *Genome* 45:1013–1024.

Neema, C., C. Lavignes, J. De Meaux, I. Cattan-Toupance, J. F. de Oliviera, A. Deville, and T. Langin. 2001. Spatial pattern for resistance to a pathogen. Theoretical approach and empirical approach at the phenotypic and molecular level. *Genetic Selection Evolution* 33: S3–S21.

Ocampo, C. H., J. P. Martin, J. M. Ortiz, M. D. Sanchez-Yelamo, O. Toro, and D. Debouck. 2002. Possible origins of common bean (*Phaseolus vulgaris* L.) cultivated in Spain in relation to the wild genetic pools of Americas. *Annual Report Bean Improvement Coop* 45: 236–237.

Ortwin-Sauer, C. 1966. *The Early Spanish Man*. University of California Press, Berkeley, CA, USA.

Papa, R., J. Acosta, A. Delgado-Salinas, and P. Gepts. In press. A genome-wide analysis of differentiation between wild and domesticated *Phaseolus vulgaris* from Mesoamerica. *Theoretical and Applied Genetics*.

Papa, R. and P. Gepts. 2003. Asymmetry of gene flow and differential geographical structure of molecular diversity in wild and domesticated common bean (*Phaseolus vulgaris* L.) from Mesoamerica. *Theoretical and Applied Genetics* 106: 239–250.

Payró de la Cruz, E., P. Gepts, P. Colunga Garcia-Marín, and D. Zizumbo Villareal. In press. Spatial distribution of genetic diversity in wild populations of *Phaseolus vulgaris* L. from Guanajuato and Michoacán, México. *Genetic Resources and Crop Evolution*.

Piergiovanni, A. R., D. Cerbino, and M. Brandi. 2000a. The common bean populations from Basilicata (southern Italy). An evaluation of their variation. *Genetic Resources & Crop Evolution* 489–495.

Piergiovanni, A. R., G. Taranto, and D. Pignone. 2000b. Diversity among common bean populations from the Abruzzo region (central Italy): A preliminary inquiry. *Genetic Resources & Crop Evolution* 47: 467–470.

Piñero, D. and L. Eguiarte. 1988. The origin and biosystematic status of *Phaseolus coccineus* subsp. *polyanthus*: Electrophoretic evidence. *Euphytica* 37: 199–203.

Provan, J., W. Powell, and P. M. Hollingsworth. 2001. Chloroplast microsatellites: New tools for studies in plant ecology and evolution. *Trends in Ecology & Evolution* 16: 142–147.

Pueyo, J. J. and A. Delgado-Salinas. 1997. Presence of α-amylase inhibitor in some members of the subtribe Phaseolinae (Phaseolae: Fabaceae). *American Journal of Botany* 84: 79–84.

Rodiño, A. P., M. Santalla, A. M. De Ron, and S. P. Singh. 2003. A core collection of common bean from the Iberian peninsula. *Euphytica* 131: 165–175.

Rodiño, A. P., M. Santalla, I. Montero, P. A. Casquero, and A. M. De Ron. 2001. Diversity in common bean (*Phaseolus vulgaris* L.) germplasm from Portugal. *Genetic Resources & Crop Evolution* 48: 409–417.

Santalla, M., A. P. Rodiño, and A. M. De Ron. 2002. Allozyme evidence supporting southwestern Europe as a secondary center of genetic diversity for common bean. *Theoretical and Applied Genetics* 104: 934–944.

Schmit, V., P. du Jardin, J. P. Baudoin, and D. G. Debouck. 1993. Use of chloroplast DNA polymorphism for the phylogenetic study of seven *Phaseolus* taxa including *P. vulgaris* and *P. coccineus*. *Theoretical and Applied Genetics* 87: 506–516.

Seo, Y. S., P. Gepts, and R. L. Gilbertson. 2004. Genetics of resistance to the geminivirus, bean dwarf mosaic virus, and the role of the hypersensitive response in common bean. *Theoretical and Applied Genetics* 108: 786–793.

Sicard, D., Y. Michalakis, M. Dron, and C. Neema. 1997a. Population structure of *C. lindemuthianum* in the three centres of diversity of the common bean. *Phytopathology* 87: 807–813.

Sicard, D., Y. Michalakis, and C. Neema. 1997b. Diversity of *Colletotrichum lindemuthianum* in wild common bean populations. *Plant Pathology* 46: 355–365.

Sicard, D., L. Nanni, O. Porfiri, D. Bulfon, and R. Papa. In press. Genetic diversity of *Phaseolus vulgaris* L. and *P. coccineus* L. landraces in central Italy. *Plant Breeding*.

Simmonds, N. W. 1976. *Evolution of Crop Plants*. Longman, London, UK.

Singh, S. P. 2001. Broadening the genetic base of common bean cultivars: A review. *Crop Science* 41: 1659–1675.

Singh, S. P., R. Nodari, and P. Gepts. 1991. Genetic diversity in cultivated common bean. I. Allozymes. *Crop Science* 31:19–23.

Snoeck, C., J. Vanderleyden, and S. Beebe. 2003. Strategies for genetic improvement of common bean and rhizobia towards efficient interactions. *Plant Breeding Reviews* 23: 21–72.

Sonnante, G., T. Stockton, R. O. Nodari, V. L. Becerra Velásquez, and P. Gepts. 1994. Evolution of genetic diversity during the domestication of common bean (*Phaseolus vulgaris* L.). *Theoretical and Applied Genetics* 89: 629–635.

Sousa, M. and A. Delgado-Salinas. 1993. Mexican Leguminosae: Phytogeography, endemism, and origin. In T. P. Ramamoorthy, R. Bye, A. Lot, and J. Fa (eds.), *Biological Diversity of Mexico: Origins and Distribution*, 459–511. Oxford University Press, Oxford, UK.

Steadman, J. R., J. Beaver, M. Boudreau, D. Coyne, J. Groth, J. Elly, M. McMillan, R. McMillan, P. Miklas, M. Pastor Corrales, H. Schwartz, and J. Stavely. 1995. Progress reported at the 2nd International Bean Rust Workshop. *Annual Report of Bean Improvement Cooperative* 38: 1–10.

Sullivan, J. G. and G. Freytag. 1986. Predicting interspecific compatibilities in beans (*Phaseolus*) by seed protein electrophoresis. *Euphytica* 35: 201–209.

Tenaillon, M. I., C. M. Sawkins, A. D. Long, R. L. Gaut, J. F. Doebley, and B. S. Gaut. 2001. Patterns of DNA sequence polymorphism along chromosome 1 of maize (*Zea mays* ssp. *mays* L.). *Proceedings of the National Academy of Science (USA)* 98: 9161–9166.

Van der Groen, J. 1669. *Den Nederlandtsen Hovenier*. Reprint 1988. Sichting Matrijs, Utrecht, The Netherlands.

Vanhouten, W. and S. MacKenzie. 1999. Construction and characterization of a common bean bacterial artificial chromosome library. *Plant Molecular Biology* 40: 977–983.

Van Tienderen, P. H., A. A. Van Haan, C. G. Van der Linden, and B. Vosman. 2002. Biodiversity assessment using markers for ecologically important traits. *Trends in Ecology & Evolution* 17: 577–582.

Vekemans, X., O. Hardy, B. Berken, B. Fofana, and J. P. Baudoin. 1998. Use of PCR-RFLP on chloroplast DNA to investigate phylogenetic relationships in the genus *Phaseolus*. *Biotechnology, Agronomy, Society and Environment* 2: 128–134.

Verdecourt, B. 1970. Studies in the Leguminosae–Papilionoideae for the flora of tropical East Africa: IV. *Kew Bulletin* 24: 507–569.

Wang, R. L., A. Stec, J. Hey, L. Lukens, and J. Doebley. 1999. The limits of selection during maize domestication. *Nature* 398: 236–239.

Wang, R. L., A. Stec, J. Hey, L. Lukens, and J. Doebley. 2001. Correction: The limits of selection during maize domestication. *Nature* 410: 718.

Yu, K., S. J. Park, V. Poysa, and P. Gepts. 2000. Integration of simple sequence repeat (SSR) markers into a molecular linkage map of common bean (*Phaseolus vulgaris* L.). *Journal of Heredity* 91: 429–434.

Mallikarjuna K. Aradhya,
Daniel Potter, and Charles J. Simon

Cladistic Biogeography of *Juglans* (Juglandaceae) Based on Chloroplast DNA Intergenic Spacer Sequences

Juglans L. is principally a New World genus within the tribe Juglandeae of the family Juglandaceae, comprising about 21 extant deciduous tree species occurring from North and South America, the West Indies, and southeastern Europe to eastern Asia and Japan (Manning, 1978). It is one of the approximately 65 genera that are known to exhibit a disjunct distributional pattern between eastern Asia and eastern North America (Manchester, 1987; Wen, 1999; Qian, 2002; figure 7.1). Four sections are commonly recognized within *Juglans,* based mainly on fruit morphology, wood anatomy, and leaf architecture (Dode, 1909a, 1909b; Manning, 1978). Section *Rhysocaryon* (black walnuts), which is endemic to the New World, comprises five North American temperate taxa: *J. californica* S. Wats., *J. hindsii* (Jeps.) Rehder, *J. nigra* L., *J. major* (Torr. ex Sitgr.) Heller, and *J. microcarpa* Berl.; three Central American subtropical taxa: *J. mollis* Engelm., *J. olanchana* Stadl. & I. O. Williams, and *J. guatemalensis* Mann.; and two South American tropical taxa, *J. neotropica* Diels and *J. australis* Griesb, mainly occurring in the highlands. They typically bear nuts that are four-chambered with thick nutshells and septa. Section *Cardiocaryon* (Asian butternuts) contains four taxa: *J. hopeiensis* Hu, *J. ailantifolia* Carr., *J. mandshurica* Maxim., and *J. cathayensis* Dode, all native to East Asia, and section *Trachycaryon* consists of the only North American butternut taxon, *J. cinerea* L. Both Asian and

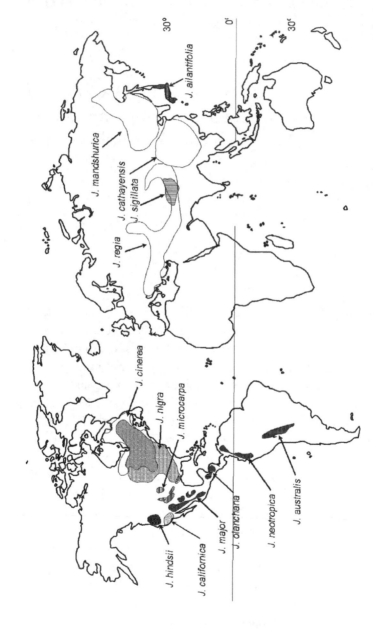

FIGURE 7.1 Geographic distribution of extant taxa of *Juglans* (Juglandaceae). The distribution of cultivated species *J. regia* extends beyond its natural home range.

American butternuts possess two-chambered nuts with thick nutshells and septa. Section *Juglans* includes two taxa: The cultivated Persian or English walnut, *J. regia* L., ranges from southeastern Europe to the Himalayas and China and bears four-chambered nuts with thin nutshells and papery septa, and the iron walnut, *J. sigillata* Dode, ranges from southern China and Tibet and has thick, rough-shelled nuts and characteristic dark-colored kernels (Dode, 1909a). The iron walnut sometimes is considered an ecotype of *J. regia,* but some botanists treat it as a separate species (Kuang et al., 1979). It is known to have been cultivated for a long time in Yunnan province of China for its oil. Complete descriptions of the morphological variation, ecological distribution, and taxonomic treatment of the genus *Juglans* are found in Manning (1957, 1960, 1978).

Plant species disjunctions have been the subject of many taxonomic and biogeographic studies. The most notable among them is the East Asian–North American disjunction, the origin of which has been studied from the paleobotanical, geological, and paleoclimatic perspectives. Various hypotheses have been proposed to explain the origin of these disjunctions, and Asa Gray's (1859, 1878) pioneering accounts in the mid-19th century of the floristic similarities of East Asia and eastern North America serve as the foundation for the modern systematic syntheses of plant species disjunctions. He proposed that many plant taxa were widely distributed throughout the Northern Hemisphere during the early Tertiary, and later disruptions by glaciation led to eastern Asian–eastern North American disjunctions. Subsequently, Chaney (1947) and Axelrod (1960) independently modified Gray's hypothesis to suggest that the floristic similarities originated as the result of range restrictions and southward migration of the homogeneous Arcto-Tertiary geoflora of the Northern Hemisphere caused by climatic changes in the late Tertiary and Quaternary. Recently, additional paleofloristic and geological discoveries have led to more complex alternative hypotheses regarding the mode and time of origin of disjunction patterns (Wolfe, 1975, 1978, 1985; Tiffney, 1985a, 1985b). However, it is now known from fossil records that deciduous woody taxa first appeared in northern latitudes as part of a mostly broad-leaved evergreen, tropical forest in the late Eocene (Wolfe, 1969, 1972). Cooling climates during the Oligocene and Miocene saw diversification and expansion of broad-leaved, deciduous taxa throughout the northern latitudes of Eurasia and North America (Wolfe, 1978, 1985), and taxa were exchanged via the Bering or North Atlantic land bridges throughout the mid-Tertiary. Continued cooling in the Pliocene produced retraction of mixed mesophytic forest from northern latitudes and greatly reduced the possibility of migration between Eurasia and North America (Wolfe, 1978, 1985; Tiffney, 1985a). Further climatic changes during the Quaternary effectively eliminated the northern mixed mesic forests, leaving eastern North America, eastern Asia, and to a much lesser extent the Balkans and Caucasus as the main refugia of many genera (Graham, 1972; Tiffney, 1985a). Others have implicated convergent adaptation to similar climatic conditions and long-distance dispersal in the development of present-day floristic disjunctions (Raven, 1972; Wolfe, 1975).

Based on fossil evidence, Manchester (1987) suggests that the origin of *Juglans*, including the initial split into black walnuts and butternuts, may have occurred sometime during the Middle Eocene in North America. Furthermore, expansion and migration between North America and Eurasia were facilitated by the presence of the Bering land bridge that connected eastern Asia with western North America throughout the mid-Tertiary and by a North Atlantic land bridge during the late Eocene, when there was a favorable climate in upper latitudes for the establishment and dispersal of deciduous and some broad-leaved evergreens. The latter were able to adapt to the Neogene cooler climate (Wolfe, 1978; Tiffney, 1985a), attaining a broad distribution extending farther south into southeastern Europe and Central and South America by the late Miocene. However, the fossil record suggests that black walnuts remained endemic to the Americas, whereas butternuts are represented by members in Asia as well as one in eastern North America. The section *Juglans* is not known in the fossil record.

The usefulness of chloroplast dna (cpdna) sequence data to estimate the rate and time of divergence between disjunct taxa is well documented (Crawford et al., 1992), but only a limited number of disjunct taxa have been examined phylogenetically using cpdna data in order to explore the biogeographic relationships, mode, and tempo of disjunction (Wen, 1999). The eastern Asian–eastern North American Tertiary disjunction in *Juglans* offers an opportunity to estimate the level and time of evolutionary divergence between vicariant groups and to compare this with the time of divergence inferred from paleobotanical evidence. Earlier molecular systematic studies based on nuclear r fl ps (Fjellstrom and Parfitt, 1995) and *matK* and internal transcribed spacer (it s) sequences (Stanford et al., 2000) support the traditional taxonomic classification of *Juglans* and are consistent with what is known about the geological history of the genus (Dode, 1909a, 1909b; Manning, 1978; Manchester, 1987).

Noncoding intergenic spacer regions of cpdna, which are presumably under less functional constraint than coding regions, are known to evolve rapidly and provide useful information to examine systematic relationships at lower taxonomic levels (Ogihara et al., 1991; Gielly and Taberlet, 1994). Recently, availability of several universal chloroplast primers to amplify noncoding regions (Taberlet et al., 1991; Demesure et al., 1995) has facilitated this effort to infer phylogenetic relationships at the generic (Gielly and Taberlet, 1994; Small et al., 1998; Cros et al., 1998; Aradhya et al., 1999; Stanford et al., 2000) and even infraspecific levels (Demesure et al., 1996; Petit et al., 1997; Mohanty et al., 2001). In the present study, we examine the utility of some of these cpdna intergenic spacer sequences for phylogenetic reconstruction and for assessing the level of evolutionary divergence within and between sections of *Juglans*. We also explore the biogeography of the genus *Juglans* based on the phylogenetic inferences and, in particular, the origin, evolution, and domestication history of the section *Juglans*, to which the cultivated walnut *J. regia* belongs.

Materials and Methods

Plant Materials, DNA Isolation, PCR Amplification, and Sequencing

Seventeen taxa representing the four sections of *Juglans* and one outgroup taxon, *Pterocarya stenoptera*, were sampled for this study (table 7.1). *Pterocarya* was chosen as the outgroup taxon because it is closely related to *Juglans* (Smith and Doyle, 1995; Manos and Stone, 2001). Total DNA was isolated using the cetyltrimethylammonium bromide method (Doyle and Doyle, 1987) and further extracted with phenol-chloroform and treated with RNAse to remove protein and RNA contaminants, respectively.

Five cpdna intergenic spacer regions: *trnT–trnF* (Hodges and Arnold, 1994), *psbA–trnH* (Sang et al., 1997), *atpB–rbcL* (Taberlet et al., 1991), *trnV–16S rRNA* (Al-Janabi et al., 1994), and *trnS–trnfM* (Demesure et al., 1995) were pcr amplified separately in a 100-μL reaction mixture containing 10 mM Tris-HCl, pH 8.3, 50 mM KCl, 2 mM $MgCl_2$ (all included in 10 μL of 10x pcr buffer), 10–20 pmol of each primer, 200 μM of each dntp, 2 U of Taq polymerase (Perkin Elmer Biosystems, ca, usa), and 50 ng of template dna. The pcr conditions were as follows: one cycle of 5 min at 94°C, 30 cycles of 45 s to 1 min at 94°C, 45 s to 1 min at 55–62°C, 2–3 min at 72°C, and one cycle of 7 min at 72°C. Amplification products were purified and concentrated using qiaquick pcr purification kit (Qiagen Inc., ca, usa)

Table 7.1 Species List, Collection Site, Geographic Origin, and GenBank Accession Numbers

			GenBank Accession Numbers				
Taxon (NCGR* Accession no.)	Collection Site	Origin	*atpB– rbcL*	*psbA– trnH*	*trnS– trnfM*	*trnT– trnF*	*trnV– 16s* rRNA
Cardiocaryon (Asian Butternut)							
J. ailantifolia (DJUG 91.4)	NCGR, Davis, CA	Japan	AY293314	AY293335	AY293365	AY293398	AY293360
J. cathayensis (DJUG 11.4)	NCGR, Davis, CA	Taiwan	AY293312	AY293334	AY293367	AY293396	AY316200
J. mandshurica (DJUG 13.1))	NCGR, Davis,CA	Korea	AY293315	AY293337	AY293364	AY293397	AY293361
J. hopeiensis (DJUG 462)	NCGR, Davis, CA	China	AY293320	AY293342	AY293371	AY293390	AY293358
Juglans (English Walnut)							
J. regia (DJUG 379.1b)	NCGR, Davis, CA	China	AY293322	AY293344	AY293369	AY293395	AY293356
J. sigillata (DJUG 528)	NCGR, Davis, ca	China	AY293317	AY293346	AY293370	AY293393	AY293357
Rhysocaryon (Black Walnut)							
J. australis (DJUG 429)	NCGR, Davis, CA	Argentina	AY293319	AY293343	AY293379	AY293391	AY293352
J. californica (DJUG 28.5)	NCGR, Davis, CA	USA	AY293323	AY293331	AY293377	AY293384	AY293359
J. microcarpa (DJUG 52.1)	NCGR, Davis, CA	USA	AY293324	AY293332	AY293372	AY293385	AY293349
J. mollis (DJUG 218.3)	NCGR, Davis, CA	USA	AY293329	AY293340	AY293375	AY293388	AY293350
J. neotropica (DJUG 330.2)	NCGR, Davis, CA	Ecuador	AY293321	AY293341	AY293368	AY293389	AY293351
J. nigra (DJUG 57.12)	NCGR, Davis, CA	USA	AY293327	AY293339	AY293366	AY293382	AY293348
J. olanchana (DJUG 212.14)	NCGR, Davis, CA	Mexico	AY293328	AY293333	AY293380	AY293387	AY293353
J. guatemalensis	UC Davis Arboretum	Guatemala	AY293316	AY293345	AY293374	AY293394	AY293354

(*continued*)

Table 7.1 (continued)

Taxon (NCGR* Accession no.)	Collection Site	Origin	atpB– rbcL	psbA– trnH	trnS– trnfM	trnT– trnF	trnV– 16s rRNA
J. hindsii (DJUG 91.4)	NCGR, Davis, CA	USA	AY293326	AY293330	AY293373	AY293383	AY293363
J. major (DJUG 78.6)	NCGR, Davis, CA	USA	AY293325	AY293338	AY293378	AY293386	AY316201
Trachycaryon (American Butternut)							
J. cinerea	UC Davis Pomology	USA	AY293318	AY293347	AY293376	AY293392	AY293355
Outgroup (Wingnut)							
Pterocarya stenoptera (DPTE 17.1)	NCGR, Davis, CA	China	AY293313	AY293336	AY293381	AY293399	AY293362

*USDA National Clonal Germplasm Repository, One Shields Avenue, University of California, Davis, CA 95616, USA.

and sequenced using an ABI PRISM 377 automated sequencer with BigDye Terminator Cycle Sequencing Kit (Perkin Elmer Biosystems).

Sequence Analyses

Alignment of DNA sequences was performed using the software Sequencher (GeneCodes Corp., Ann Arbor, MI, USA) and subsequently manually adjusted. Indels were coded as binary characters regardless of length, and all characters were equally weighted and unordered. Congruence of intergenic spacer sequences was examined with the incongruence length difference (ILD; Farris et al., 1994, 1995) test as implemented in PAUP* 4.0b10 (partition homogeneity test) (Swofford, 2002). Invariant sites were removed from the test, and 100 replications were performed.

Phylogenetic analyses were performed with PAUP* using the maximum parsimony (MP), maximum likelihood (ML), and minimum evolution (ME) methods. MP analysis was performed using the branch-and-bound algorithm with MulTrees activated and the addition of sequence set to Furthest (character optimization accelerated transformation and tree bisection and reconnection [TBR] branch swapping options) to find most parsimonious trees. Bootstrap analysis (100 replicates) using a heuristic search with the

TBR branch swapping option was performed to assess relative support for different clades. Decay values (Bremer, 1988), the number of extra steps needed to collapse a clade, were computed by examining trees longer than the MP solutions, in which strict consensus trees for all topologies that were up to five steps longer than the MP trees generated using branch-and-bound approach were evaluated. An ME tree was constructed using the Kimura (1980) two-parameter distance with ML estimates of gamma and proportion of invariable sites, and 100 bootstrap replications were used to estimate the support for different nodes. The ML analysis was performed using the best evolutionary models identified by the hierarchical likelihood ratio test and Akaike information criterion method provided in the program Modeltest version 3.06 (Posada and Crandall, 1998) with a Jukes–Cantor tree as the starting tree and indel characters excluded from the analysis. A heuristic search with 10 replications of random addition sequence and TBR branch-swapping options was used. One hundred bootstrap replications were performed under the same conditions.

The sequence divergence between two sister lineages was estimated as the average of all pairwise divergence values between species from the two different clades (Xiang et al., 2000). Evolutionary rates were estimated based on the fossil record, and the time of evolutionary divergence was estimated by dividing the pairwise sequence divergence by twice the rate of nucleotide substitution. The molecular clock hypothesis (Zuckerkandl and Pauling, 1965) was tested by computing the difference in the log likelihood scores between ML trees with and without a molecular clock assumption ($2\Delta = \log L_{\text{no clock}} - \log L_{\text{clock}}$), which follows a chi-square distribution with $n - 2$ degrees of freedom where n is the number of sequences or taxa. A likelihood ratio test (Muse and Weir, 1992), which allows for different transversion and transition rates, was used to test the equality of evolutionary rates along different paths of descent leading to two species, using *Pterocaya* as the reference taxon.

Results

Sequence Characteristics and Divergence

More than 3.8 kb of cpDNA sequence from five spacer regions was assembled for each of the 17 ingroup and 1 outgroup taxa. Although potentially parsimony informative characters were found in all five regions, the variation within individual regions was insufficient to obtain a reasonable level of phylogenetic resolution. The ILD test to examine the null hypothesis that the five data sets were homogenous with respect to phylogenetic information suggested that pairwise combinations and combination of all five data partitions did not result in significant incongruence ($p = .01$). The sequence data therefore were combined to obtain a composite data matrix to perform the phylogenetic analyses. There were 112 (2.9%) variable sites among 3834 total characters within *Juglans*, of which 40 (1.04%) were potentially parsimony informative. Eight indels out of a total of 19 observed were potentially informative. Alignment of the *trnT–trnF* region required one 18-bp deletion for the sections *Rhysocaryon* and *Trachycaryon* and a 9-bp insertion for *J. microcarpa* within *Rhysocaryon*, and the rest of the indels, including the remaining four spacer regions, were 1–5 nucleotides long. The GC content ranged from 30.1% for the *atpB–rbcL* region to 47.2% for the *trnV–16S* *r*RNA region, with an overall average of 31.7%, which is typical for plastomes (Palmer, 1991). The ti/tv ratio for pairwise comparisons between taxa ranged from 0 to 3.0, and, surprisingly, most comparisons showed a bias favoring transversion. In general, pairwise sequence divergence was extremely low within and between the sections of *Juglans* (table 7.2). Within the section *Rhysocaryon*, sequence divergence ranged from 0.08% between the two Central American taxa, *J. mollis* and *J. guatemalensis*, to 0.51% between the Central American walnut, *J. olanchana*, and northern California walnut, *J. hindsii*. Among the four Asian butternuts, divergence ranged from 0.159% between *J. ailantifolia* and *J. mandshurica* to 0.635% between *J. cathayensis* and *J. hopeiensis*. Surprisingly, the degree of divergence between American butternut *J. cinerea* and the black walnuts (0.26%) was lesser than to its Asian counterparts (0.717%). The Persian walnut *J. regia* (section *Juglans*) was found to be more similar to the Asian butternuts (0.773% divergence) than to black walnuts (0.818% divergence).

Phylogenetic Reconstruction

Parsimony analysis of the combined data matrix using a branch-and-bound search generated three equally most parsimonious trees of 146 steps (including autapomorphies) with a consistency index of 0.795 (0.595 excluding autapomorphies) and retention index of 0.762. The trees differ only in relative positions of *J. microcarpa* and *J. guatemalensis*. Three major clades are apparent in the strict consensus tree corresponding to the sections *Juglans* (J clade), *Cardiocaryon* (c clade), and *Rhysocaryon–Trachycaryon* (RT clade) (figure 7.2). The single butternut species, *J. cinerea*, native to eastern

Table 7.2 Estimates of Pairwise Distance Between Taxa: Absolute Distance (Above Diagonal) and Kimura 2-Parameter Distance (Below Diagonal)

Taxon	1	2	3	4	5	6	7	8	9	10	11	12	13	14	15	16	17	18
1 *J. nigra*		11	13	12	7	14	6	11	9	6	8	30	22	20	36	27	24	38
2 *J. hindsii*	0.0016		14	15	8	19	11	14	12	11	11	35	25	23	41	32	29	41
3 *J. californica*	0.0019	0.0030		16	10	17	13	12	12	11	13	34	23	19	37	28	27	39
4 *J. microcarpa*	0.0019	0.0030	0.0029		13	17	8	13	13	7	10	31	24	23	38	29	28	39
5 *J. major*	0.0005	0.0016	0.0019	0.0019		15	7	8	8	7	9	31	21	19	37	28	25	39
6 *J. olanchana*	0.0024	0.0035	0.0038	0.0035	0.0024		12	15	13	10	14	32	26	25	41	31	32	40
7 *J. mollis*	0.0011	0.0021	0.0024	0.0013	0.0011	0.0024		9	9	3	8	26	21	20	35	26	23	37
8 *J. neotropica*	0.0011	0.0021	0.0024	0.0019	0.0011	0.0030	0.0011		8	7	11	33	24	21	38	29	28	40
9 *J. australis*	0.0011	0.0021	0.0024	0.0024	0.0011	0.0030	0.0016	0.0005		7	9	32	19	21	37	26	27	40
10 *J. guatemalensis*	0.0008	0.0019	0.0021	0.0013	0.0008	0.0022	0.0005	0.0008	0.0013		5	24	18	18	31	23	23	33
11 *J. cinerea*	0.0011	0.0021	0.0024	0.0024	0.0011	0.0030	0.0016	0.0016	0.0016	0.0011		30	22	20	35	27	26	34
12 *J. regia*	0.0062	0.0073	0.0070	0.0065	0.0062	0.0070	0.0056	0.0064	0.0070	0.0051	0.0064		13	30	35	25	27	40
13 *J. sigillata*	0.0040	0.0051	0.0049	0.0048	0.0040	0.0060	0.0049	0.0043	0.0046	0.0038	0.0046	0.0024		18	27	17	18	34
14 *J. hopeiensis*	0.0038	0.0048	0.0040	0.0048	0.0038	0.0054	0.0043	0.0043	0.0043	0.0040	0.0043	0.0064	0.004		24	15	12	37
15 *J. cathayensis*	0.0083	0.0094	0.0086	0.0091	0.0083	0.0100	0.0086	0.0086	0.0089	0.0078	0.0086	0.0086	0.0067	0.0056		8	14	46
16 *J. mandshurica*	0.0057	0.0067	0.0059	0.0065	0.0057	0.0076	0.0059	0.0059	0.0062	0.0054	0.0062	0.0059	0.0043	0.0029	0.0019		6	38
17 *J. ailantifolia*	0.0051	0.0062	0.0054	0.0059	0.0051	0.0070	0.0054	0.0054	0.0056	0.0051	0.0056	0.0062	0.0037	0.0024	0.0032	0.0008		39
18 *Pterocarya*	0.0078	0.0089	0.0081	0.0089	0.0078	0.0087	0.0081	0.0081	0.0086	0.0073	0.0078	0.0083	0.0070	0.0080	0.0107	0.0083	0.0083	

North America, representing the section *Trachycaryon,* is placed within the black walnut (*Rhysocaryon*) clade. Clades J and RT are strongly supported (bootstrap > 90%, decay = 4), whereas support for C clade, including *J. hopeiensis,* is somewhat lower (bootstrap = 69%), and the sister relationship between the C and RT clades is only weakly supported. However, there is strong support for *J. ailantifolia, J. mandshurica,* and *J. cathayensis* (bootstrap = 83%, decay = 3) within the C clade. The J clade, weakly supported as a sister group to the C and RT clades, is itself strongly supported (bootstrap = 97%, decay = 4) with four unique synapomorphies. Within the clade, the English walnut has seven unique autapomorphies, whereas its sister taxon, *J. sigillata,* possesses one unique mutation. The ME tree (figure 7.2) is basically concordant with the MP analysis, and there is strong bootstrap support for all three major clades.

Modeltest found two optimum models of sequence evolution: the F81+ I+G model (I = 0.8957; α = 0.9144; base frequencies: A = 0.3491, C = 0.1465, G = 0.1722, T = 0.3322; Felsenstein, 1981) based on the likelihood ratio test, and the K81uf+I model (R[A$\leftrightarrow$C] = 1, R[A$\leftrightarrow$G] = 0.8245, R[A$\leftrightarrow$T] = 0.1759, R[C$\leftrightarrow$G] = 0.9378, R[C$\leftrightarrow$T] = 0.8245, R[G$\leftrightarrow$T] = 1; I = 0.9378; Kimura, 1981) based on the Akaike information criterion. However, both F81+I+G (–Ln = 5760.87) and K81uf+I (–Ln = 5748.94) models resolved trees with a topology identical to the mp and me analyses (figure 7.2) and strong bootstrap support, estimated based on the analysis using K81uf+I model, to the sections *Juglans, Cardiocaryon,* and *Rhysocaryon–Trachycaryon.*

There is some evidence for differentiation within the black walnut clade in all three analyses (mp, me, and ml), indicating biogeographic assemblages representing North American temperate, Central American subtropical, and South American tropical highland black walnuts. However, these affinities are weakly supported except for the South American group comprising *J. neotropica* and *J. australis,* which is supported by two unique synapomorphies. Surprisingly, southern California black walnut, *J. californica,* which is considered a conspecific variant of *J. hindsii,* is placed as sister to the rest of the section, *Rhysocaryon.*

Rate of Divergence

The cpDNA intergenic spacer sequence divergence rates for *Juglans* are unknown. However, one can use estimates of time since divergence based on fossil records to compute the rates of sequence evolution. The average overall rate was calculated by dividing the Kimura 2-parameter distances by twice the

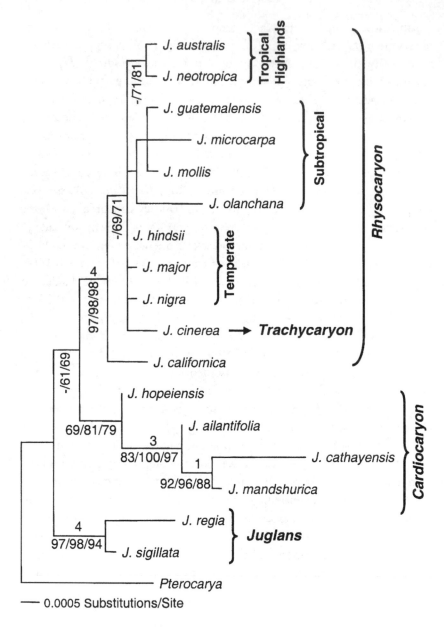

FIGURE 7.2 Phylogram of the genus *Juglans* inferred from maximum likelihood method using K81uf+I model of sequence evolution. Numbers above branches are decay indices, and numbers below are bootstrap support (>50%) based on the following analyses: MP/ME/ML (K81uf+I).

time since divergence. The two landmark divergence events in the evolutionary history of *Juglans* documented in the fossil record were used to compute overall nucleotide substitution rates: the late Paleocene/early Eocene time frame for the divergence of *Pterocarya* and *Juglans* (~54 mya), which yields an average overall rate of sequence divergence of 0.772×10^{-10} substitutions per site per year; and the middle Eocene time frame for the divergence of *Rhysocaryon* and *Cardiocaryon* (~45 mya), as proposed by Manchester (1987), which yields a divergence rate of 0.69×10^{-10} substitutions per site per year. If one of these rates or the average rate (0.731×10^{-10}) is used to compute the time since divergence of different sections within *Juglans,* the results contradict the evolutionary hypothesis based on the fossil history. To address this discrepancy, the test of relative overall nucleotide substitution rates (Muse and Weir, 1992) was used. Using *Pterocarya* as the reference taxon, rates along different paths of descent leading to two ingroup taxa indicated that the section *Juglans,* especially the cultivated walnut *J. regia,* and some taxa in the section *Cardiocaryon* seem to have evolved at significantly different rates than the taxa in the section *Rhysocaryon* (table 7.3). This rate heterogeneity demonstrates that either the ~50-million-year-old *Juglans* lineage is not adequately represented by the extant taxa included in the study, or many taxa at the basal and intermediate nodes might have undergone extinction.

Discussion

Sequence Evolution

Noncoding regions of the chloroplast genome have been suggested to be potentially informative in reconstructing phylogenetic relationships at lower taxonomic levels (Taberlet et al., 1991; Demesure et al., 1995). Nevertheless, the five intergenic spacer sequences (*trnT–trnF, psbA–trnH, atpB–rbcL, trnV–16S rRNA,* and *trnS–trnfM*) used in our study provided little resolution within the major clades, especially among the New World black walnuts and butternut (RT clade). Such low resolution often is seen among taxa that have undergone radiation recently, or it may be result from reticulate evolution within the clade. Despite variation in the information content between different intergenic spacers, the region-specific analysis indicated that the overall phylogenetic structure is conserved across the spacer regions, which was further confirmed by the ILD test. Among the substitutions, transversions were more prevalent than transitions except for the region *psbA–trnH* located within the inverted repeat region of the cpDNA. Although intergenic

Table 7.3 Likelihood Ratio Between Taxa Pairs for Comparing Rates of Evolutionary Change, with *Pterocarya* Used as a Reference Taxon

Taxon	1	2	3	4	5	6	7	8	9	10	11	12	13	14	15	16	17
J. nigra												***			*		
J. hindsii	4.473									*		***			*		
J. californica	1.390	3.107										***					
J. microcarpa	3.612	2.347	0.547							*		***					
J. major	0.000	4.472	1.389	3.610								***			*		
J. olanchana	3.326	0.289	1.934	0.995	3.325							***			*		
J. mollis	1.352	1.289	1.800	4.291	1.351	1.025						***			*		
J. neotropica	1.357	1.278	1.803	2.080	1.357	0.758	0.000					***			*		
J. australis	2.766	0.484	2.869	2.752	2.769	0.475	0.417	1.852				***			**		*
J. guatemalensis	2.197	7.554	5.363	7.216	2.197	3.715	2.816	4.141	5.087			***			**		
J. cinerea	2.766	0.671	3.214	3.386	2.765	0.797	0.510	0.505	0.017	3.340		***			**		*
J. regia	3307.380	3272.150	3299.350	3277.770	3307.410	3280.030	3300.140	3298.290	3283.570	3320.240	3306.360		***	***	***	***	***
J. sigillata	0.926	2.858	0.600	1.955	0.925	2.257	0.887	0.888	2.139	0.506	2.252	3345.540			**	***	
J. hopeiensis	1.558	2.243	0.117	0.482	1.557	1.135	1.729	1.729	2.413	5.790	2.773	3309.320	0.501		*		**
J. cathayensis	8.757	7.739	5.175	2.607	8.755	6.497	6.814	6.816	8.708	11.182	12.330	3238.370	12.816	6.479			
J. mandshurica	4.349	4.742	1.554	0.537	4.348	3.107	2.922	2.923	5.197	4.246	6.127	3295.320	4.248	2.056	5.645		
J. ailantifolia	4.816	5.812	1.987	1.100	4.814	4.156	3.476	3.477	6.076	5.144	6.868	3303.260	3.775	4.082	10.045	3.883	

Above diagonal: Taxa pair, with significance at ***$p < .001$, **$p < .01$, and *$p < .05$.

spacers are considered to be under fewer functional constraints and expected to evolve more rapidly than coding sequences (Wolfe et al., 1987; Zurawski and Clegg, 1987), surprisingly, the level of within-clade resolution observed is far lower than the divergence levels reported for the cpDNA *matK* gene and nuclear ITS spacer sequences for the genus *Juglans* (Stanford et al., 2000). Two possibilities could explain the low rate: Either the rate of substitution is inherently low for *Juglans,* or the extant species may not represent the entire ˜50 million years of evolutionary history but represent a more recent divergence or a part of it, indicating past extinctions.

Molecular Phylogeny and Cladogenesis

The cladograms from the three analyses (MP, ML, and ME) are concordant with each other and contain three well-supported, monophyletic clades corresponding to the sections *Juglans, Cardiocaryon,* and *Rhysocaryon–Trachycaryon* described within the genus *Juglans.* The clades exhibit a high degree of differentiation and differ significantly in leaf architecture, wood anatomy, and pollen and fruit morphology (Manchester, 1987). However, monophyly of the genus was not evident, probably because of past extinctions obscuring the evolutionary history.

The low consistency index apparently indicates that the spacer regions have been subjected to a moderate level of homoplasy across the lineages during the evolution and diversification of *Juglans.* Previous molecular systematic studies generally supported two major groups, one corresponding to section *Rhysocaryon* (black walnuts) and the second including the members of sections *Cardiocaryon* (Asian butternuts), *Trachycaryon* (North American butternut), and *Juglans* (Fjellstrom and Parfitt, 1995; Stanford et al., 2000). In a recent study, Manos and Stone (2001) found section *Juglans* as the sister group to the black walnuts, suggesting a second biogeographic disjunction within the genus *Juglans.* The single North American butternut species, *J. cinerea,* with nut characteristics (two-chambered nuts with four-ribbed husks) resembling the members of section *Cardiocaryon,* is placed within the *Rhysocaryon* clade, members of which are characterized by four-chambered nuts with indehiscent hulls. The placement of *J. cinerea* within *Rhysocaryon* was supported in a recent phylogenetic study based on the chloroplast *matK* sequences, whereas the phylogeny based on the nuclear it s sequences, nuclear genome r fl ps, and the combined data set placed *J. cinera* sister to *Cardiocaryon* (Fjellstrom and Parfitt, 1995; Stanford et al., 2000). This controversial placement of butternut into the black walnut clade by cpd na, with five unique synapomorphies, strong bootstrap support, and decay index = 4, suggests historical introgression of *Rhysocaryon* chloroplast into an ancestral member of section *Cardiocaryon*, which later may have given rise to the North American butternut, *Trachycaryon*. The introgression may have occurred during range reduction and selective extinction of juglandaceous taxa in general and of *Juglans* in particular in northern latitudes, including some of the ancestral butternuts in North America in the early Neogene. Fossil records indicate that butternuts were widely distributed throughout the northern latitudes during the late Eocene and Oligocene. Chloroplast capturing has been documented in several plant groups, perhaps the best studied of which are in cotton (Wendel et al., 1991). The present-day *Trachycaryon* is represented by a single taxon, *J. cinerea*, found only in eastern North America and sympatric with *Rhysocaryon*.

Members of the section *Rhysocaryon* are not well resolved; however, in the mp and ml analyses, they are segregated into three biogeographic groups reflecting specific adaptations to the temperate, subtropical, and tropical highland environments in which they are found (figure 7.2). The clade as a whole is well supported, with five unique synapomorphies and a bootstrap value and decay index of 97% and 4, respectively. Many of these taxa have accumulated a number of autapomorphic mutations along with some homoplasious ones shared mostly within and to a less extant between different clades. The basal placement of southern California black walnut, *J. californica*, within the r t clade, well separated from its putative close relatives *J. hindsii* and *J. major*, was surprising because *J. hindsii* has often been treated as a conspecific variant within *J. californica* (Wilken, 1993), and a sister relationship between these two taxa has been reported in other studies (Fjellstrom and Parfitt, 1995; Stanford et al., 2000). The basal placement of *J. californica* probably results from two substitutions that it shares with the section *Cardiocaryon*, which may represent convergence. Lower resolution within the black walnut section probably indicates recent diversification, possibly in the upper Miocene; reticulate evolution within the section; and persistence of ancestral polymorphisms through speciation. This is contrary to the fossil evidence that suggests that the earliest evolutionary split within *Juglans* during the middle Eocene involved the origin of black walnut and butternut sections and thus these two sections would have had enough time for intersectional and intrasectional diversification.

Section *Cardiocaryon* is well supported and resolved as a monophyletic lineage. Within *Cardiocaryon*, *J. hopeiensis* is moderately supported as sister to the remaining three Asian butternuts, *J. ailantifolia, J. cathayensis,* and

J. mandshurica, which are well supported as a clade in all three analyses. In overall tree morphology, *J. hopeiensis* closely resembles the Persian walnut, *J. regia,* but the nut characters are similar to *J. mandshurica,* and it has been considered as either an interspecific hybrid between *J. regia* and *J. mandshurica* (Rehder, 1940) or as a subspecies of *J. mandshurica* (Kuang et al., 1979). In contrast to earlier studies that placed *J. mandshurica* as sister to *J. ailantifolia* and *J. cathayensis* (Stanford et al., 2000; Fjellstrom and Parfitt, 1995), in our study *J. cathayensis* and *J. mandshurica* are closely united with five unique synapomorphies.

The Persian walnut, *J. regia,* and its sister taxon, *J. sigillata* (section *Juglans*), form a distinct clade sister to both *Cardiocaryon* and *Rhysocaryon–Trachycaryon* in all three of our analyses. This was in contrast to earlier studies, which placed the cultivated walnut *J. regia* either within *Cardiocaryon* (Fjellstrom and Parfitt, 1995; Stanford et al., 2000) or within *Rhysocaryon* (Manos and Stone, 2001). The early evolutionary split of this clade within the genus *Juglans* contradicts the traditional taxonomic treatments and fossil evidence, both of which supported the almost simultaneous ancient divergence of sections *Cardiocaryon* and *Rhysocaryon,* and the origin of the genus in the middle Eocene (Manchester, 1987). Within the section *Juglans,* the cultivated species *J. regia* accumulated seven unique autapomorphies with unique nut characteristics (thin-shelled four chambered nuts) and is differentiated from its sister taxon *J. sigillata,* which contains one unique mutation and retains many primitive nut characteristics such as thick rough-shelled nuts with dark kernels (Dode, 1909a). *J. sigillata* may represent a semidomesticated form within the section. It is known to have been cultivated in southern China for its oil and wood. Furthermore, early Chinese records suggest that domestication and selection of walnut occurred in the southern Tibetan and Yunnan regions, and better varieties were brought to the north during the Han dynasty (de Candolle, 1967).

Biogeography

The extant species of *Juglans* show an intercontinental disjunction with the modern distributions of sections *Juglans* and *Cardiocaryon* limited to Eurasia and section *Rhysocaryon* endemic to the Americas. A single butternut species, *J. cinerea,* with modern distribution in eastern North America, is generally considered to be a disjunct of *Cardiocaryon* (Asian butternuts) (Manchester, 1987). Recently, Manos and Stone (2001) proposed a sister group relationship between the cultivated walnut, *J. regia,* and section *Rhysocaryon,* suggesting the possibility of a second disjunction within *Juglans*. These disjunctions could have arisen as a result of either a vicariance event disrupting the geographic continuity of ancestral populations that once spanned from Eurasia to North America or a long-distance dispersal from one region to the other. The vicariance hypothesis is favored over the long-distance dispersal theory because of the large fruit size in *Juglans*, which does not appear to have great dispersal ability.

It is likely that the ancestral populations of *Juglans* were widely distributed throughout the middle and upper latitudes of the Northern Hemisphere during the early Tertiary up until the late Miocene, when the climate was generally warm enough (Wolfe and Upchurch, 1987) for the successful establishment and periodic exchange of broad-leaved deciduous taxa across the Bering and North Atlantic land bridges connecting Asia, North America, and Europe (McKenna, 1983; Tiffney, 1985b; Ziegler, 1988). The gradual cooling during the Neogene produced range contraction and greatly reduced the migration between Eurasian and North American floras by the mid-Pliocene (Wolfe, 1978; Tiffney, 1985a). Further climatic changes during the Quaternary eliminated mixed mesophytic forests in the northern latitudes, leaving eastern North America, eastern Asia, and to a much lesser extent the Balkans and Caucasus as the main refugia of many genera (Tiffney, 1985a).

Based on fossil evidence, Manchester (1987) proposed that the divergence of *Pterocarya* and *Juglans* may have occurred sometime during the late Paleocene or early Eocene (~54 mya) and that the initial split of sections *Rhysocaryon* and *Cardiocaryon* probably occurred during the middle Eocene (45 mya) in North America, but the two sections were clearly resolved only in the early Oligocene (38 mya). However, based on extensive analysis of nut specimens of a fossil walnut, *J. eocinerea* from the Beaufort Formation (Tertiary), southwestern Banks Island, arctic Canada, Hills et al. (1974) concluded that it is closely related and probably ancestral to fossil *J. tephrodes* from early Pliocene Germany and the extant *J. cinerea* from the eastern United States. Furthermore, they argued that butternuts may have evolved independently in the Arctic, attaining a broad distribution in the upper latitudes of the Northern Hemisphere by the Miocene, and that subsequent geoclimatic changes (Wolfe and Leopold, 1967; Axelrod and Bailey, 1969; Wolfe, 1971) resulted in the southward movement of the floras across the Bering Strait. However, the early Pleistocene glaciations have completely eliminated butternuts from Europe and northwestern parts of North America, leaving small disjunct populations in eastern Asia to evolve into three major present-day taxa, *J. cathayensis, J. mandshurica,* and *J. ailantifolia,* and one south of the glacial limit in North America to evolve to its present form, *J. cinerea.* The geographic and stratigraphic fossil distribution strongly supports the hypothesis that butternuts may have originated and radiated from high northern latitudes. At about the same time, black walnuts spanned throughout North America and extended into the Southern Hemisphere, reaching Ecuador by the late Neogene, and remained endemic to the Americas throughout their evolutionary history.

One can argue that if butternuts and black walnuts diverged from a common ancestor in North America during the middle Eocene, as suggested by Manchester (1987), there would have been ample opportunity for both groups to become established in both Asia and North America because both the Bering and North Atlantic land bridges were in continuous existence from the middle Eocene through the late Miocene, when there was a favorable climate in upper latitudes for the establishment and dispersal of broad-leaved deciduous taxa (Wolfe, 1972, 1978; Tiffney, 1985b). However, the distributional range of the Tertiary fossils of butternuts and black walnuts does not overlap except in the northwestern parts of the United States around 40°N latitude, strongly suggesting that they may have evolved independently, as suggested by Hills et al. (1974). The weak support for the sister relationship between these two groups observed in our phylogenetic analysis further substantiates this point and also suggests that they may not share an immediate common ancestor.

An analysis of the comparative rates of molecular evolution along the branches of the cladogram indicated that the rates did not conform to the expectation of the molecular clock hypothesis (Zuckerkandl and Pauling, 1965). Relative rates of sequence evolution based on overall substitutions, estimated using *Pterocarya* (outgroup) as the reference, indicated that the differences between species pairs are mostly insignificant except for combinations involving *J. regia* and a few members of *Cardiocaryon,* especially *J. cathayensis* (table 7.3). The differential rates of divergence associated with these Eurasian taxa and their basal placement in the cladograms could indicate their ancient and distinct origin or the fact that extant taxa may not reflect the entire evolutionary history of *Juglans.* The range reduction, local extinctions, and geographic isolation during the late Tertiary and early quaternary glaciations and the subsequent expansion into central Asia and southeastern Europe might have played an important role in the evolution and diversification of sections *Juglans* and *Cardiocaryon.* Influence of both natural and human selection and introgression during domestication may have further altered the rate and direction of evolution of the cultivated walnut.

Estimates of time since divergence may be obtained from fossil evidence or from computations assuming a molecular clock. For *Juglans,* the sequence divergence rates for the five intergenic cpdna regions used in this study are unknown, and the estimation of divergence times relies strictly on fossil records. Therefore, the accuracy of fossil records and the variation of molecular evolutionary rate and patterns of extinction in a clade affect the estimations. Nevertheless, the estimations of nucleotide substitution rates or time since divergence using the molecular clock hypothesis, although based on uncertain assumptions and approximate values, are helpful in understanding the tempos of evolution and plant historical geographies (Parks and Wendel, 1990; Crawford et al., 1992; Wendel and Albert, 1992).

Paleobotanical evidence suggests two major landmarks in the evolution and diversification of *Juglans,* the first corresponding to the divergence of *Pterocarya* and *Juglans* (early Eocene, ˜54 mya) and the second corresponding to the early split between sections *Rhysocaryon* and *Cardiocaryon* (mid-Eocene, ˜45 mya) (Manchester, 1987). Based on these events, the rates of divergence between the outgroup taxon *Pterocarya* and the ingroup *Juglans,* and between the sections *Rhysocaryon* and *Cardiocaryon* within *Juglans,* were estimated to be approximately 0.772×10^{-10} and 0.69×10^{-10} nucleotide sites per year, respectively. These estimates were much lower than the earlier reports between *Pterocarya* and *Juglans* (3.36×10^{-10}) based on the cpdna rflps (Smith and Doyle, 1995) and between the sections *Cardiocaryon* and *Rhysocaryon* (1.17×10^{-9}) based on nuclear genome rflps (Fjellstrom and Parfitt, 1995). The nonparametric rate smoothing method (Sanderson, 1997), which combines likelihood and the nonparametric penalty function to estimate ages for different nodes based on fossil calibration, has resulted in inconsistent estimation of age for different nodes with large variances.

Given the many caveats mentioned earlier, we proceeded with caution in calculating the time since divergence for some of the other major bifurcations observed in the phylogenetic analyses. The time since divergence between clades provides a rough estimate of the time since isolation between them. If an overall divergence rate of 0.772×10^{-10} substitutions per site per year, estimated from the time since divergence between the outgroup taxon, *Pterocarya,* and the ingroup *Juglans* (54 mya) as a whole, is used, then the divergence times between sections *Rhysocaryon* and *Juglans, Rhysocaryon* and *Cardiocaryon,* and *Cardiocaryon* and *Juglans* are estimated to be 41.6, 40.2, and 43.8 mya, respectively. However, if it is based on 0.69×10^{-10} nucleotide sites per year, estimated using the Middle

Eocene as the time frame for divergence between sections *Rhysocaryon* and *Cardiocaryon* (45 mya) (Manchester, 1987), the divergence times between section *Rhysocaryon* and *Juglans* and section *Cardiocaryon* and *Juglans* are estimated to be 46.5 and 50 mya. Based on sequence data, estimated divergence times for different lineages within *Juglans* range from the early to late Eocene, which coincide roughly with the divergence times proposed by Manchester (1987), but the sequence of divergence events contradicts the fossil evidence. Contrary to fossil evidence, which suggests the split between black walnuts and butternuts as the earliest evolutionary event, our analyses suggest that the divergence of section *Juglans* is the first splitting of the lineage to have occurred within the genus *Juglans*.

Origin and Domestication of Cultivated Walnut, *J. regia*

One of the puzzling biogeographic questions in *Juglans* is the presence of a Eurasian section comprising two taxa, *J. regia* and *J. sigillata*, with four-chambered nuts similar to *Rhysocaryon*, which is endemic to the New World. The nutshell thickness of these taxa may vary from extremely thick, as in black walnuts in the case of *J. sigillata*, to paper-thin, as in *J. regia*, whereas the other Asian section, *Cardiocaryon*, strictly possesses two-chambered, thick-shelled nuts. The placement of the cultivated species *J. regia* has been problematic in earlier phylogenetic studies, and recent studies place it as sister to either butternuts (Stanford et al., 2000) or black walnuts (Manos and Stone, 2001). Our data strongly support the section *Juglans* as an independent clade basal to the remaining three sections within the genus *Juglans*. It evolved at a significantly higher rate than section *Rhysocaryon* and some taxa of section *Cardiocaryon*. However, the evolutionary history of the section *Juglans* may have been confounded by widespread extinctions, geographic isolation, and bottlenecks during the Pleistocene glaciations, when the ancestral forms were in refugia in central Asia and southeastern Europe. Subsequent expansion, human selection, and introgression among isolated diverse populations during the post-Pleistocene glaciations may have rapidly changed the genetic structure and differentiation patterns within the section *Juglans* (Popov, 1929; Beug, 1975; Huntley and Birks, 1983). *J. regia* is a highly domesticated and economically important walnut species, occurring mostly under cultivation in both the Old and New World, whereas its sister taxon, *J. sigillata*, with primitive nut characteristics, may represent a semidomesticated or primitive form within the section restricted to parts of southern China.

It is appropriate here to provide some details on the domestication history and development of cultivated walnut. All walnut species bear edible nuts, but the Persian or English walnut (*J. regia*) is the most delicious, economically important, and successfully cultivated throughout the temperate regions of the world. Although its origin is obscure, it has been thought to be indigenous to the mountainous regions of central Asia extending from the Balkan region across Turkey, the Caucasus, Iraq, Iran, and Afghanistan, parts of Kazakhstan, Uzbekistan, and southern Russia to northern India (Dode, 1909b; Forde, 1975; McGranahan and Leslie, 1991). However, the pollen data (Bottema, 1980) suggest that *J. regia* went into extinction in southeastern Europe and southwestern Turkey during the glacial period but survived in the Pontic and Hyrcanic refugia and reappeared there around 2000 bc (Zohary and Hopf, 1993). If true, this evidence strongly points to the Caucasus and northern Iran as the most plausible area of walnut domestication. The walnuts have been found in prehistoric deposits in Europe dating back to the Iron Age and were also prevalent in Palestine and Lebanon during that period (Rosengarten, 1984). At present, natural populations of Persian walnut, some as good as modern cultivars, exist in many parts of Central Asia from the Caucasus to the mountains of Tien-Shan. They represent the natural range of diversity, probably as a consequence of complex interactions of natural and human selection after postglacial expansion and domestication (Takhtajan, 1986; Vavilov, 1992). However, *J. regia* found in the flora of the Khasi-Manipur province belong to the eastern Asiatic elements tied to floras of the eastern Himalayas, upper Burma, and eastern China. This region represents one of the most important centers of the Tertiary flora of eastern Asia (Bor, 1942). Furthermore, it is suggested that the mountainous regions of central and western China and adjacent lowlands along with west Asia and Asia Minor are areas of diversity for walnut. The Chinese center of diversity is further supported by the ancient walnut fossils and archaeological material found in the ruins at Cishan Hebei and the walnut pollen dating back to 4000–5000 bc found in the spore pollen analysis of Banpo Xian (Rong-Ting, 1990).

Further support for the Eurasian origin of cultivated walnuts comes from the fact that the Tertiary relict flora comprising mostly deciduous and some evergreen woody taxa survived in the regions of equable climate in southeastern Europe, the Caucasus, and southwestern and eastern Asian refugia during the late Miocene to Pliocene cooling and Quaternary glaciations (Tiffney, 1985a, 1985b; Wen, 1999; Xiang et al., 2000), where perhaps small remnant populations of ancestral walnuts may have survived. Expansion of these relict floras into the central European regions comprising Balkan, Carpathian, and Euxinian provinces and south into Asia Minor, northern parts of Iran, Afghanistan, Turkmenistan, Uzbekistan, north into Tien-Shan mountains, and the Himalayas started at the end of the glacial period and the beginning of the Holocene (Beug, 1975; Davis, 1982; Takhtajan, 1978). There is evidence of a floristic connection between some Tertiary relict species from the south central European refugia, which migrated via the southern route of the North Atlantic land bridge, and the East Asian relicts including eastern China and some regions in the Himalayas, derived predominantly through migration across the Bering land bridge. The East Asian refugia may have included some of the ancestral forms of butternuts and cultivated walnut, *J. regia,* which may have gradually evolved into the modern Asian butternut clade (Wen, 1999, 2001; Milne and Abbott, 2002). According to Rong-Ting (1990), the native populations of walnut in China exhibit a wide range of variation for all discernible characters, with 6000–7000 years of evolutionary and domestication history, extending across a wide range of environments. Dode (1909b) described the section *Juglans* by recognizing six species in addition to *J. regia* with distribution extending from central to East Asia including China and the Himalayan region, which others have not accepted but which could be treated as ecotypes within *J. regia.*

In summary, the cladogenesis within *Juglans* based on cpdna intergenic sequence analyses does not fully corroborate the evolutionary hypothesis based on the fossil history and biogeographic evidence. Neither the fossil nor molecular phylogenetic evidence strongly supports the monophyletic origin of *Juglans.* If Eocene North America is considered the center of origin and diversification of *Juglans,* as suggested by Manchester (1987), there would have been sufficient opportunity for members of different sections to become distributed in both North America and Eurasia because land bridges across the Bering Sea and North Atlantic Ocean were in continuous existence from the middle Eocene through the late Miocene (Tiffney, 1985b). On the contrary, the Tertiary fossil evidence suggests that section *Rhysocaryon* remained endemic to the Americas throughout its evolutionary history, and the section *Juglans* was not represented in the fossil records from North America. Furthermore, the results allow for some generalizations on the origin and evolution of the genus *Juglans*: The cpdna intergenic spacer sequence divergence levels observed within and between different sections of *Juglans* are low; basal placement of the section *Juglans* in the phylogenetic analyses suggests its ancient origin contrary to fossil evidence, which suggests the earliest origin of sections *Rhysocaryon* and *Cardiocaryon;* the two Asian sections, *Juglans* and *Cardiocaryon*, evolved at different rates than *Rhysocaryon*; and the extant taxa may not adequately represent the entire evolutionary history of the genus.

Acknowledgments

This study was funded by the U.S. Department of Agriculture, Agricultural Research Service (Project No. 5306-21000-015-00D). We thank Clay Weeks, Warren Roberts, and Chuck Leslie for contributing to the collection of samples and many helpful suggestions.

References

Al-Janabi, M., M. McClelland, C. Petersen, and W.S. Sobral. 1994. Phylogenetic analysis of organellar DNA sequences in the Andropogoneae: Saccharinae. *Theoretical and Applied Genetics* 88: 933–944.

Aradhya, M.K., R.M. Manshardt, F. Zee, and C.W. Morden. 1999. A phylogenetic analysis of the genus *Carica* L. (Caricaceae) based on restriction fragment length variation in a cpDNA intergenic spacer region. *Genetic Resources and Crop Evolution* 46: 579–586.

Axelrod, D.I. 1960. The evolution of flowering plants. In S. Tax (ed.), *Evolution After Darwin*, Vol. 1, 227–305. Chicago University Press, Chicago, IL, USA.

Axelrod, D.I. and H.P. Bailey. 1969. Paleotemperature analysis of Tertiary floras. *Palaeogeography, Palaeoclimatology, Palaeoecology* 6: 163–195.

Beug, H.-J. 1975. Man as a factor in the vegetational history of the Balkan Peninsula. In D. Jordanov, I. Bondev, S. Kozuharov, B. Kuzmanov, and E. Palamarev (eds.), *Problems of Balkan Flora and Vegetation. Proceedings of the First International Symposium on Balkan Flora and Vegetation, Varna, June 7–14, 1973*, 72–78. Publishing House of the Bulgarian Academy of Sciences, Sofia, Bulgaria.

Bor, N.L. 1942. The relict vegetation of the Shillong Plateau Assam, India. *Forestry Records* 3: 152–195.

Bottema, S. 1980. On the history of the walnut in southeastern Europe. *Acta Botanica Neerlandica* 29: 343–349.

Bremer, K. 1988. The limits of amino acid sequence data in angiosperm phylogenetic reconstruction. *Evolution* 42: 795–803.

Chaney, R.W. 1947. Tertiary centers and migration routes. *Ecological Monographs* 17: 139–148.

Crawford, D.J., M.S. Lee, and T.F. Stuessy. 1992. Plant species disjunctions: Perspectives from molecular data. *Aliso* 13: 395–409.

Cros, J., M.C. Combes, P. Trouslot, F. Anthony, S. Hamon, A. Charrier, and P. Lashermes. 1998. Phylogenetic analysis of chloroplast DNA variation in *Coffea* L. *Molecular Phylogenetics and Evolution* 9: 109–117.

Davis, P.H. 1982. *Flora of Turkey and the East Aegean Islands*, Vol. 7. University Press, Edinburgh, UK.

de Candolle, A. 1967. *Origin of Cultivated Plants*. Hafner Publishing Company, New York, NY, USA.

Demesure, B., B. Comps, and R.J. Petit. 1996. Chloroplast DNA phylogeography of the common beech (*Fagus sylvatica* L.) in Europe. *Evolution* 50: 2515–2520.

Demesure, B., N. Sodzi, and R. J. Petit. 1995. A set of universal primers for amplification of polymorphic non-coding regions of mitochondrial and chloroplast DNA in plants. *Molecular Ecology* 4: 129–131.

Dode, L. A. 1909a. Contribution to the study of the genus *Juglans* (English translation by R. E. Cuendett). *Bulletin of the Society of Dendrology, France* 11: 22–90.

Dode, L. A. 1909b. Contribution to the study of the genus *Juglans* (English translation by R. E. Cuendett). *Bulletin of the Society of Dendrology, France* 12: 165–215.

Doyle, J. J. and J. L. Doyle. 1987. A rapid DNA isolation procedure for small quantities of fresh leaf tissue. *Phytochemical Bulletin* 19: 11–15.

Farris, J. S., M. Kallersjo, A. G. Kluge, and C. Bult. 1994. Testing significance of incongruence. *Cladistics* 10: 315–319.

Farris, J. S., M. Kallersjo, A. G. Kluge, and C. Bult. 1995. Constructing a significance test for incongruence. *Systematic Biology* 44: 570–572.

Felsenstein, J. 1981. Evolutionary trees from DNA sequences: A maximum likelihood approach. *Journal of Molecular Evolution* 17: 368–376.

Fjellstrom, R. G. and D. E. Parfitt. 1995. Phylogenetic analysis and evolution of the genus *Juglans* (Juglandaceae) as determined from nuclear genome RFLPs. *Plant Systematics and Evolution* 197: 19–32.

Forde, H. I. 1975. Walnuts. In J. Janick and J. N. Moore (eds.), *Advances in Fruit Breeding*, 439–455. Purdue University Press, West Lafayette, IN, USA.

Gielly, L. and P. Taberlet. 1994. Chloroplast DNA polymorphism at the intrageneric level: Implications for the establishment of plant phylogenies. *Comptes Rendus de l'Académie des Sciences Life Science* 317: 685–692.

Graham, A. 1972. Outline of the origin and historical recognition of floristic affinities between Asia and eastern North America. In A. Graham (ed.), *Floristics and Paleofloristics of Asia and Eastern North America*, 1–18. Elsevier, Amsterdam, The Netherlands.

Gray, A. 1859. Diagnostic characters of phanerogamous plants, collected in Japan by Charles Wright, botanist of the U.S. North Pacific Exploring Expedition, with observations upon the relationship of the Japanese flora to that of North America and of other parts of northern temperate zone. *Memoirs of the American Academy of Arts* 6: 377–453.

Gray, A. 1878. Forest geography and archaeology. A lecture delivered before the Harvard University Natural History Society. *American Journal of Science and Arts*, Series 3 & 16: 85–94 & 183–196.

Hills, L. V., J. E. Klovan, and A. R. Sweet. 1974. *Juglans eocinerea* n. sp., Beaufort Formation (Tertiary), southwestern Banks Inland, Arctic Canada. *Canadian Journal of Botany* 52: 65–90.

Hodges, S. A. and M. L. Arnold. 1994. Columbines: A geographic widespread species flock. *Proceedings of the National Academy of Sciences* 91: 5129–5132.

Huntley, B. and H. J. B. Birks. 1983. *An Atlas of Past and Present Pollen Maps for Europe: 0–13,000 Years Ago.* Cambridge University Press, New York, NY, USA.

Kimura, M. 1980. A simple method for estimating evolutionary rates of base substitutions through comparable studies of nucleotide sequences. *Journal of Molecular Evolution* 16: 111–120.

Kimura, M. 1981. Estimation of evolutionary distances between homologous nucleotide sequences. *Proceedings of the National Academy of Sciences*, USA 78: 454–458.

Kuang, K., S. Cheng, P. Li, and P. Lu. 1979. Juglandaceae (in Chinese, unpublished translation by W. E. Manning). In K.-Z. Kuang and C. P. Li (eds.), *Flora Republicae Popularis Sinicae*, Vol. 21, 8–42. Institutum Botanicum Academiae Sinicae, Peking, China.

Manchester, S. R. 1987. The fossil history of Juglandaceae. *Missouri Botanical Garden Monograph* 21: 1–137.

Manning, W. E. 1957. The genus *Juglans* in Mexico and Central America. *Journal of Arnold Arboretum* 38: 121–150.

Manning, W. E. 1960. The genus *Juglans* in South America and West Indies. *Brittonia* 12: 1–26.

Manning, W. E. 1978. The classification within the Juglandaceae. *Annals of Missouri Botanical Garden* 65: 1058–1087.

Manos, P. S. and D. E. Stone. 2001. Evolution, phylogeny, and systematics of the Juglandaceae. *Annals of Missouri Botanical Garden* 88: 231–269.

McGranahan, G. and C. Leslie. 1991. Walnuts (*Juglans*). In J. N. Moore and J. R. Ballington Jr. (eds.), *Genetic Resources of Temperate Fruit and Nut Crops*, Vol. 2, 907–951. International Society of Horticultural Sciences, Wageningen, The Netherlands.

McKenna, M. C. 1983. Cenozoic paleogeography of North Atlantic land bridge. In M. H. P. Bott, S. Saxov, M. Talwani, and J. Thiede (eds.), *Structure and Development of the Greenland-Scotland Ridge*, 351–399. Plenum, New York, NY, USA.

Milne, R. I. and R. J. Abbott. 2002. The origin and evolution of tertiary relict floras. *Advances in Botanical Research* 38: 281–314.

Mohanty, A., J. P. Martin, and I. Aguinagalde. 2001. Chloroplast DNA study in wild populations and some cultivars of *Prunus avium* L. *Theoretical and Applied Genetics* 103: 112–117.

Muse, S. V. and B. S. Weir. 1992. Testing for equality of evolutionary rates. *Genetics* 132: 269–276.

Ogihara, Y., T. Terachi, and T. Sasakuma. 1991. Molecular analysis of the hot spot region related to length mutations in wheat chloroplast DNAs. I. Nucleotide divergence of genes and intergenic spacer regions located in the hot spot region. *Genetics* 129: 873–884.

Palmer, J. D. 1991. Plastid chromosome: Structure and evolution. In L. Bogorad and I. K. Vasil (eds.), *The Molecular Biology of Plastids*, 5–52. Academic Press, San Diego, CA, USA.

Parks, C. R. and J. F. Wendel. 1990. Molecular divergence between Asian and North American species of *Liriodendron* (Magnoliaceae) with implications of fossil floras. *American Journal of Botany* 77: 1243–1256.

Petit, R. J., E. Pineau, B. Demesure, R. Bacilieri, A. Ducousso, and A. Kremer. 1997. Chloroplast DNA footprints of postglacial recolonisation by oaks. *Proceedings of National Academy of Sciences* 94: 9996–10001.

Popov, M. G. 1929. Wild growing fruit trees and shrubs of Asia Minor (in Russian). *Bulletin of Applied Botany and Plant Breeding* 22: 241–483.

Posada, D. and K. A. Crandall. 1998. Modeltest: Testing the model of DNA substitution. *Bioinformatics* 14: 817–818.

Qian, H. 2002. Floristic relationships between eastern Asia and North America: Test of Gray's hypothesis. *American Naturalist* 160: 317–332.

Raven, P. H. 1972. Plant species disjunctions: A summary. *Annals of Missouri Botanical Garden* 59: 234–146.

Rehder, A. 1940. *Manual of Cultivated Trees and Shrubs in North America*. Macmillan, New York, NY, USA.

Rong-Ting, X. 1990. Discussion on the origin of walnut in China. *Acta Horticulturae* 284: 353–361.

Rosengarten, F. 1984. *The Book of Edible Nuts*. Walker, New York, NY, USA.

Sanderson, M. J. 1997. A nonparametric approach to estimating divergence times in the absence of rate consistency. *Molecular Biology and Evolution* 14: 1218–1231.

Sang, T., D. Crawford, and T. Stuessy. 1997. Chloroplast DNA phylogeny, reticulate evolution, and biogeography of *Paeonia* (Paeoniaceae). *American Journal of Botany* 84: 1120–1136.

Small, R. L., J. A. Ryburn, R. C. Cronn, T. Seelanan, and J. F. Wendel. 1998. The tortoise and the hare: Choosing between noncoding plastome and nuclear ADH sequences for phylogeny reconstruction in a recently diverged plant group. *American Journal of Botany* 85: 1301–1315.

Smith, J. F. and J. J. Doyle. 1995. A cladistic analysis of chloroplast DNA restriction site variation and morphology for the genera of the Juglandaceae. *American Journal of Botany* 82: 1163–1172.

Stanford, A. M., R. Harden, and C. R. Parks. 2000. Phylogeny and biogeography of *Juglans* (Juglandaceae) based on *mat*K and ITS sequence data. *American Journal of Botany* 87: 872–882.

Swofford, D. L. 2002. PAUP*. *Phylogenetic Analysis Using Parsimony (* and Other Methods)*, Version 4. Sinauer Associates, Sunderland, MA, USA.

Taberlet, P., L. Gielly, G. Pautou, and J. Bouvet. 1991. Universal primers for amplification of three non-coding regions of chloroplast DNA. *Plant Molecular Biology* 17: 1105–1109.

Takhtajan, A. 1978. *Floristic Regions of the World*. Nauka, Leningrad, USSR.

Takhtajan, A. 1986. *Floristic Regions of the World* (translated by T. J. Crovello). University of California Press, Berkeley, CA, USA.

Tiffney, B. H. 1985a. The Eocene North Atlantic land bridge: Its importance in Tertiary and modern phytogeography of the Northern Hemisphere. *Journal of Arnold Arboretum* 66: 243–273.

Tiffney, B. H. 1985b. Perspectives on the origin of the floristic similarity between eastern Asia and eastern North America. *Journal of Arnold Arboretum* 66: 73–94.

Vavilov, N. I. 1992. *Origin of Cultivated Plants* (translated by D. Love). Cambridge University Press, Cambridge, UK.

Wen, J. 1999. Evolution of eastern Asian and eastern North American disjunct distributions in flowering plants. *Annual Review of Ecology and Systematics* 30: 421–455.

Wen, J. 2001. Evolution of eastern Asian and eastern North American biogeographic disjunctions: A few additional issues. *International Journal of Plant Science* 162: S117–S122.

Wendel, J. F. and V. A. Albert. 1992. Phylogenetics of the cotton genus (*Gossypium*): Character-state weighted parsimony analysis of chloroplast—DNA restriction site data and its systematic and biogeographic implications. *Systematic Botany* 17: 115–143.

Wendel, J. F., J. M. Stewart, and J. H. Rettig. 1991. Molecular evidence of homoploid reticulate evolution among Australian species of *Gossypium*. *Evolution* 45: 694–711.

Wilken, D. H. 1993. Juglandaceae. In J. C. Hickman (ed.), *The Jepson Manual: Higher Plants of California*, 709. University of California Press, Berkeley, CA, USA.

Wolfe, J. A. 1969. Neogene floristic and vegetational history of the Pacific northwest. *Madrano* 20: 83–110.

Wolfe, J. A. 1971. Tertiary climatic fluctuations and methods of analysis of Tertiary floras. *Palaeogeography, Palaeoclimatology, Palaeoecology* 9: 27–57.

Wolfe, J. A. 1972. An interpretation of Alaskan Tertiary floras. In A. Graham (ed.), *Floristics and Paleofloristics of Asia and Eastern North America*, 201–233. Elsevier, Amsterdam, The Netherlands.

Wolfe, J. A. 1975. Some aspects of plant geography of the Northern Hemisphere during the late Cretaceous and Tertiary. *Annals of Missouri Botanical Garden* 62: 264–279.

Wolfe, J. A. 1978. A paleobotanical interpretation of Tertiary climates in the Northern Hemisphere. *American Scientist* 66: 694–703.

Wolfe, J. A. 1985. Distribution of major vegetational types during the Tertiary. In E. T. Sundquist and W. S. Broecker (eds.), *The Carbon Cycle and the Atmospheric CO_2: Natural Variations Archean to*

Present, American Geophysical Monograph 32: 357–375. American Geophysical Union, Washington, DC, USA.

Wolfe, J. A. and E. B. Leopold. 1967. Neogene and early Quaternary vegetation of North America and northeastern Asia. In D. M. Hopkins (ed.), *The Bering Land Bridge,* 193–206. Stanford University Press, Stanford, CA, USA.

Wolfe, J. A. and G. R. Upchurch Jr. 1987. North American nonmarine climates and vegetation during the late Cretaceous. *Palaeogeology, Palaeoclimatology, Palaeoecology* 61: 33–78.

Wolfe, K. H., W.-H. Li, and P. Sharp. 1987. Rates of nucleotide substitution vary greatly among plant mitochondria, chloroplast, nuclear DNAs. *Proceedings of National Academy of Science* 84: 9054–9058.

Xiang, Q., D. E. Soltis, P. S. Soltis, S. R. Manchester, and D. L. Crawford. 2000. Timing the eastern Asian–eastern North American floristic disjunction: Molecular clock corroborates paleontological estimates. *Molecular Phylogenetics and Evolution* 15: 462–472.

Ziegler, P. A. 1988. Evolution of the Arctic–North Atlantic and the western Tethys. *American Association of Petroleum Geologists (AAPG) Memoir* 43: 164–196.

Zohary, D. and M. Hopf. 1993. *Domestication of Plants in the Old World.* Clarendon Press, Oxford, UK.

Zuckerkandl, E. and L. Pauling. 1965. Evolutionary divergence and convergence in proteins. In V. Bryson (ed.), *Evolving Genes and Proteins,* 97–106. Academic Press, New York, NY, USA.

Zurawaski, G. and M. T. Clegg. 1987. Evolution of higher-plant chloroplast DNA-encoded genes: Implications for structure function and phylogenetic studies. *Annual Review of Plant Physiology* 38: 391–418.

Hugh Cross, Rafael Lira Saade,
and Timothy J. Motley

CHAPTER 8

Origin and Diversification of Chayote

The habitat and life history of a plant species will influence how individuals are selected and used by humans. For weedy climbers, such as members of the cucumber and squash family, Cucurbitaceae, little manipulation of the natural genetic stock has been necessary apart from selection for improved fruit size and taste. Once edible individuals were discovered and propagated, over time an enormous diversity of fruit size and shape arose by selection and dispersal. One potential source for this expansion of phenotypic diversity could be the gene pool that includes the crop's wild relatives (Harlan, 1992). It is possible that this has been a factor for the morphological diversification seen among chayote cultivars as well. This chapter uses molecular data to determine the origin of chayote and the role of wild relatives in the subsequent diversification of the crop.

Background

Chayote (*Sechium edule* (Jacq.) Swartz) is a crop grown primarily for its fruits, although the tubers, leaves, and shoots are also consumed (Lira, 1996). Like many other Cucurbitaceae, chayote is a vigorously growing vine that produces tendrils to pull the plant onto and above other vegetation. In cultivation these tendrils are trained onto trellises, from which the fruits are harvested from below as they mature (Lira, 1995; Newstrom, 1989). Chayote plants are perennial, and in ideal climates (such as the growing areas of the Mexican states of Veracruz and Jalisco) multiple harvests can be achieved in a single year (Lira, 1996). This high productivity makes chayote economically important in several Latin American countries. Mexico and Costa Rica are by far the largest producers and exporters of the fruit, followed by the Dominican Republic, Peru, and Brazil (Lira, 1996). Chayote has been dispersed all over the world and is now grown in many tropical and subtropical regions. In most of the areas outside Mexico and Central America there is little phenotypic variation among the chayotes (Cross and Motley, 2002). The large monocultures that produce most of the fruit for export create problems for farmers because the low genetic diversity makes them more susceptible to diseases (Lira, 1995). Often disease-resistant varieties are found among landraces and wild relatives of a crop species (Brush, 1989). Despite this, little is known about the variability of chayote landraces in southern Mexico and Central America, where their diversity is greatest.

In contrast to the genetically uniform, high-production orchards, the house gardens and smaller orchards of Veracruz and Oaxaca, Mexico—chayote's center of diversity—represent a great reservoir of genetic diversity for the crop (Lira, 1995; Newstrom, 1989; Cross, 2003). Landraces of chayote from these regions demonstrate high variability in fruit shape, size, and color (figure 8.1a). The fruits are generally round to pear-shaped, varying from 2 to more than 30 cm in length. The fruits are white (the smaller, oval fruits of this color class are called *chayote papa* for their resemblance to white potatoes), to light green (the most common color, and the one most commonly exported), to very dark green (distinctive from the lighter green and called *negrito* in Mexico). Fruits can also be prickly or glabrous, and when prickles are present they cover the fruit in varying degrees (i.e., covering the entire fruit, confined along the ridges, or at the apex). In addition, there are detectable differences in taste and texture of the fruit flesh between varieties and landraces. Yet for all its diversity in Mexico, few of these varieties are known outside this country.

Perhaps the most distinctive characteristic of the chayote is the opening or cleft at the tip of the fruit (figure 8.1b). The seedling and primary root emerge from the cleft of ripe fruits. The nutrients are provided to the seedling by the fruit, which shrinks as the plant grows. Anatomical studies of chayote fruits (Giusti et al., 1978) have shown that the vascular tissue in the endosperm has been rerouted to the seed, and in this way the fruits

FIGURE 8.1 Fruits of chayote (*Sechium edule* ssp. *edule*) and related wild taxa: **(A)** Chayote varieties from Oaxaca, Mexico; **(B)** maturing fruit of chayote showing the seedling emerging from the apical cleft; **(C)** wild subspecies *Sechium edule* ssp. *sylvestre*; **(D)** *Sechium chinantlense*, showing the apical cleft; and **(E)** *Sechium compositum*. Scale bars are equal to 1 cm.

of chayote act as nutrient reservoirs for the seed. This feature represents a fundamental shift in the function of the fruit and has consequences for how the crop is grown, distributed, and maintained in collections. In some instances vivipary has also been observed, in which seedlings may emerge from the fruit while it is still on the vine, although this has never been observed in the wild (Lira, 1996). The consequence of this biology for germplasm conservation is that neither seeds nor fruits can be stored for long periods: The seed needs the fruit to germinate, and the fruit either germinates or rots if left in storage. This means that chayote landraces must be kept in living collections, which entails labor-intensive management and large land areas that has severely limited the capacity and effectiveness of these efforts.

Despite these limitations, there have been some efforts in recent years to catalog and conserve the diversity of chayote landraces. However, limited financial resources have hampered these efforts. In contrast to other major crops, such as corn, rice, and potato, gene bank conservation programs for minor or underused crops are more difficult to fund and establish. Nevertheless, two chayote gene banks have been established in the past 20 years, one in Nepal and the other at the National University in Costa Rica (Sharma et al., 1995). Unfortunately, these gene banks have encountered problems maintaining and storing their collections. The Nepalese collection contains only locally adapted varieties and because of space and money limitations has had to give up some accessions that have not been as useful in their regional breeding program (L. Newstrom, pers. comm., 2001). The Costa Rican collection has also lost accessions over the last 2–3 years and is primarily a repository of Costa Rican varieties, but accessions from Mexico have been added when available (A. Brenes, pers. comm., 2001; Sharma et al., 1995). Because these gene banks were established to serve the agricultural needs of the individual countries, the prevalence of locally adapted varieties in their collections is understandable. However, as a consequence there is currently no gene bank that represents the entire spectrum of chayote diversity.

Historical Evidence of the Origin of Chayote

The origin of chayote has been obscured by its spread around the world over the last several centuries. Native populations have been reported in many countries, including Mexico, Puerto Rico, and Venezuela (Lira, 1996; Newstrom, 1990, 1991). Many of these reports probably are of naturalized escapes from

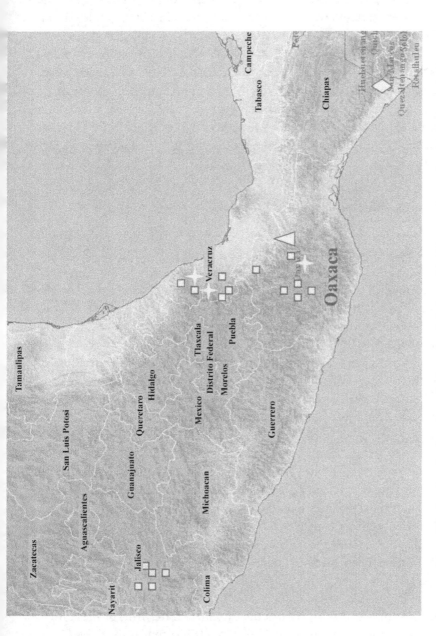

FIGURE 8.2 Map of Mexico showing locations of selected populations of *Sechium* species used in this study. White squares indicate chayote (*S. edule* ssp. *edule*), white stars indicate *S. edule* ssp. *sylvestre*. The white triangle indicates the population of *S. chinantlense* in Oaxaca, and the white diamond shows the population of *S. compositum* in Chiapas.

cultivation. The mountainous region stretching across the Mexican states of Veracruz, Puebla, and Oaxaca is the likely area of origin of chayote because this is where its closest wild relatives are indigenous (figure 8.2) (Newstrom, 1991; Lira, 1995), and it represents the center of morphological diversity of the crop (Lira, 1996; Cross and Motley, 2002). Furthermore, based on linguistic evidence, Newstrom (1991) contends that the name *chayote* comes from the Nahua word *chayojtli,* from Mexico, and that the South American names for the crop are derived from this word (e.g., *cho cho* and *xuxu* in Brazil and *chayota* in Colombia). Direct historical evidence of the chayote's history is scarce because archaeological remains of the nonligneous seeds are rare.

Wild Relatives of Chayote

The most compelling argument for locating the natural origin of a crop is the geographic distribution of its closest wild relatives (figure 8.2). The taxa most closely related to chayote are found primarily in the mountainous regions of central and southern Mexico and northern Guatemala. Taxonomically, these wild populations comprise two species (*S. chinantlense* Lira & Chiang and *S. compositum* J.D. Smith) and a subspecies of the cultivated *S. edule* (*S. edule* ssp. *sylvestre* Lira & Castrejón) (figure 8.1c–e). Despite their classification as species, the possible genetic contribution of these wild taxa to the origin and spread of chayote has not been clearly determined. However, before reviewing recent evidence for these hypotheses, background information on chayote's wild relatives (both conspecific and congeneric) is needed.

The wild subspecies *S. edule* ssp. *sylvestre* is distributed in the narrow strip of montane rainforest between 500–1700 m, which stretches north to south from the Mexican states of Hidalgo to Oaxaca, occurring primarily in damp areas around ravines, waterfalls, and rivers (Newstrom, 1990; Lira, 1995). Generally the fruits of the wild subspecies are smaller than chayote, densely prickly and bitter (figure 8.1c). However, some fruit variation has been reported among fruits of some free-living plants in Veracruz (Newstrom, 1989; Lira, 1995), although these may be escapes from cultivation or hybrids. The most important difference between the wild and cultivated subspecies is the extremely bitter fruit of the former. These were the criteria used by Lira et al. (1999) to classify *S. edule* into the two subspecies, *S. edule* ssp. *edule* for cultivated types and *S. edule* ssp. *sylvestre* for the wild forms.

Sechium chinantlense is found in the state of Oaxaca, where it is endemic to the lower foothills and valleys (20–800 m elevation) of the Chinantla range of the Sierra Madre de Oaxaca (figure 8.2). This species has medium-sized (6–9 cm), ovoid fruits similar to many varieties of chayote and also possesses an apical cleft on the fruit from which the seedling germinates (figure 8.1d). This is the only other species of *Sechium* besides *S. edule* to possess this character (Lira, 1995). Newstrom (1989) originally described this lowland species as another wild type of chayote (wild type III in her classification). However, the stamens of *S. chinantlense* are distinct in that the pollen thecae are confined to the underside of the anther, whereas in *S. edule* the thecae are distributed around the entire apex of the anther (Lira and Chiang, 1992; Lira, 1995). This distinctive flower morphology, along with reported reproductive incompatibility with *S. edule* (both wild and cultivated), led Lira and Chiang (1992) to classify this as a new species.

Sechium compositum is the third species of the chayote species complex. It is found in southernmost Mexico in the Motozintla range of Chiapas state and adjacent Guatemala. The fruits of this species are medium-sized (6–9 cm), usually with longitudinal ridges containing prickles, and extremely bitter (figure 8.1e). Although it differs in physical aspects from *S. edule* (e.g., it lacks an apical cleft), there are anecdotal reports of hybrids between the two species. Two other *Sechium* species in Mexico, *S. hintonii* P. G. Wilson and *S. mexicanum* Lira & Nee, are thought to be more distantly related to chayote (Lira et al., 1997a, 1997b). *S. hintonii* is very rare, known from only two localities in central Mexico (the states of Guerrero and Mexico); therefore its relationship to chayote is enigmatic (Lira and Soto, 1991). The recently described *S. mexicanum* (Lira and Nee, 1999) is distinct from the other *Sechium* species in Mexico and placed in a separate section of the genus (Lira, 1995; Lira and Nee, 1999).

Tacaco (*S. tacaco* Pittier), the other domesticated species in the genus, is very similar to chayote in habitat and morphology and is also grown for its fruit. However, unlike chayote, it is little known outside Costa Rica. Even in its native country it is essentially an heirloom crop, found mostly in private gardens (Lira, 1995; A. Brenes, pers. comm., 2001). In contrast to chayote, tacaco shows little variability, having only one or two named varieties. The origin of tacaco is less well known than that of chayote because no wild populations of *S. tacaco* have been found; the closest wild species is thought to be *S. talamancense* Wunderlin, primarily because it is the only other *Sechium* species in Costa Rica lacking the distinctive pouchlike covering over the floral nectaries. *S. talamancense* is endemic to the higher elevations (more than 2000 m) in the southern Talamanca range of Costa Rica. Its distribution does not appear to overlap with that of tacaco, which is grown at lower elevations in mountain valleys of central Costa Rica.

Despite the discovery of wild taxa and advances in the taxonomy and systematics of *Sechium*, questions remain about the nature of the divergence of chayote from its respective wild relatives and the evolutionary relationships in the genus. Additional data are needed to determine whether chayote was derived from wild *S. edule* or is of hybrid origin from a cross between two Mexican *Sechium* species. Recently, molecular sequence and marker data have become available, and in the following sections we will discuss what each data set reveals about the origin and diversification of chayote.

Molecular Systematics of *Sechium*

Molecular sequence data were used to evaluate hypotheses of chayote's origin in a phylogenetic context. Species were sampled from all genera of the single-seeded cucurbits (subtribe Sicyinae), including the largest and most widespread genus, *Sicyos* (table 8.1). A phylogeny of subtribe Sicyinae based on molecular sequence data was obtained from two gene regions, the nuclear ribosomal internal transcribed spacer (ITS) and external transcribed spacer (ETS) using maximum parsimony and heuristic methods as implemented in the software program PAUP* (Swofford, 1998). These are neutral markers, essentially free from selection, and therefore can provide enough variability to detect differences between closely related species (Baldwin, 1992; Baldwin and Markos, 1998).

The results of these analyses are congruent with those from chloroplast molecular sequence data (Cross, 2003) and provide a different picture of the relationships in Sicyinae than those proposed by previous, morphology-based taxonomy (Lira et al., 1997a, 1997b) (figure 8.3). Many genera, including *Sechium*, do not appear to represent monophyletic lineages. The relationships between the two *Sechium* clades represent a geographic division: The species of *Sechium* from Mexico, with the exception of *S. mexicanum*, form a single clade (hereafter *Sechium* sensu stricto), and the species from Costa Rica and Central America form another clade (hereafter Central American *Sechium*). The Central American clade is quite divergent from other *Sechium* and forms a strongly supported clade at the base of the single-seeded cucurbits.

Table 8.1 Taxa Included in the Combined ITS–ETS Phylogenetic Analysis

Taxon (accession #)	Voucher Data	Locality
Rytidostylis carthaginensis	A. K. Neill 3560 (NY)	Ecuador
Microsechium helleri	H. Cross 58 (NY)	Veracruz, Mexico
Parasicyos dieterleae	R. Lira 1103 (MEXU)	Veracruz, Mexico
Sechiopsis distincta	R. Torres 13828 (NY)	Motozintla Range, Chiapas, Mexico
Sechiopsis laciniatus	R. Lira 1530 (MEXU)	Motozintla Range, Chiapas, Mexico
Sechiopsis triquetra	T. Andres 38 (NY)	Michoacan, Mexico
Sechium chinantlense (H15)	J. Castrejon 86 (NY)	Sierra Chinantla, Oaxaca, Mexico
Sechium chinantlense (H344)	H. Cross 108 (NY)	Sierra Chinantla, Oaxaca, Mexico
Sechium chinantlense (H359)	H. Cross 123 (NY)	Sierra Chinantla, Oaxaca, Mexico
Sechium compositum (H370)	J. Cadena s.n. (Chapingo)	Motozintla Range, Chiapas, Mexico
S. edule ssp. *edule* (H54)	H. Cross 54 (NY)	Orizaba, Veracruz, Mexico
S. edule ssp. *edule* (H62)	H. Cross R2 (NY)	Oaxaca City, Oaxaca, Mexico
S. edule ssp. *edule* (H67)	H. Cross D4 (NY)	Oaxaca City, Oaxaca, Mexico
S. edule ssp. *edule* (H264)	H. Cross 150 (NY)	Zaachila, Oaxaca, Mexico
S. edule ssp. *edule* (H279)	H. Cross 177 (NY)	Chocaman, Veracruz, Mexico
S. edule ssp. *sylvestre* (H25)	R. Lira 1370 (NY)	Queretaro, Mexico
S. edule ssp. *sylvestre* (H55)	H. Cross 57 (NY)	Veracruz, Mexico
S. edule ssp. *sylvestre* (H250)	H. Cross 136 (NY)	Oaxaca, Mexico
S. edule ssp. *sylvestre* (H292)	H. Cross 191 (NY)	Xico, Veracruz, Mexico
S. edule ssp. *sylvestre* (H294)	H. Cross 193 (NY)	Xico, Veracruz, Mexico
S. edule ssp. *sylvestre* (H302)	J. Cadena 408 (Chapingo)	Ixtac, Veracruz, Mexico
S. edule ssp. *sylvestre* (H304)	J. Cadena 408 (Chapingo)	Ixtac, Veracruz, Mexico
S. edule ssp. *sylvestre* (S8)	T. Andres 171 (NY)	Tabasco, Mexico
S. edule ssp. *sylvestre* (S16)	R. Fernandez 3173 (NY)	Queretaro, Mexico
S. edule ssp. *sylvestre* (S18)	L. E. Newstrom 1473 (NY)	Veracruz, Mexico
Sechium hintonii	J. Castrejon 1226 (NY)	Guerrero, Mexico
Sechium mexicanum	R. Lira 1368 (MEXU)	Veracruz, Mexico
Sechium pittieri	H. Cross 68 (NY)	Talamanca Range, Costa Rica
Sechium tacaco (H160)	M. Murrell sn (NY)	Heredia, Costa Rica
Sechium tacaco (H176)	H. Cross 92 (NY)	Cartago, Costa Rica
Sechium talamancense (H173)	H. Cross 79 (NY)	Talamanca Range, Costa Rica
Sechium talamancense (H174)	H. Cross 80 (NY)	Talamanca Range, Costa Rica
Sechium villosum	H. Cross 97 (NY)	Volcan Poas, Costa Rica
Sicyos angulata	H. Cross 43 (NY)	New York, New York, USA

(continued)

Table 8.1 (continued)

Taxon (accession #)	Voucher Data	Locality
Sicyos alba	S. Perlman 15666 (BISH)	Hawaii, USA
Sicyos guatemalensis	I. Rodriguez 260 (MEXU)	Ixtlan, Oaxaca, Mexico
Sicyos hispidus	W. Takeuchi 8517 (BISH)	Hawaii, USA
Sicyos microphyllus	I. Rodriguez 253 (MEXU)	Michoacan, Mexico
Sicyos motozintlensis	R. Lira 951 (MEXU)	Motozintla Range, Chiapas, Mexico
Sicyos parviflorus	I. Rodriguez 234 (MEXU)	Mexico State, Mexico
Sicyos polyacanthus	Mulgara 1745 (NY)	Parana, Brazil
Sicyosperma gracile	V. W. Steinman 961 (NY)	Sonora, Mexico

S. mexicanum is allied with other species of *Sicyos* and does not appear to be very closely related to the other species of *Sechium*.

With respect to the origin of chayote and its relationships to the wild taxa, the phylogeny obtained from the sequence data do not resolve the relationships within *Sechium* s.s., indicating that these species are genetically very similar. Sequences from ITS and ETS were obtained from many individuals of both subspecies of *S. edule* to represent the maximum geographic range and morphological variation of the species, yet little sequence variation was observed between individuals. Furthermore, very little interspecific sequence variation between *S. edule, S. compositum,* and *S. chinantlense* was evident. Therefore, from the sequence data it is not possible to determine whether these species are recently diverged or perhaps belong to a single, highly variable species. The possibility of gene flow between these taxa may have also obscured the phylogenetic signal (Rieseberg and Soltis, 1991). This is especially relevant for a crop such as chayote that has been brought into cultivation in close proximity to its wild relatives (see also chapter 15, this volume).

The results of the phylogenetic analysis suggest three (not entirely mutually exclusive) possibilities regarding the evolution of the *Sechium* s.s. clade: Either *Sechium* s.s. is a single species with many morphological variants, there is extensive gene flow between these taxa, or speciation among *S. edule, S. chinantlense, S. compositum,* and *S. hintonii* occurred very recently. Regarding this last hypothesis, there is additional evidence that supports the recognition of four distinct species. This includes the differing chromosome numbers between taxa (Mercado et al., 1993), the

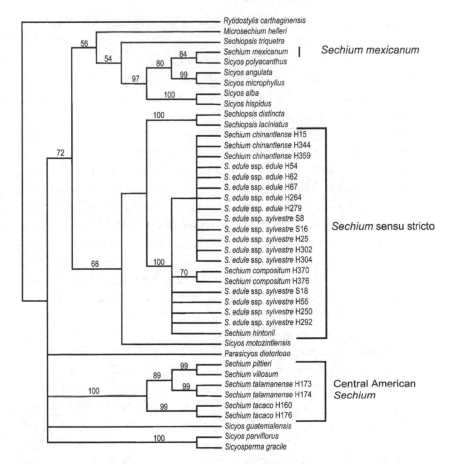

FIGURE 8.3 Strict consensus of more than 10,000 most parsimonious trees from an analysis of the combined external and internal transcribed spacer data sets. Length = 748, C.I. = 0.66, R.I. = 0.85. Numbers above branches indicate bootstrap support (500 replicates). Individual collections of *Sechium* species (including both subspecies of *S. edule*) are indicated by their accession number. See table 8.1 for collection details.

inability to produce successful hybrid crosses (Castrejón and Lira, 1992; Lira, unpublished data), distinctive morphological characters, and biogeographic distributions.

Evidence from Molecular Marker Data

The lack of resolution among *Sechium* s.s. in the phylogenetic analysis necessitates a different approach. Amplified fragment length polymorphism

(AFLP) (Vos et al., 1995) is a polymerase chain reaction–based technique that provides fragment length differences based on single–base pair polymorphisms from across the plant's genomes. It has been shown to be useful at differentiating between individuals and very closely related species (Milbourne et al., 1997; Cervera et al., 1998) and has greater resolving power than molecular sequence data can usually provide.

A total of 453 markers from five primer pairs were obtained for 178 individuals of *S. edule* (both subspecies; 127 individuals of chayote and 21 individuals of ssp. *sylvestre*), *S. chinantlense* (20 individuals), and *S. compositum* (10 individuals) from Mexico, with additional chayote accessions from Costa Rica. The majority of the collections were of chayote because more individuals were available and because we wanted to represent the morphological diversity of chayote in its native range and from each major growing region. The chayotes from Costa Rica were largely from the germplasm collection of the National University of Costa Rica. Both neighbor joining (NJ) analysis (figure 8.4) and principal component analysis (PCA) (figure 8.5) were conducted on the AFLP data. These two genetic distance analyses provide different perspectives on the same data, but both tell essentially the same story. The results suggest that the species delineations within *Sechium* s.s. represent very closely related, distinct taxonomic entities. The NJ analysis (figure 8.4) reveals three main clusters in accordance with morphology-based specific circumscriptions of Lira (1995). However, the wild subspecies of *S. edule* does not form a monophyletic group. The populations of *S. edule* ssp. *sylvestre* from Oaxaca are sister to all other *S. edule,* and individuals of the wild subspecies from Veracruz are sister to the chayote cluster. The *S. edule* ssp. *sylvestre* from Oaxaca are distinct from the other clusters of *S. edule* based on genetic distance (represented in figure 8.4 by branch length) and have a position between *S. chinantlense* and the remaining groups of *S. edule.* Morphologically, these populations represent *S. edule* ssp. *sylvestre,* but their position on the tree is somewhat ambiguous and may indicate gene flow between these wild taxa.

The PCA analysis also provides evidence for the recognition of three distinct species, although it also shows ambiguity among the Oaxacan *S. edule* ssp. *sylvestre,* the *S. chinantlense,* and the remaining *S. edule* individuals. In the PCA analysis both *S. compositum* and *S. chinantlense* form distinct clusters, but the individuals of *S. edule* cluster into three groups conforming to the three main branches of the NJ analysis. The population of *S. edule* ssp. *sylvestre* from Oaxaca appears genetically intermediate between *S. chinantlense* and

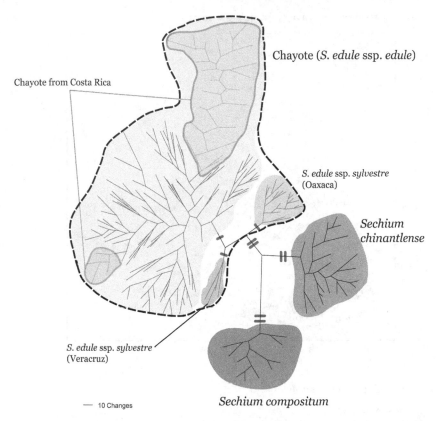

Chayote from Costa Rica

Chayote (*S. edule* ssp. *edule*)

S. edule ssp. *sylvestre* (Oaxaca)

Sechium chinantlense

S. edule ssp. *sylvestre* (Veracruz)

— 10 Changes

Sechium compositum

FIGURE 8.4 Unrooted phylogram from neighbor-joining analysis of AFLP data of *Sechium* species. Double bars on branches indicate species delimitations; single bars indicate subspecific delimitations. *S. edule* (both subspecies) is encircled by a dashed line. Shaded areas nested within chayote (*S. edule* ssp. *edule*) indicate the accessions of Costa Rica.

the remaining *S. edule*. The populations of *S. chinantlense* and *S. edule* ssp. *sylvestre* from Oaxaca were collected only a few kilometers apart, along the same road. Two AFLP markers were present only in members of *S. chinantlense* and the Oaxacan population of *S. edule* ssp. *sylvestre*. These markers were common in the population of *S. chinantlense* and rare in *S. edule* ssp. *sylvestre* (only 3 of 15 individuals), a situation that can suggest interspecific gene flow (Arias and Rieseberg, 1995). There is also evidence of gene flow between chayote and *S. edule* ssp. *sylvestre* in the same region: Two markers are found exclusively in these same populations of *S. edule* ssp. *sylvestre* in Oaxaca and chayotes collected from a farm a few kilometers away.

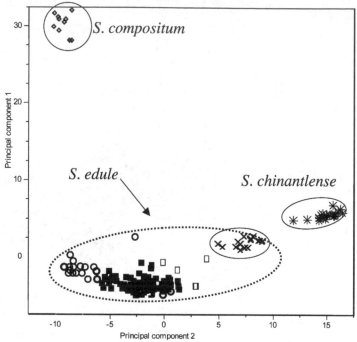

FIGURE 8.5 Graph of principal component analysis (PCA) of *Sechium* AFLP data set. The first component (14.93% of the variation) is displayed along the *y*-axis, and the second component (10.61% of the variation) is displayed along the *x*-axis. Lines are drawn around each species.

However, other evidence suggests that there are significant reproductive barriers to hybridization among these taxa. As mentioned earlier, chromosome numbers vary between the two species and even between the subspecies of *S. edule*. For chayote (*S. edule* ssp. *edule*), counts of n = 12, 13, 2n = 22, 24, 26, and 28, have been reported (Giusti et al., 1978; Goldblatt, 1981, 1984, 1990; Palacios, 1987; Singh, 1990; Sobti and Singh, 1961;

Sugiura, 1938, 1940). For *S. edule* ssp. *sylvestre,* 2n = 24 was reported for the populations from Veracruz (Palacios, 1987) and n = 13 for the populations from Oaxaca (Mercado et al., 1993; Mercado and Lira, 1994). For *S. chinantlense,* a count of 2n = 30 was reported (Mercado et al., 1993). Furthermore, crosses attempted between *S. edule* and *S. chinantlense* yielded no viable progeny; fruits were produced in only 2 of 52 attempts when crossed with chayote and 2 of 29 when crossed with *S. edule* ssp. *sylvestre* (Castrejón and Lira, 1992; Lira, unpublished data). Only one of the fruits in those four cases germinated (in a cross of *S. chinantlense* and chayote) and quickly died (Castrejón and Lira, 1992; Lira, unpublished data). Given the lack of viable hybrid offspring, differences in karyology, and the genetic differentiation between species seen in the distance analyses, any gene flow between these species probably would be rare.

Based on the AFLP data presented here, *Sechium* s.s. appears to represent four distinct species: *S. edule, S. compositum, S. chinantlense,* and *S. hintonii.* Although no specimens of *S. hintonii* were available for the AFLP study, it is morphologically distinct from the other species (Lira, 1995). Each of these four species can be easily differentiated with morphological characters. The central question left unresolved is the actual genetic contribution of these species to the gene pool of chayote. Although reproductive barriers exist, the AFLP data suggest (at least in some cases) that they may not be completely insurmountable. It is possible that introgression from wild taxa, though rare, could have been sufficient to enhance the chayote gene pool and contribute to its diversification. In other words, there has not been sufficient gene flow between these taxa to blur the species boundaries, yet occasional introgression from the wild taxa could have contributed to the genetic diversity of the crop. In addition to the AFLP data, the observation that the fruit characteristics of chayote overlap those of each of the wild taxa (e.g., some chayote fruits have characteristics of the fruits of *S. compositum,* some are like *S. chinantlense*) provides further compelling if circumstantial evidence. Another interpretation of these data is that there existed extensive genotypic variability in the ancestral *Sechium* species before its divergence into the four species recognized today, and this diversity was retained in enough chayote varieties (and populations of the wild subspecies from Veracruz, with which it is interfertile) to have been available to the chayote gene pool as the crop expanded into new environments. These two interpretations are not mutually exclusive, and it seems likely that a combination of these factors has been responsible for maintaining the diversity of the crop. Additional data are needed to determine the extent of introgression from wild species and their role in the origin and diversification of chayote.

Evolution and Diversification Within *S. edule*

The large number of chayotes sampled for the AFLP analysis allows a more detailed evaluation of the evolutionary patterns within the domesticated species and what they reveal about the crop's diversification. The results show that a good proportion of the variability among the molecular marker data is accounted for by differences between the cultivated varieties of chayote. According to an analysis of molecular variance of the AFLP data set (table 8.2), 57% of the variation is found between populations and 36% of the variation was distributed within populations (Cross, 2003). This is also reflected in the PCA analysis, in which a major portion of the second principal component (10.61% of the variation, the x-axis in figure 8.5) resulted from variation between chayote individuals. The NJ analysis also shows large genetic distances between the chayote individuals (figure 8.4). Given the large variation in fruit morphology and greater geographic distribution in chayote than in the wild populations, this was not unexpected.

One factor that may have contributed to the expansion of genetic diversity within chayote is the geographic expansion of the crop beyond its native range and habitat. It is clear that chayotes from Costa Rica represent a generally distinct collection because most of the samples from this country form a single lineage within the larger cluster of chayote (figure 8.4). Other Costa Rican chayotes not in this cluster probably represent more recent Mexican accessions in the chayote breeding program (Sharma et al., 1995). The variability of the Costa Rican samples suggests that this country is a secondary center of diversity for chayote. Chayote was an early introduction into Costa Rica (at least pre-Columbian; Newstrom, 1991), and locally adapted

Table 8.2 Analysis of Molecular Variance Based on 453 AFLP Loci

Source of Variation	df	Sum of Squares	Variance Components	% Variance	*P* Value
Between species	3	1283.0338	4.35537	6.73	<.05
Between populations	4	2532.2989	36.9172	57.03	<.05
Within populations	7	1407.4364	23.45	36.24	<.05

genotypes and phenotypes have been selected over the last several centuries. Although Costa Rica does not contain as many morphotypes as Mexico, it is home to several local varieties. Most of the collections from Costa Rica included in this study are from the National University of Costa Rica gene bank, and they were selected to represent chayote variation from all growing areas of the country, including high valleys, cloud forests, and dry areas in the northwest of the country. Therefore, the genetic diversity observed in this collection (as illustrated by the long branch lengths in this cluster; figure 8.4) is not surprising.

Origin of Chayote

Comparing patterns of domestication and evolution between closely related crop species can be informative in understanding the domestication process. Many factors influence the domestication and diversification of a species, including differences in climate, biology, life history, and genetic preadaptation to domestication. When comparing domestication within a well-defined group of species (such as the single-seeded cucurbits) or even across a single family, it is possible to control for some of these factors. The Costa Rican crop tacaco provides a point of comparison for examining significant factors that may have contributed to the origin and diversification of chayote. There are some parallels between the crops in morphology, plant parts used, cultivation practices, and habitat, but there are also important differences. Exploring these differences and the possible causes may help shed light on the evolutionary mechanisms involved in the origins of both crops.

The molecular sequence data suggest that the mode of domestication of tacaco was different than that of chayote. The molecular phylogeny (figure 8.3) and genetic distances based on pairwise sequence comparisons (data not shown) demonstrate that tacaco, unlike chayote and its relatives, is very distinct genetically from other *Sechium* in Central America. The Central American clade of *Sechium* is well resolved and strongly supported, with tacaco sister to the other species (figure 8.3). The species most similar morphologically to tacaco, *S. talamancense,* appears more closely related to two other species, *S. pittieri* and *S. villosum,* which, like *S. talamancense,* are found in the high elevations of Central America. No species in the Central American clade of *Sechium* is sympatric with *S. tacaco,* and with no known conspecific wild relative, the origin of tacaco remains enigmatic. There are two other *Sechium* species in Central America, *S. venosum* and

S. panamense, which were not available for this study, so it is possible that they could provide more clues to the origin of tacaco. However, they are morphologically more similar to *S. pittieri* and do not overlap in distribution with tacaco.

Given its genetic isolation and lack of close wild relatives, it is possible that tacaco is a semidomesticated species (Harlan, 1992) in which natural selection for nonbitter fruits may have occurred long before humans noticed them. By this scenario, the species was simply brought into cultivation under less selective pressure. The high valleys in Costa Rica where tacaco is grown have been continuously populated for many millennia, and by now any natural populations of the species probably have been destroyed. However, it is possible that wild relatives of tacaco may yet be identified.

The much greater variation of chayote compared with tacaco may be a result of the species originating from a much broader gene pool of closely related taxa. A similar pattern can be observed in other domesticated species in the family. For example, the genus *Cucurbita* has several domesticated species, and the most diverse of these is *C. pepo.* Similar to *S. edule, C. pepo* has many wild conspecific taxa (three subspecies) and other species that are closely related (Nee, 1990; Sanjur et al., 2002). Furthermore, *C. pepo* exhibits incredible diversity in its fruit types (e.g., zucchini, pumpkin, spaghetti squash, and acorn squash). Another domesticated species in the genus, *C. ficifolia,* mirrors tacaco in being very monomorphic, with no known wild relatives, and is phylogenetically isolated from the other species of *Cucurbita* (Nee, 1990; Sanjur et al., 2002). More research is needed both across and within families to determine the degree to which the broader gene pool influences the intraspecific diversity of crop species.

With molecular sequence data it is now possible to infer time to the branching points on a phylogenetic tree (Arbogast et al., 2002). By estimating these divergence dates, it is possible to speculate on geological factors that may have contributed to the divergence of *Sechium* and related genera and also provide another means of comparing the divergent paths that the crops chayote and tacaco have taken. Dates of molecular divergence were calculated using a penalized likelihood approach as implemented in r8s (Sanderson, 2002, 2003) on a subset of the nuclear ribosomal DNA sequence data for Sicyinae (figure 8.6), using a Miocene fossil of *Sicyos* as the single calibration point at the root of the tree. Because only one fossil was available and its age is not precise (from 15 to 5 mya; Li-Jianqiang, 1997), the molecular clock analysis was run three times, using

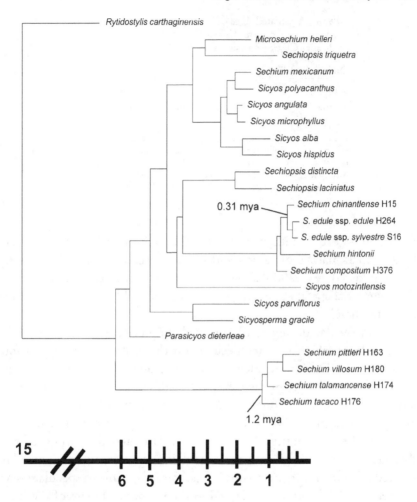

FIGURE 8.6 Maximum likelihood chronogram derived from ETS and ITS sequence data. Branch lengths represent estimated divergence times of nodes as calculated in r8s, based on a maximum age of divergence of 15 million years before present (mya). Bar on bottom indicates scale in mya. Scale disjunction is indicated by slashed bars. The divergence times for the crop species and their closest wild relatives are indicated on the figure.

the oldest, median, and youngest dates of the fossil (15, 10, and 5 mya, respectively). Based on these analyses, the timing of divergence of *S. edule* from *S. chinantlense* was estimated to have occurred from the middle to late Pleistocene (200,000–500,000 years ago), depending on the calibration date used. By contrast, the divergence times estimated for tacaco from the other Central American *Sechium* species were much earlier,

Table 8.3 Estimated Ages of Divergence Between the Crops Chayote and Tacaco and Their Wild Relatives, Based on a Molecular Clock Analysis Using Penalized Likelihood

Calibration Point	Chayote	Tacaco
5 mya	0.20 ± 0.0017	0.46 ± 0.003
10 mya	0.32 ± 0.0062	0.83 ± 0.048
15 mya	0.40 ± 0.0080	1.18 ± 0.015

All ages are in millions of years before present (mya). For chayote, the divergence time is between *S. edule* and *S. chinantlense;* for tacaco, divergence time is between *S. tacaco* and the other Central American *Sechium* species (*S. talamancense, S. pittieri,* and *S. villosum*).

from 500,000 to 1 million years BP (table 8.3). This is also supported by a similar analysis of tacaco and related species in Costa Rica using AFLP markers (Cross, 2003) showing tacaco to be more genetically distant from the other species than chayote is from any wild species in the study presented here.

The molecular dating results, though imprecise, provide a possible scenario from which we can speculate on the evolution and speciation of *Sechium* s.s. During the Pleistocene even tropical Mexico was affected by the climate changes of the ice age, though not as severely as at higher latitudes. The pine–oak forests of Mexico, the primary habitat of *S. edule* ssp. *sylvestre,* expanded to lower elevations during periods of colder climates in the Pleistocene (Toledo, 1982). Although the timing of the divergence of *S. chinantlense* from *S. edule* is open to interpretation and is in need of much more precise measures, it is reasonable to speculate that ancestors of *S. chinantlense* adapted to the warming climate of the lowland tropics as the pine–oak forests retreated to higher elevations during interglacial periods of the Pleistocene. The subsequent uphill movement of this habitat during the interglacial also could have isolated populations of *S. edule* in the mountains to the north in Veracruz and Hidalgo, further contributing to genetic differentiation between the populations of *S. edule.*

Conclusions

A synthesis of studies of morphological and molecular variation provides a picture of the evolution of *Sechium* at both the interspecific and intraspecific levels. At the interspecific level, the genus *Sechium* s.s. is quite distinct genetically from its sister taxon. However, the phylogenetic signal of the sequence data does not definitively demarcate the species boundaries within *Sechium* s.s., suggesting very recent speciation events, during the middle to late Pleistocene (table 8.3). The shifting climate of the last ice age seems to have provided a sufficient mechanism to drive isolation of these populations and their subsequent speciation. AFLP markers provided ample genetic signal at and below the species level, resolving three discrete clusters corresponding to the recognized species and revealing geographic differentiation within the *S. edule* cluster. Based on the AFLP results and the clear morphological differences between these taxa, they appear to be distinct species. The subsequent domestication and spread of chayote in the Holocene brought these populations back in contact, and despite the physical and reproductive barriers it is possible that at least some interspecific gene flow has occurred. In addition, the very recent divergence of chayote from a widespread, diverse *Sechium* s.s. species complex seems to have provided a genetic reservoir from which the chayote diversified into the many varieties known today.

The origin and diversification of chayote based on molecular evidence demonstrate that knowledge of the wild populations and closely related species of a crop can be very important for understanding its evolution both before and after domestication. This is especially important if, as in the case of chayote, the barriers to gene flow are porous. These results suggest that for the continued improvement of a crop, the conservation of wild relatives is crucial. In the case of chayote, the populations of all the wild taxa are endangered by deforestation and human encroachment into their habitat. Currently no habitats of wild *Sechium* species are protected in reserves in Mexico (Lira et al., 1999). For chayote, the future success of the crop may be closely linked to the conservation of its wild relatives.

Acknowledgments

The authors would like to thank several people and institutions that helped make this research possible. We are grateful to Jorge Cadena for the use of his photograph in figure 8.1e. We thank Jorge Cadena, Ismael Calzada, Javier Castrejón, the Fuentes family of Oaxaca and Jalisco, Patricia Fuentes-Cross, Alvaro Gutíerrez, and Isela Rodriguez, who assisted with collecting in Mexico. We also thank Abdenago Brenes, Jorge Madriz, and Manfred Murrell, who assisted with collecting in Costa Rica.

We would like to acknowledge Kobinah Abdul-Salim, Bill Hahn, Fabian Michelangeli, Susan Pell, and Nyree Zerega for assistance with laboratory work and analyses. We are also grateful to anonymous reviewers for their invaluable comments on the manuscript. Finally, the field and laboratory work was completed with funding assistance from the New York Botanical Garden, the Lewis B. & Dorothy Cullman Foundation, the Center for Environmental Research and Conservation, Columbia University, the Torrey Botanical Society, and the Explorers Club.

References

Arbogast, B. S., S. V. Edwards, J. Wakeley, P. Beerli, and J. B. Slowinski. 2002. Estimating divergence times from molecular data on phylogenetic and population genetic timescales. *Annual Review of Ecology and Systematics* 33: 707–740.

Arias, D. M. and L. H. Rieseberg. 1995. Genetic relationships among domesticated and wild sunflowers (*Helianthus annuus*, Asteraceae). *Economic Botany* 49: 239–248.

Baldwin, B. G. 1992. Phylogenetic utility of the internal transcribed spacers of nuclear ribosomal DNA in plants: An example from the Compositae. *Molecular Phylogenetics and Evolution* 1: 3–16.

Baldwin, B. G., and S. Markos. 1998. Phylogenetic utility of the external transcribed spacer (ETS) of 18S–26S rDNA: Congruence of ETS and ITS trees of *Calycadenia* (Compositae). *Molecular Phylogenetics and Evolution* 10: 449–463.

Brush, S. B. 1989. Rethinking crop genetic resources conservation. *Conservation Biology* 3: 19–29.

Castrejón, J. and R. Lira. 1992. Contribución al conocimiento de la relación silvestre-cultivo en el "chayote" *Sechium edule* (Jacq.) Swartz (Cucurbitaceae). In *Resúmenes Simposio Etnobotánica* 92: 345. Jardines Botánicos de Córdoba, Córdoba, Spain.

Cervera, M. T., J. A. Cabezas, J. C. Sancha, F. Martínez de Toda, and J. M. Martínez-Apater. 1998. Application of AFLPs to the characterization of grapevine *Vitis vinifera* L. genetic resources. A case study with accessions from Rioja (Spain). *Theoretical and Applied Genetics* 97: 51–59.

Cross, H. B. 2003. *Evolution, Systematics, and Domestication in* Sechium *and Related Genera (Sicyeae, Cucurbitaceae)*. Ph.D. dissertation, Columbia University, New York, NY, USA.

Cross, H. B. and T. J. Motley. 2002. Phenotypic and genetic diversity of chayote germplasm. *Proceedings of Cucurbitaceae* 2002: 138–143.

Giusti, L., M. Resnik, T. Del V. Ruiz, and A. Grau. 1978. Notas acerca de la biología de *Sechium edule* (Jacq.) Swartz (Cucurbitaceae). *Lilloa* 35: 5–13.

Goldblatt, P. (ed.). 1981. Index to plant chromosome numbers (1975–1978). *Monographs in Systematic Botany 5*. Missouri Botanical Garden, St. Louis, MO, USA.

Goldblatt, P. (ed.). 1984. Index to plant chromosome numbers (1979–1981). *Monographs in Systematic Botany 8*. Missouri Botanical Garden, St. Louis, MO, USA.

Goldblatt, P. (ed.). 1990. Index to plant chromosome numbers (1988–1989). *Monographs in Systematic Botany 40*. Missouri Botanical Garden, St. Louis, MO, USA.

Harlan, J. R. 1992. *Crops and Man*. American Society of Agronomy, Madison, WI, USA.

Li-Jianqiang. 1997. On the systematics of *Thladiantha* (Cucurbitaceae). *Acta Botanica Yunnanica* 19: 103–127.

Lira, R. 1995. *Estudios Taxonómicos y Ecogeográficos de las Cucurbitaceae Latinoamericanas de Importancia Económica:* Cucurbita, Sechium, Sicana, y Cyclanthera. *Systematic and Ecogeographic Studies on Crop Genepools. 9.* International Plant Genetic Resources Institute/Instituto de Biologia, Universidad Nacional Autonoma de Mexico, Rome, Italy.

Lira, R. 1996. *Chayote,* Sechiumedule *(Jacq.) Sw. Promoting the Conservation and Use of Underutilized and Neglected Crops. 8.* Institute für Pflanzengenetik und Kulturpflanzenforschung Gatersleben/ International Plant Genetic Resources Institute, Rome, Italy.

Lira, R., J. Caballero, and P. Dávila. 1997a. A contribution to the generic delimitation of *Sechium* (Cucurbitaceae, Sicyinae). *Taxon* 46: 269–282.

Lira, R., J. Castrejón, S. Zamudio, and C. Rojas-Zenteno. 1999. Propuesta de ubicación taxonómica para los chayotes silvestres (*Sechium edule,* Cucurbitaceae) de Mexico. *Acta Botanica Mexicana* 49: 47–61.

Lira, R. and F. Chiang. 1992. Two new combinations in *Sechium* (Cucurbitaceae) from Central America, and a new species from Oaxaca, Mexico. *Novon* 2: 227–231.

Lira, R. and M. Nee. 1999. A new species of *Sechium* sect. *Frantzia* (Cucurbitaceae, Sicyeae, Sicyinae) from Mexico. *Brittonia* 51: 204–209.

Lira, R. and J.C. Soto. 1991. *Sechium hintonii* (P.G. Wilson) C. Jeffrey (Cucurbitaceae): Rediscovery and observations. *FAO/IBPGR Plant Genetic Resources Newsletter* 87: 5–10.

Lira, R., J.L. Villasenor, and P.D. Davila. 1997b. A cladistic analysis of the subtribe Sicyinae (Cucurbitaceae). *Systematic Botany* 22: 415–425.

Mercado, P. and R. Lira. 1994. Contribución al conocimiento de los números cromosómicos de los géneros *Sicana* Naudin y *Sechium* P. Br. (Cucurbitaceae). *Acta Botanica Mexicana* 27: 7–13.

Mercado, P., R. Lira, and J. Castrejón. 1993. Estudios cromosómicos en *Sechium* P. Br. y *Sicana* Naudin (Cucurbitaceae). In *XII Congreso Mexicano de Botanica,* p. 176. Sociedad Botanica de Mexico, Merida, Yucatan, Mexico.

Milbourne, D.R., R. Meyer, J. Bradshaw, E. Baird, N. Bonar, J. Provan, W. Powell, and R. Waugh. 1997. Comparison of PCR-based marker systems for the analysis of genetic relationships in cultivated potato. *Molecular Breeding* 3: 127–136.

Nee, M. 1990. The domestication of *Cucurbita. Economic Botany Supplement* 44: 56–68.

Newstrom, L.E. 1989. Reproductive biology and evolution of the cultivated chayote *Sechium edule:* Cucurbitaceae. In J.H. Bock and Y.B. Linhart (eds.), *The Evolutionary Ecology of Plants,* 491–509. Westview Press, Boulder, CO, USA.

Newstrom, L.E. 1990. Origin and evolution of chayote, *Sechium edule.* In D.M. Bates, W.R. Robinson, and C. Jeffrey (eds.), *Biology and Utilization of the Cucurbitaceae,* 141–149. Cornell University Press, Ithaca, NY, USA.

Newstrom, L.E. 1991. Evidence for the origin of chayote, *Sechium edule* (Cucurbitaceae). *Economic Botany* 45: 410–428.

Palacios, R. 1987. *Estudio Exploratorio del Número Cromosómico del chayote* Sechium edule *Sw.* Licenciado en Ciencia Agronómica Universidad Veracruzana, Vera Cruz, Mexico.

Rieseberg, L.H. and D.E. Soltis. 1991. Phylogenetic consequences of cytoplasmic gene flow in plants. *Evolutionary Trends in Plants* 5: 65–84.

Sanderson, M.J. 2002. Estimating absolute rates of molecular evolution and divergence times: A penalized likelihood approach. *Molecular Biology and Evolution* 19: 101–109.

Sanderson, M.J. 2003. *r8s,* Version 1.60, *Unix for Mac OSX.* University of California, Davis, CA, USA.

Sanjur, O.I., D.R. Piperno, T.C. Andres, and L. Wessel-Beaver. 2002. Phylogenetic relationships among domesticated and wild species of *Cucurbita* (Cucurbitaceae) inferred from a mitochondrial gene: Implications for crop plant evolution and areas of origin. *Proceedings of the National Academy of Sciences* 99: 535–540.

Sharma, M.D., L. Newstrom-Lloyd, and K.R. Neupane. 1995. Nepal's new chayote gene bank offers great potential for food production in marginal lands. *Diversity* 11: 7–8.

Singh, A.K. 1990. Cytogenetics and evolution in the Cucurbitaceae. In D.M. Bates, W.R. Robinson, and C. Jeffrey (eds.), *Biology and Utilization of the Cucurbitaceae,* 10–28. Cornell University Press, Ithaca, NY, USA.

Sobti, S.N. and S.D. Singh. 1961. A chromosome survey of Indian medicinal plants. Part I. *Proceedings of the Indian Academy of Sciences* 54: 138–144.

Sugiura, T. 1938. A list of chromosome numbers in angiosperm plants. V. *Proceedings of the Imperial Academy of Japan* 14: 391–392.

Sugiura, T. 1940. Studies on the chromosome numbers in higher vascular plants. *Citologia* 10: 363–370.

Swofford, D.L. 1998. *PAUP*. Phylogenetic Analysis Using Parsimony (* and Other Methods),* Version 4. Sinauer Associates, Sunderland, MA, USA.

Toledo, V.M. 1982. Pleistocene changes of vegetation in tropical Mexico. In G.T. Prance (ed.), *Biological Diversification in the Tropics,* 93–111. Columbia University Press, New York, NY, USA.

Vos, P., R. Hogers, M. Bleeker, M. Rijans, T. Van de Lee, M. Hornes, A. Frijters, J. Pot, J. Peleman, M. Kuiper, and M. Zabeau. 1995. AFLP: A new technique for DNA fingerprinting. *Nucleic Acids Research* 23: 4407–4414.

PART 3

THE DESCENT OF MAN

Human History and Crop Evolution

Terence A. Brown, Sarah Lindsay,
and Robin G. Allaby

CHAPTER 9

Using Modern Landraces of Wheat to Study the Origins of European Agriculture

Origins of European Agriculture

Agriculture began independently in China, Mesoamerica, and the Fertile Crescent of Southwest Asia, a region comprising the plains of Mesopotamia, parts of Syria and Palestine, and some of the mountainous areas to the east of Anatolia (Diamond, 2002). In Southwest Asia, cereals were among the first plants to be domesticated, with einkorn wheat (*Triticum monococcum* L.), emmer wheat (*T. dicoccum* Schübl.), and barley (*Hordeum vulgare* L.) present at farming sites dating to the 9th millennium BC (Bell, 1987; Kislev, 1992; Zohary and Hopf, 2000). After some 1500 years, cereal cultivation began to expand out of Southwest Asia into Europe, Central Asia, and northeast Africa, with emmer in particular becoming a widespread feature of prehistoric agriculture across much of the Old World and not being substantially replaced by hexaploid bread wheat (*T. aestivum* L.) until 2000 years ago (Zohary and Hopf, 2000). Agriculture first appeared in the Balkans at about 6500 BC and during the next 3000 years spread into Europe by two principal routes, one following the Danube and Rhine valleys through central Europe and into the north European plain, and the second taking a coastal route through Italy and Iberia to northwestern Europe (Barker, 1985; van Zeist et al., 1991; Price, 2000).

Archaeological Questions Concerning the Origins of European Agriculture

There has been much debate about the processes responsible for the origin of agriculture in Southwest Asia and for its subsequent spread into Europe. Blumler (1992) describes two models for agricultural origins: stimulus diffusion, in which agriculture has a very localized start point, and independent invention, in which agriculture has a dispersed geographic origin. When applied to Southwest Asia these models have important implications: Stimulus diffusion at one extreme suggests an almost heroic breakthrough by a small group of humans whose activities resulted in assembly of "a balanced package of domesticates meeting all of humanity's basic needs" (Diamond, 1997:1243), and dispersed origins at the other extreme indicates a transition to agriculture that may have been driven not by human ingenuity but solely or largely by the climatic and other environmental changes occurring across Southwest Asia at the end of the last major glaciation (Sherratt, 1997). Distinguishing between these possibilities has been a goal of archaeologists for the last 20 years, as stated by Harris (1996:6): "If it can be determined that a particular plant . . . was domesticated once only, or several times in different areas, we can gain important insights into the early history of agriculture and pastoralism. . . . This must continue to be a major part of the research agenda for the study of 'agricultural origins.'"

Equally important questions surround the factors responsible for the spread of agriculture into Europe. The application of human genetics to this problem has polarized views between the migrationist and indigenist positions, the former supported by the detailed analysis of nuclear DNA markers (Cavalli-Sforza et al., 1994) and the latter promoted by mitochondrial DNA studies (Richards, 2003). The migrationist view holds that the primary force responsible for agricultural spread into Europe was the immigration of farmers from Southwest Asia, possibly driven by population growth brought about by farming itself, resulting in the displacement of the hunting–gathering communities of preagricultural Europe. The indigenist position is that agriculture spread primarily through contact between frontier populations and subsequent acculturation (Zvelebil, 2000). This debate has now become sterile, with a general consensus that 20–30% of the modern European population arrived on the continent at the same time as farming, so the human dynamic was neither migrationist nor indigenist. In reality, the attention of archaeologists has moved forward and is no longer focused on these simplistic interpretations of agricultural spread. Interest is now centered on the more detailed and complex issues relating to the precise trajectories followed by agricultural spread within and between localized geographic regions and on the nature of the factors responsible for the initial establishment and subsequent development of agriculture in these regions (Zvelebil, 2000). These factors include not only the contact between the preagricultural foragers and the first farmers but also the ecological pressures placed on the crops and the genetic responses of the crops to these pressures. The issues are exemplified by the debates regarding the stop–go pattern of agricultural spread. Agriculture spread rapidly into Greece and the Balkans but apparently slowed down when it reached southeast Hungary (Halstead, 1989; Zvelebil and Lillie, 2000) before again spreading rapidly through the Danube and Rhine valleys. This and other delays have been ascribed either to human factors, agriculture being an unattractive alternative to a successful hunter–gatherer lifestyle in an environment rich in wild resources or to genetic factors, the delay being the time needed for crops to adapt to alien climatic conditions (Zvelebil and Rowley-Conwy, 1986; Halstead, 1989; Bogucki, 1996; Zvelebil and Lillie, 2000).

Plant Genetics and the Origins of Agriculture

Plant genetics has the potential to play a key role in addressing the questions described in this chapter, but so far this potential has been exploited only with regard to the origin of agriculture in Southwest Asia, not with respect to its spread into Europe. Before 1997, a substantial body of disparate information had been accumulated about the genetics of the founder crops of Southwest Asian agriculture (summarized by Zohary, 1996). The discovery that a key domestication trait in cultivated barley, the nonbrittle phenotype characterized by retention rather than shedding of the grain when the ears become mature, is coded by two different mutations, with some cultivars having one mutation and some having the other (Takahashi, 1964, 1972), led to the view that barley was taken into cultivation at least twice. With einkorn and emmer, however, the absence of evidence to the contrary was taken as indicating that these crops were both taken into cultivation just once (Zohary, 1996).

Since 1997, this area of research has been revolutionized through the acquisition of large amplified fragment length polymorphism (AFLP) data sets that have been analyzed in a phylogeographic manner not only to determine whether a crop is monophyletic or polyphyletic but also, through comparisons with wild populations, to infer the geographic location of the initial cultivations (Salamini et al., 2002). The first analysis, involving

288 AFLP loci in 388 accessions of einkorn, concluded that cultivated einkorn is monophyletic and originates from the Karacadag region of southeast Turkey, the area in which the most similar populations of the wild progenitor, *Triticum boeoticum,* are found today (Heun et al., 1997). Subsequent projects using the same methods assigned monophyletic origins to tetraploid wheats and to barley, the former also originating from southeast Turkey and the latter from the Israel–Jordan area (Badr et al., 2000; Özkan et al., 2002). These studies have been looked on as strong support for the stimulus diffusion model for agricultural origins, but doubts are now being raised about the veracity of the analyses and, if correct, the meaning of the results. One problem is that there is a contradiction between the apparent monophyly of cultivated barley, as shown by the AFLP analysis, and the presence of two separate mutations for the nonbrittle ear phenotype (highlighted by Abbo et al., 2001). A possible explanation is that one of the two mutations arose in the cultivated crop after domestication (Salamini et al., 2004). A second question surrounds the support that the archaeological record provides for the identification of the Karacadag region as a birthplace for agriculture (Jones et al., 1998), but this debate is inconclusive because of the incompleteness of the archaeological record, especially with regard to the identification of domesticated grains at early Southwest Asian sites. Equally difficult to assess, because of a lack of solid evidence, is the possibility that the wild phylogeography has changed in the period since the plants were taken into cultivation. If this has happened then the geographic location of the wild population most related to the crop will not necessarily indicate where that crop was first cultivated. All of these issues raise questions about the interpretations of the AFLP studies, but none provides conclusive evidence against those interpretations. More critical is the demonstration that the method used to analyze the AFLP data sets is not sufficiently robust to enable a monophyletic crop to be distinguished from a diphyletic one under all circumstances. If the markers being studied do not display tight genetic linkage (as may be the case with AFLPs), then the neighbor-joining algorithm that was used in three of the studies previously described (Heun et al., 1997; Badr et al., 2000; Özkan et al., 2002) may combine the members of a diphyletic crop into a single, apparently monophyletic grouping (Allaby and Brown, 2003). The question of whether the AFLPs used in the einkorn, tetraploid wheat, and barley studies display sufficient linkage for neighbor-joining analysis to be valid has not been established (Allaby and Brown, 2003, 2004). Reanalysis of the data using principal coordinate analysis, a more appropriate method, does not contradict

the monophyletic inferences (Salamini et al., 2004) but does not provide conclusive support for them (Allaby and Brown, 2004).

Even if the conclusions of the AFLP projects are correct, it is not reasonable to extrapolate from the demonstration of monophyly for a crop to the assumption that the crop was taken into cultivation just once. Part of the problem is that concepts such as monophyly, which have clear meanings and implications when the evolution of several species is studied, become much less determinative when applied to populations of a single species. A modern crop could appear to be monophyletic because it originated from a single domestication event, but monophyly could equally well result from events occurring after the initial cultivations. Salamini et al. (2002) point out that there are inconsistencies between the apparent monophyly displayed by the key founder crops of Southwest Asian agriculture and the gradual transition from gathering to cultivation to domestication that is apparent in the archaeological record for at least some of these crops. They suggest that genetic monophyly might arise after multiple domestications have taken place if for each crop a superior landrace emerges from the variety of forms generated by the initial cultivations, and this superior landrace subsequently spreads and becomes the progenitor of all the modern landraces and cultivars sampled. Considerations such as this show that there is difficulty in linking studies of the genetics of modern crops to archaeological questions regarding agricultural origins. In this particular example, it cannot be assumed that the superior landrace is descended from the first wild plants to be cultivated, and it may not even be the first cultivated population to become domesticated. The geographic origin of this superior landrace therefore cannot identify the location of the farming communities that first took the wild plants into cultivation, nor can it identify the location of the possibly different communities whose cultivated forms first became transformed into domesticated varieties.

Wheat Glutenin Loci

Large data sets have a seductive charm simply because of their size: After all, it must be better to study many loci rather than just one. However, each marker in an AFLP or similar data set is, in effect, a point mutation, and therefore a similar amount of information can be obtained by studying a single locus with many polymorphic sites. The single locus has the added, major advantage that the tight linkage between the informative sites enables evolutionary models to be constructed, tested, and applied to broader questions regarding the evolution of the organism in which the locus is found. Even when the number of polymorphic sites at a locus is few, the potential information that can be obtained is arguably greater than is possible with data sets of dispersed markers, which can be analyzed only by methods based on similarity matrices. The potential of single-locus studies is illustrated by work that we have carried out with the high-molecular weight (HMW) glutenin loci of wheat.

The HMW glutenins are a complex group of seed storage proteins coded by a pair of tightly linked multiallelic loci, *Glu-1-1* and *Glu-1-2,* on homologous chromosome 1 (Payne et al., 1982). We have carried out an extensive phylogenetic analysis of the *Glu* genes in order to understand the long-term evolution of these genes and of the A, B, D, and G genomes of wild and cultivated wheats (Allaby et al., 1999). One observation arising from this work is that cultivated emmers and their descendants can be divided into two genetic lineages according to the allele type present at the *Glu-B1-1* locus (the x-type *Glu* gene on the B chromosome set). We refer to these two lineages as α and β (figure 9.1) and have dated their divergence to

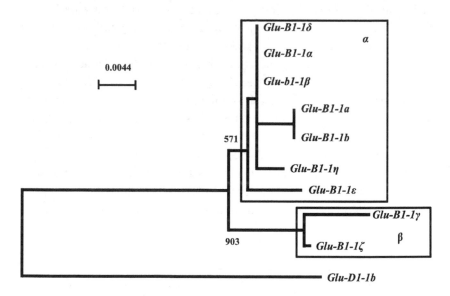

FIGURE 9.1 Neighbor-joining tree of the nine known *Glu-B1-1* alleles, all of which are present in cultivated emmers or emmer descendants (e.g., *T. aestivum* L.), based on multiple-sequence alignment of a 241- to 243-bp region immediately upstream of the open reading frame (Allaby et al., 1999). The *Glu-D1-1b* allele was used as the outgroup, and the robustness of the branching order was tested by creating 1000 bootstrap replicate trees using the CLUSTAL W program. The bootstrap values are the numbers above the branches of the tree. The α and β allele groups within the *Glu-B1-1* clade are highlighted.

1.4–2.0 million years ago by application of the appropriate molecular clock (Wolfe et al., 1989). The date is clearly many millennia before the origins of emmer cultivation, which could indicate that this crop was domesticated twice, once from a population of wild plants belonging to the α lineage and once from plants belonging to the β lineage. The heterogeneity could also have arisen from a single domestication of a mixed population of α and β plants or by introgression of α (or β) alleles into a crop domesticated on a single occasion from a wild population of β (or α) plants. The data are also consistent with many domestications of emmer (rather than just two) because if different assumptions are made about the earlier evolution of the *Glu* loci, the phylogenetic analysis that results in identification of the α and β lineages could be interpreted as indicating the presence in cultivated emmers of ancient lineages additional to α and β. As we state in Allaby et al. (1999), the data do not enable a distinction to be made between different scenarios for emmer domestication.

Phylogeography of Glutenin Alleles

To gain further information on the α and β subclades, we determined the lineage affiliation for a total of 185 cultivated emmers (table 9.1), spanning the full range of the expansion of emmer cultivation from Southwest Asia into Europe, Asia, and Africa. Alpha alleles were more common than β alleles among these cultivated wheats (78% α, 22% β), and the geographic distributions of the two *Glu-B1-1* allele types among cultivated emmers were different (figure 9.2). The more common α alleles were present in all areas from which accessions were obtained, whereas β alleles were found only in cultivated emmers from Turkey, the Balkans, southeastern and central Europe, and Italy.

We also examined *Glu-B1-1* alleles in 59 wild emmer wheats (*T. dicoccoides* (Korn) Schweinf.) (table 9.1). Most of the wild emmers came from the two regions of the Fertile Crescent in Southwest Asia that have been highlighted as possible locations for crop domestication: the southern Levant and the border between southeast Turkey and northern Syria (Jones et al., 1998; Nesbitt and Samuel, 1998). The 36 southern accessions came from Jordan, Israel, Lebanon, and the D'ara region of south Syria, and the 20 northern specimens were from the north Syrian borderlands and the Gaziantep region of southeastern Turkey. The collection also included two accessions from Iran and one from Iraq, within the eastern arm of the Fertile Crescent, outside the postulated domestication centers. The α and β allele frequencies were

Table 9.1 *Glu-B1-1* Allele Types in Wild and Cultivated Emmer Wheats

Wheat and Country of Origin	Number of Accessions Containing	
	α Alleles	β Alleles
Cultivated Emmer		
Armenia	6	0
Bulgaria	1	1
Czech Republic	12	20
Ethiopia	20	0
Georgia	3	0
Germany	15	1
Greece	1	1
Hungary	1	0
India	3	0
Iran	6	0
Israel	1	0
Italy, north and central	21	12
Italy, south	24	1
Kuwait	1	0
Montenegro	2	0
Morocco	2	0
Romania	1	0
Serbia	3	2
Slovenia	1	0
Spain	8	0
Switzerland	5	2
Turkey	3	1
USSR	4	0
All cultivated emmers	144	41
Wild Emmer		
Iran	0	2
Iraq	1	0
Israel	13	3
Jordan	3	4
Lebanon	3	3
Syria (north)	2	2
Syria (south)	4	3
Turkey	4	12
All wild emmers	30	29

Wheats were obtained from the John Innes Centre, Norwich, UK; Institut für Pflanzengenetik und Kulturpflanzenforschung, Gatersleben, Germany; the International Centre for Agricultural Research in the Dry Areas, Aleppo, Syria; and private collections.

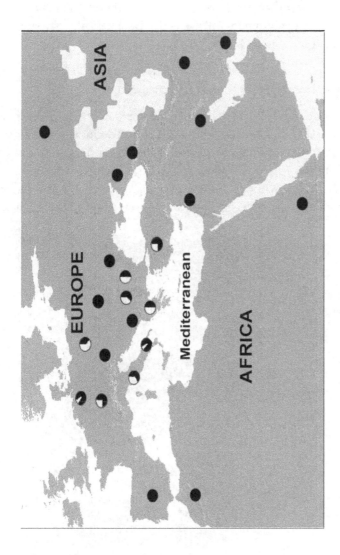

FIGURE 9.2 Geographic distribution of *Glu-B1-1* subclades in cultivated emmers. The pie charts show the proportion of α (*black*) and β (*white*) alleles in each geographic region, using the data listed in table 9.1.

similar among the wild emmers as a whole (50.8% α, 49.2% β), but there were distinct geographic biases, with α alleles common in Israel (81.3% α, 18.7% β) and β alleles common in Turkey (25% α, 75% β). Largely as a result of these biases, the allele frequency in the southern domestication region was higher for α than for β (64% α, 36% β) whereas in the northern region β alleles were predominant (30% α, 70% β). These phylogeographic data are illustrated in figure 9.3.

Limitations of a Phylogeographic Approach with Cultivated Wheats

The objective of this analysis was to obtain a broad picture of the geographic distribution of *Glu-B1-1* alleles in wild and cultivated emmers and to see whether these distributions can be related to the expansion of cultivated wheats from Southwest Asia. For cultivated wheats, the current phylogeography of *Glu-B1-1* alleles will reflect the phylogeography established during the expansion phase if there was no significant movement of wheats or alleles

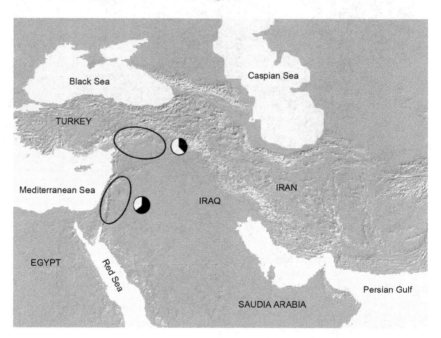

FIGURE 9.3 Map of Southwest Asia showing the northern and southern regions from which most of the 59 wild emmers we studied were collected. The southern region includes part of the Levant, and the northern region is located on the border between southeast Turkey and northern Syria. The pie charts show the proportion of α (*black*) and β (*white*) alleles in each geographic region, using the data listed in table 9.1.

during the millennia since the initial phylogeography was set up. Movement of cultivated wheats requires human agency because these plants, lacking a fragile rachis, cannot shed their seeds without human intervention and therefore cannot move from one geographic region to another unless physically transported by humans. Similarly, extensive movement of alleles requires human agency because cultivated wheats are predominantly self-fertilizing, limiting the opportunities for gene flow in the absence of directed cross-fertilization by humans. Modern phylogeographies therefore reflect ancient events only if wheats and alleles have not been moved extensively by human activity during the millennia since the initial expansion of agriculture from Southwest Asia. Extensive movement of both wheats (by trade and exchange of seed corn) and alleles (through breeding programs) has occurred in the last 150 years, but we made particular efforts to use accessions thought by the suppliers to be genuine landraces associated with a specified geographic region so that the results would be affected as little as possible by these recent events.

The question of whether the resulting phylogeographies have been significantly affected by human activities in premodern agricultural periods therefore is an open one. However, studies we have made of microsatellite genotypes in emmer accessions from Italy suggest that at least some landraces retain a Neolithic phylogeography. Isaac et al. (submitted) genotyped five microsatellite loci in 52 landraces of Italian emmer wheat. Each of the five loci was polymorphic, with 43 allele combinations identified in the 52 accessions. The allele combinations fell into two evolutionarily distinct groups, the larger of these comprising 27 genotypes found in 42 accessions, with a significant correlation between geographic and genetic distance matrices ($r = .393$, $p = .003$). Using a model that predicts the point of origin of crop cultivation within a geographic region by comparing the genetic and geographic distances between accessions, we identified a point on the coastline of northern Puglia as the most likely origin for this group of wheats. This phylogeographically determined origin corresponds closely with the location of the earliest agricultural sites in Italy; radiocarbon dating shows that they occur at 6100–5900 BC in northern Puglia and eastern Basilicata, in a geographically distinct region known as the Tavoliere. The coincidence between the origin predicted by the genetic analysis and the actual origin as revealed by archaeology lends strong support to the hypothesis that at least some emmer landraces have remained geographically static since their original introduction into Europe, so phylogeographic analysis of these modern plants can provide information on events occurring as agriculture spread into Europe.

Implications of the Glutenin Phylogeographies

Two Expansions of Cultivated Emmer into Europe

Among cultivated emmers, the phylogeographic distributions of the α and β alleles are markedly different: α alleles are ubiquitous, but β alleles are restricted mainly to central and southern Europe. A possible explanation of the dissimilar distributions is that α alleles have a selective advantage over β alleles throughout the greater part of the geographic range of cultivated emmers. It is difficult to imagine what the nature of this selective advantage might be because the nucleotide differences between the α and β alleles appear to be neutral: They lie upstream of the *Glu-B1-1* open reading frame, and the variations between the α and β sequences do not affect motifs thought to be involved in transcription or translation initiation (Allaby et al., 1999). In the absence of selection it is unlikely that the two allele subclades achieved their modern distributions via a single agricultural expansion. Therefore the data suggest that there have been at least two independent expansions of emmer cultivation into Europe, one involving plants carrying α alleles and the other involving plants with β alleles.

The possibility that there were two independent expansions of emmer cultivation correlates with evidence from other sources. The archaeological record contains direct evidence of two trajectories of spread of agriculture into Europe, one following the Mediterranean coast to Western Europe and the other following the major river valleys through the Balkans to northern Europe (reviewed by Bell, 1987). Similarly, the expansion of Indo-European languages into Europe, thought to be associated with the expansion of agriculture, involves two language groups: the Slavo-Germanic branch, which gave rise to the Slavic and Germanic languages of central and eastern Europe, and the Greco-Italic-Celtic branch, from which the Romance and Celtic languages of Western Europe are derived (Renfrew, 1989). One explanation of the glutenin phylogeography is that β alleles were underrepresented or even absent among plants that followed the Mediterranean trajectory.

The archaeological evidence appears to indicate that the two expansions of agriculture occurred at different times. The geographical ubiquity of the α subclade, not only in Europe but also in Asia and northern Africa, could be taken as evidence that it was associated with the primary expansion of wheat farming out of Southwest Asia and that expansion of plants with the β subclade was a secondary phenomenon. The archaeological record can also accommodate a more localized expansion of β plants during the early Neolithic, covered over by a later α expansion with a global impact. Although our current data do not allow us to distinguish between these alternatives, it should be possible to address the question by examining ancient DNA from charred wheats, using techniques that are now well established for genetic analysis of this type of material (Brown, 1999). Whichever scenario is correct, the implication is that the human events that led to the initial expansion of agriculture from Southwest Asia during the period 6500–3500 BC were not unique and recurred on at least one occasion.

Origins of the *Glu-B1-1* Allele Subclades

The presence of the α and β allele subclades in cultivated wheats can be explained by multiple domestication of emmer, single domestication of a highly divergent wild population, or introgression of novel alleles after domestication (Allaby et al., 1999). The second of these possibilities is unlikely because single domestication of a population of wheats containing both α and β alleles would be expected, after expansion, to give a phylogeography in which α and β alleles are fairly equally distributed, presuming that, as argued earlier, there is no differential selection between alleles of the two subclades. The distributions of α and β alleles in wild emmers do not preclude multiple domestication; a possible scenario is that the α subclade entered the cultivated gene pool via domestication of an emmer population from Israel, where α alleles are common, and the β subclade originates from a domestication in the Gaziantep region of southeastern Turkey, where β alleles predominate. Both areas contain some of the earliest farming villages and therefore are possible locations for crop domestication according to the archaeological record (Jones et al., 1998; Nesbitt and Samuel, 1998). However, the *Glu-B1-1* phylogeographies are equally consistent with a single domestication of emmer, in either the south or north of the western arm of the Fertile Crescent, followed by acquisition of the other allele subclade by introgression from nonancestral wild wheats. Introgression could have been by direct cross-hybridization between wild and cultivated emmers or by hybridization between a wild emmer and a cultivated hexaploid, the latter resulting in a pentaploid intermediate whose segregation products could include a tetraploid with domestication traits inherited from the hexaploid parent and *Glu-B1-1* alleles from the wild emmer. Introgression of one form or another is supported by other results, based on 5S rDNA comparisons, that suggest that wild emmers from several parts of the western Fertile Crescent have contributed to the gene pool of domesticated wheat (Allaby and Brown, unpublished results).

Conclusions

Both the origins of agriculture in Southwest Asia and its spread into Europe are accessible to examination by genetic analysis. Although to date the large AFLP data sets obtained for einkorn, emmer, and barley have not been analyzed in a convincing manner, these data sets and others like them have the potential to provide extensive information on the development of early crops in Southwest Asia. Genetic studies of crops throughout Europe are beginning to show that some landraces have remained geographically static since their first introduction into the continent, and more detailed phylogeographic analysis of these will tell us much about the trajectories followed by the spread of agriculture. Through examination of selective markers, it may be possible to assess the impact of environmental factors on the spread of cereal cultivation from the Fertile Crescent into the less hospitable regions of northern Europe. The great challenge for the next decade is to link the findings of plant genetics with archaeological evidence so that the former can contribute to the debates about the human dynamics underlying the transition from hunting–gathering to agriculture in Southwest Asia and across Europe.

Acknowledgments

We thank the John Innes Centre, Norwich, UK, the Institut für Pflanzengenetik und Kulturpflanzenforschung, Gatersleben, Germany, and the International Centre for Agricultural Research in the Dry Areas, Aleppo, Syria, for providing wheats. We thank Martin Jones and Chris Howe (University of Cambridge, UK), Glynis Jones (University of Sheffield, UK), and Keri Brown (UMIST, UK) for productive discussions on the origins and spread of agriculture. Our work on the glutenin loci was supported by the UK Natural Environment Research Council.

References

Abbo, S., S. Lev-Yadun, and G. Ladizinsky. 2001. Tracing the wild genetics stocks of crop plants. *Genome* 44: 309–310.

Allaby, R. G., and T. A. Brown. 2003. AFLP data and the origins of domesticated crops. *Genome* 46: 448–453.

Allaby, R. G., and T. A. Brown. 2004. On DNA markers, phylogenetic trees and mode of origin of crops: Response to Salamini et al. *Genome* 47: 621–622.

Allaby, R. G., M. Banerjee, and T. A. Brown. 1999. Evolution of the high–molecular-weight glutenin loci of the A, B, D and G genomes of wheat. *Genome* 42: 296–307.

Badr, A., K. Müller, R. Schäfer-Pregl, H. El Rabey, S. Effgen, H. H. Ibrahim, C. Pozzi, W. Rohde, and F. Salamini. 2000. On the origin and domestication history of barley (*Hordeum vulgare*). *Molecular Biology and Evolution* 17: 499–510.

Barker, G. 1985. *Prehistoric Farming in Europe.* Cambridge University Press, Cambridge, UK.

Bell, G. D. H. 1987. The history of wheat cultivation. In F. G. H. Lupton (ed.), *Wheat Breeding: Its Scientific Basis,* 31–49. Chapman & Hall, London, UK.

Blumler, M. A. 1992. Independent inventionism and recent genetic evidence on plant domestication. *Economic Botany* 46: 98–111.

Bogucki, P. 1996. The spread of early farming in Europe. *American Scientist* 84: 242–253.

Brown, T. A. 1999. How ancient DNA may help in understanding the origin and spread of agriculture. *Philosophical Transactions of the Royal Society of London,* series B 354: 89–98.

Cavalli-Sforza, L. L., P. Menozzi, and A. Piazza. 1994. *The History and Geography of Human Genes.* Princeton University Press, Princeton, NJ, USA.

Diamond, J. 1997. Location, location, location: The first farmers. *Science* 278: 1243–1244.

Diamond, J. 2002. Evolution, consequences and future of plant and animal domestication. *Nature* 418: 700–707.

Halstead, P. 1989. Like rising damp? An ecological approach to the spread of farming in south east and central Europe. In A. Milles, D. Williams, and N. Gardner (eds.), *The Beginnings of Agriculture* (BAR International Series 496), 23–53. BAR, London, UK.

Harris, D. R. 1996. Introduction: Themes and concepts in the study of early agriculture. In D. R. Harris (ed.), *The Origins and Spread of Agriculture and Pastoralism in Eurasia,* 1–9. UCL Press, London, UK.

Heun, M., R. Schäfer-Pregl, D. Klawan, R. Castagna, M. Accerbi, B. Borghi, and F. Salamini. 1997. Site of einkorn wheat domestication identified by DNA fingerprinting. *Science* 278: 1312–1314.

Isaac, A. D., L. Nencioni, M. Muldoon, K. A. Brown, and T. A. Brown. Submitted. Phylogeographical analysis of wheat landraces reveals the point of origin of Italian agriculture. *Molecular Ecology.*

Jones, M. K., R. G. Allaby, and T. A. Brown. 1998. Wheat domestication. *Science* 279: 302–303.

Kislev, M. E. 1992. Agriculture in the Near East in VIIth millennium BC. In P. C. Anderson (ed.), *Préhistoire de l'Agriculture: Nouvelles Approches Expérimentales et Ethnographiques* (Monographie 6, Centre de Recherches Archéologiques), 87–93. Editions du CNRS, Paris, France.

Nesbitt, M. and D. Samuel. 1998. Wheat domestication: Archaeobotanical evidence. *Science* 279: 1433.

Özkan, H., A. Brandolini, R. Schäfer-Pregl, and F. Salamini, F. 2002. AFLP analysis of a collection of tetraploid wheats indicates the origin of emmer and hard wheat domestication in southeast Turkey. *Molecular Biology and Evolution* 19: 1797–1801.

Payne, P. I., L. M. Holt, A. J. Worland, and C. N. Law. 1982. Structural and genetic studies on the high–molecular-weight subunits of wheat glutenin. 3. Telocentric mapping of the subunit genes on the long arms of the homeologous group 1 chromosomes. *Theoretical and Applied Genetics* 63: 129–138.

Price, T. D. 2000. *Europe's First Farmers.* Cambridge University Press, Cambridge, UK.

Renfrew, C. 1989. The origins of Indo-European languages. *Scientific American* 261(4): 82–90.

Richards, M. 2003. The Neolithic invasion of Europe. *Annual Review of Anthropology* 32: 135–162.

Salamini, F., M. Heun, A. Brandolini, H. Özkan, and J. Wunder. 2004. On DNA markers, phylogenetic trees and mode of origin of crops. *Genome* 47: 615–620.

Salamini, F., H. Özkan, A. Brandolini, R. Schäfer-Pregl, and W. Martin. 2002. Genetics and geography of wild cereal domestication in the Near East. *Nature Reviews Genetics* 3: 429–441.

Sherratt, A. 1997. Climatic cycles and behavioural revolutions: The emergence of modern humans and the beginning of farming. *Antiquity* 71: 271–287.

Takahashi, R. 1964. Further studies on the phylogenetic differentiation of cultivated barley. *Barley Genetics* 1: 19–26.

Takahashi, R. 1972. Non-brittle rachis 1 and non-brittle rachis 2. *Barley Genetics Newsletter* 2: 181–182.

van Zeist, W., K. Wasylikowa, and K.E. Behre. 1991. *Progress in Old World Palaeoethnobotany: A Retrospective View on the Occasion of the International Work Group for Palaeoethnobotany.* Balkema, Rotterdam, The Netherlands.

Wolfe, K., P.M. Sharp, and W.–H. Li. 1989. Rates of synonymous substitution in plant nuclear genes. *Journal of Molecular Evolution* 29: 208–211.

Zohary, D. 1996. The mode of domestication of the founder crops of Southwest Asian agriculture. In D.R. Harris (ed.), *The Origins and Spread of Agriculture and Pastoralism in Eurasia,* 142–158. UCL Press, London, UK.

Zohary, D. and M. Hopf. 2000. *Domestication of Plants in the Old World,* 3rd ed. Clarendon Press, Oxford, UK.

Zvelebil, M. 2000. The social context of the agricultural transition in Europe. In C. Renfrew and K. Boyle (eds.), *Archaeogenetics,* 57–79. McDonald Institute Monographs, Cambridge, UK.

Zvelebil, M. and M. Lillie. 2000. Transition to agriculture in eastern Europe. In T.D. Price (ed.), *Europe's First Farmers,* 57–92. Cambridge University Press, Cambridge, UK.

Zvelebil, M. and P. Rowley-Conwy. 1986. Foragers and farmers in Atlantic Europe. In M. Zvelebil (ed.), *Hunters in Transition,* 67–93. Cambridge University Press, Cambridge, UK.

Nyree Zerega, Diane Ragone, and
Timothy J. Motley

CHAPTER 10

Breadfruit Origins, Diversity, and Human-Facilitated Distribution

I received the seeds of the bread tree.... One service of this kind rendered to a nation, is worth more to them than all the victories of the most splendid pages of their history, and becomes a source of exalted pleasure to those who have been instrumental in it.
—Letter from Thomas Jefferson to M. Giraud (1797)

Background

Breadfruit (*Artocarpus altilis* (Parkinson) Fosberg, Moraceae) is a staple crop in Oceania, where it was originally domesticated. It is a versatile tree crop with many uses including construction, medicine, animal feed, and insect repellent. However, it is principally grown as a source of carbohydrates and is an important component of agroforestry systems. Unlike many herbaceous starch crops harvested for their vegetative storage tissues, breadfruit is a large tree grown for its fruit (technically an infructescence, as the breadfruit is a syncarp made up of many small fruitlets fused together) (figure 10.1). Many cultivars have no seeds, just tiny aborted ovules (these will be called seedless cultivars), whereas others may have few to many seeds. Breadfruit typically is harvested when it is slightly immature and still firm, and seedless cultivars are prepared in much the same way as potatoes: baked, boiled, steamed, roasted, or fried. Ripe fruits are sweet and used in desserts. In seeded cultivars, seeds are chestnut-like in both size and taste and are boiled or roasted.

Although breadfruit yields vary between individual trees and cultivars, productivity typically is quite high. A commonly cited figure for seedless

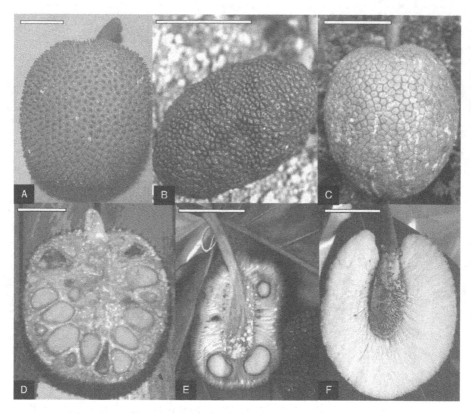

FIGURE 10.1 Breadfruit and wild relatives. **(A–C)** Syncarp surfaces of **(A)** *Artocarpus camansi,* Zerega 88; **(B)** *A. mariannensis,* Zerega 107; and **(C)** *A. altilis* cultivar Mei uhp from Pohnpei, Zerega 172. **(D–F)** Cross-sections of **(D)** *A. camansi,* Zerega 88; **(E)** *A. mariannensis,* Zerega 146; and **(F)** seedless *A. altilis* cultivar Lemae from Rota, Mariana Islands, Zerega 142. Scale bar = 5 cm.

breadfruit is 700 fruits per tree per year, with each fruit averaging 1–4 kg (Purseglove, 1968). In a specific case study of an agroforestry system in Pohnpei of the Federated States of Micronesia, average yields for five cultivars ranged from 93 to 219 fruits per tree per season. In-depth yield studies for more than 100 cultivars growing in a common location are being conducted by Diane Ragone. Breadfruit is a seasonal crop, and because trees produce large quantities of highly perishable fruit, various methods of preservation have been developed for long-term storage. Some traditional preservation methods include fermentation in underground pits (Atchley and Cox, 1985; Aalbersberg et al., 1988) and the production of a starchy, sun-dried paste (Coenen and Barrau, 1961). A limited number of studies have examined breadfruit's nutritional value. Compared with other starch crops it provides comparable levels of carbohydrates and is a better source of protein than cassava and equivalent to banana and sweet potato (Graham and Negron de Bravo, 1981).

Breadfruit Biology

Breadfruit plants are monoecious with separate pistillate and staminate inflorescences borne in the leaf axils of a single tree. The pistillate inflorescence is typically globose to subglobose, whereas the staminate inflorescence is cylindrical. Both inflorescences consist of hundreds of tiny flowers, which are tightly packed together and sit on a fleshy receptacle. The staminate flowers of fertile cultivars produce copious amounts of viable pollen, whereas few-seeded and seedless cultivars produce little or no viable pollen (Sunarto, 1981; Ragone, 2001). It has been demonstrated that fruit development in seedless breadfruit is parthenocarpic and does not require pollen to be initiated (Hasan and Razak, 1992). As the pistillate inflorescence develops, the fleshy perianths of the individual flowers expand and provide the edible starchy portion of the syncarpous fruit (figure 10.1). Little is known about pollination in seeded cultivars or wild relatives of breadfruit, although both wind (Jarrett, 1959a) and insect pollination (Brantjes, 1981; Momose et al., 1998; Sakai et al., 2000) have been suggested for various *Artocarpus* species.

Thousands of years of breadfruit cultivation and human selection in Oceania have given rise to a tremendous amount of morphological diversity, including variation in the number of seeds per fruit. Cultivars in Melanesia typically produce viable, edible seeds and are propagated by seed. Other cultivars, especially in Polynesia and Micronesia commonly produce few to no seeds and must be propagated vegetatively. This is usually accomplished by planting of root suckers, through air layering, or by grafting. The loss of fertility is caused by triploidy ($2n = 3x = {}^-84$) or is the result of hybridization in the case of sterile diploids ($2n = 2x = 56$) (Ragone, 2001; Zerega et al., 2004).

Breadfruit Distribution

At the end of the sixteenth century, European explorers and naturalists traveling to Oceania quickly recognized the potential of breadfruit as a highly productive, cheap source of nutrition and introduced a limited number of cultivars to their tropical colonies (Ragone, 1997). The most famous of these attempts was led by William Bligh and culminated in the mutiny aboard the *H.M.S. Bounty* (Bligh, 1792). Today breadfruit is grown throughout the tropics but is especially important in Oceania and the Caribbean. Breadfruit historically has had little commercial value outside the Pacific islands, where it has served primarily as a subsistence crop. However, in the last few decades, the Caribbean Islands have become the primary exporter of fresh breadfruit to Europe and North America (Marte, 1986; Andrews, 1990), and the Fijian Ministry of Agriculture reported breadfruit as one of Fiji's top four agricultural exports to New Zealand in the *Pacific Business News* in December 2001. Additionally, promising methods of preservation that could increase the export market for breadfruit include fried breadfruit chips, freeze drying, flour, canning, and extracting starch for use in the textile industry (Roberts-Nkrumah, 1993; Ragone, 1997).

Breadfruit Diversity and Conservation in Oceania

Over millennia, Pacific Islanders have selected and named hundreds of traditional cultivars based on fruiting season, fruit shape, color and texture of the flesh and skin, absence or presence of seeds, flavor, cooking and storage qualities, leaf shape, and horticultural needs (Wilder, 1928; Ragone, 1997). These cultivars have adapted to local climates and soils, including the harsh saline soils of coral atolls, and many of them are endemic to a single island group. However, the use of breadfruit has been declining since World War II with the introduction and convenience of a western-style diet, causing some cultivars to be neglected and knowledge about fruit storage and preparation to be lost. Climate change and cyclones also contribute to the loss of cultivars. To help conserve and study breadfruit, many germplasm collections have been assembled throughout the tropics, especially in the Pacific islands, over the last several decades. Because most cultivars are seedless, and even when seeds are present they are recalcitrant and cannot be dried or stored, collections must be maintained as living trees in field gene banks. This is a time-consuming and expensive task. For this reason, many collections are no longer being maintained (Ragone, 1997). A noteworthy exception is the Breadfruit Institute at the National Tropical Botanical Garden in Hawaii. This collection, with 120 cultivars and 192 accessions from 18 Pacific island groups, Indonesia, the Philippines, Papua New Guinea, and the Seychelles, represents a broad range of diploid, triploid, and hybrid cultivars and accessions of breadfruit's wild progenitors and has become an important genetic repository for conservation and research. Several other important collections representing primarily local cultivars are being maintained in various Pacific and Caribbean islands.

Breadfruit's Closest Relatives

Breadfruit belongs to the genus *Artocarpus* in the Moraceae family. This family also includes other important members such as figs, mulberries, and jackfruit. The wild species of *Artocarpus* are restricted to Southeast Asia and the Indo-Pacific and comprise nearly 60 species divided into two subgenera, four sections, and eight series based on leaf and inflorescence morphology and anatomy (Jarrett, 1959a). Recent phylogenetic analyses of morphological and DNA sequence data from the nuclear ribosomal internal transcribed spacers (ITS) and the chloroplast *trnL-F* region for 38 *Artocarpus* species representing each of the 8 series and 13 Moraceae outgroup taxa indicated that *A. camansi* Blanco and *A. mariannensis* Trécul form a very highly supported monophyletic lineage with *A. altilis,* and they are breadfruit's closest relatives (figure 10.2; Zerega, 2003). *Artocarpus camansi* (figure 10.1), commonly called breadnut, is native to New Guinea and possibly the Moluccas (Jarrett, 1959b). It has been introduced for its edible seeds to other tropical locations outside Oceania and is especially common in the Caribbean and South America. *Artocarpus mariannensis* (figure 10.1) is native to the Mariana Islands and Palau and has been introduced to a limited number of Micronesian and Polynesian islands for its edible fruits and seeds (Ragone, 1997, 2001). Both species are diploid (2n = 2x = 56) (Ragone, 2001) and produce viable seeds.

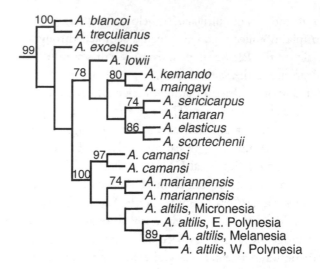

FIGURE 10.2 Strict consensus tree of 20 most parsimonious trees derived from ITS and *trnL-F* DNA sequence and morphological data. Jackknife support values are indicated above the branches. Breadfruit and its putative progenitors form a strongly supported clade. (Modified from Zerega, 2003.)

Zerega et al. (2004) explored the origins of breadfruit using amplified fragment length polymorphism (AFLP) (Vos et al., 1995) and found that both *A. camansi* and *A. mariannensis* played roles in the origins of breadfruit to varying degrees in different regions of the Pacific. These data will be summarized and elaborated on here in combination with additional isozyme data (Ragone, 1991) in order to identify the role of wild progenitors in breadfruit origins, assess genetic diversity and relationships between wild relatives and breadfruit cultivars throughout Oceania, and trace historical human-mediated breadfruit movement through Oceania.

Origins of Breadfruit

In order to discuss the regions of Oceania, the geographic classification originally proposed by French voyager Dumont d'Urville (1832) is followed. Although the regions do not necessarily reflect cultural or historical unity, they are a commonly used, practical way in which to describe the islands of the Pacific Basin. The regions are Melanesia (included in this study: Papua New Guinea, Solomon Islands, Vanuatu, Fiji, and Rotuma), western Polynesia (included in this study: Samoa), eastern Polynesia (included in this study: Cook Islands, Society Islands, Hawaii, and the Marquesas), and Micronesia (included in this study: Mariana Islands, Chuuk, Yap, Palau, Kiribati, Kosrae, and Pohnpei).

The roles that *A. camansi* and *A. mariannensis* may have played in breadfruit origins throughout Oceania were explored using AFLP data. Using three different primer pair combinations, AFLP data were collected from a total of 254 individuals. These samples came from accessions in the Breadfruit Institute or from field collections deposited at the New York Botanical Garden (NY). Samples comprised 24 *A. mariannensis*, 30 *A. camansi*, and 200 Pacific breadfruit cultivars from the island groups of Fiji (9), the Solomon Islands (7), Vanuatu (7), Rotuma (8), Papua New Guinea (3), Chuuk (9), Palau (6), the Mariana Islands (21), Pohnpei (47), Yap (2), Kiribati (2), the Society Islands (45), the Cook Islands (11), the Marquesas (9), Hawaii (1), and Samoa (13) (accession information is listed in Zerega et al., 2004). The AFLP data were collected and scored as a binary matrix to indicate the presence or absence of each AFLP fragment. Three AFLP primer pair combinations yielded 149 polymorphic markers across all 254 individuals (52 markers from *Eco*RI-ACA/*Mse*I-CTC, 44 markers from *Eco*RI-ACA/*Mse*I-CAT, and 53 markers from *Eco*RI-AAG/*Mse*I-CTG).

To better understand the relationships between breadfruit and wild relatives, the AFLP data were analyzed using several methods. First, unweighted pair group method with arithmetic mean (UPGMA) dendrograms were drawn using Nei's (1978) unbiased genetic identity and distance based on AFLP data in Popgene version 1.31 (Yeh et al., 1999). Cultivars from the same island group were treated as a population, and *A. camansi* and *A. mariannensis* samples were treated as separate populations. The cultivar from Hawaii was not included in this analysis because only one individual was available. To further investigate relationships, the AFLP data were also analyzed using principal component analysis (PCA) on a square symmetric matrix of covariances in the software package JMP (SAS Institute, Cary, NC, USA). Finally, in examining the AFLP data from breadfruit's progenitors, four markers were found that were diagnostic and constant, one in *A. camansi* and three in *A. mariannensis*. That is, one marker was present in all *A. camansi* individuals and never in *A. mariannensis*, and three markers were present in all *A. mariannensis* individuals and never in *A. camansi*. These diagnostic markers are distributed variously throughout breadfruit cultivars and play a role in the discussions of breadfruit origins and human-mediated dispersal (figure 10.3).

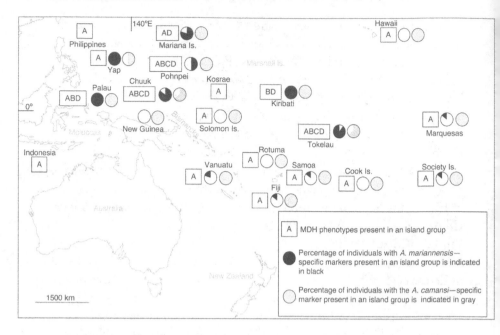

FIGURE 10.3 Map of Oceania indicating the distributions among breadfruit cultivars of malic dehydrogenase (MDH) isozyme phenotypes and *A. camansi–* and *A. mariannensis*–specific AFLP markers. Letters in the boxes refer to the four different MDH isozyme phenotypes present in an island group. The percentage of individuals within an island group with *A. mariannensis*–specific AFLP markers is indicated by the black portion of the pie chart on the left for each island group. White portions of the pie chart indicate the percentage of individuals with no *A. mariannensis* markers present. The percentage of individuals in an island group with an *A. camansi*–specific marker is indicated by the gray portion of the pie chart on the right for each island group. The percentage of individuals with no *A. camansi*–specific marker is indicated by the white portion of the pie chart.

In the UPGMA dendrogram, all of the island groups in Polynesia cluster together, as do most of the Melanesian islands (Fiji, Rotuma, Solomon Islands, and Vanuatu), and the cultivars from both Polynesia and Melanesia share a higher genetic similarity with *A. camansi* than with *A. mariannensis* (figure 10.4). Interestingly, the cultivars collected in Papua New Guinea are sister to Polynesian rather than to other Melanesian cultivars. This is not surprising because they are seedless cultivars that are believed to have been introduced from elsewhere. The AFLP data suggest they were brought

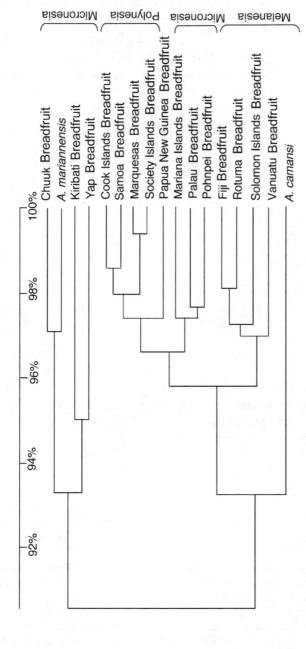

FIGURE 10.4 UPGMA dendrogram based on AFLP data of breadfruit cultivars from various island groups in Oceania and progenitor species, *A. camansi* and *A. mariannensis.*

from Polynesia. Among cultivars in Micronesia, some (Mariana Islands, Palau, and Pohnpei) share a higher genetic similarity with *A. camansi,* whereas others (Chuuk, Kiribati, and Yap) are more similar to *A. mariannensis.* The results from the PCA analysis demonstrate a similar pattern (figure 10.5a). Cultivars from Melanesia and Polynesia cluster with one another and with *A. camansi,* and Micronesian cultivars cluster between *A. mariannensis* and the Polynesian and Melanesian breadfruit. These results

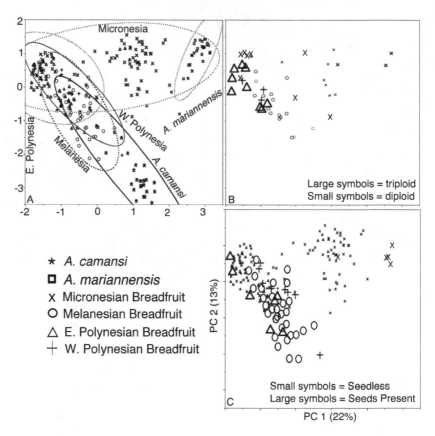

FIGURE 10.5 Principal component analysis (PCA) of 200 breadfruit cultivars and wild relatives (24 *A. camansi* and 30 *A. mariannensis*) based on 149 AFLP markers. **(A)** Bivariate normal ellipses with *p* = .95 are drawn around *A. camansi, A. mariannensis,* Melanesian breadfruit, western Polynesian breadfruit, eastern Polynesian breadfruit, and Micronesian breadfruit. **(B)** The same PCA, showing only breadfruit cultivars whose ploidy level has been tested (Ragone, 2001). **(C)** The same PCA analysis with seedless cultivars and seeded cultivars indicated by small and large symbols, respectively.

suggest that Melanesian and Polynesian breadfruit cultivars may share similar origins, whereas many of the Micronesian cultivars have a different evolutionary history.

Melanesian and Polynesian breadfruit cultivars are more closely related to *A. camansi* than *A. mariannensis* and may have been derived from the former species. The distribution of both *A. camansi* and *A. mariannensis* diagnostic markers further illustrates this point (figure 10.3). All Melanesian and Polynesian cultivars have the *A. camansi* marker present, whereas very few have *A. mariannensis* markers. The rare presence of *A. mariannensis* markers in these regions is discussed in more detail in the section about human-mediated dispersal of breadfruit later in this chapter.

Micronesian cultivars are closely related to Polynesian and Melanesian *A. camansi*–derived breadfruit and to *A. mariannensis*, as revealed by UPGMA (figure 10.4) and PCA (figure 10.5a) analyses. This suggests that many Micronesian cultivars may be the product of hybridization between *A. camansi*–derived breadfruit and *A. mariannensis* and subsequent introgression. This is illustrated by the prevalence of both *A. camansi* and *A. mariannensis* diagnostic AFLP markers throughout individual Micronesian breadfruit cultivars and by additional evidence from isozyme data (figure 10.3).

Diversity in Breadfruit and Its Closest Relatives

Despite the fact that many breadfruit cultivars are vegetatively propagated, a great deal of morphological diversity has been selected for by humans. This is evident in gross fruit and leaf morphology, the number of cultivar names that exist, and the various environments in which breadfruit can thrive (Wilder, 1928; Ragone, 1988, 1997). However, little is known about the underlying genetic diversity of breadfruit. Here we examine diversity in breadfruit and its wild relatives using protein and DNA techniques.

Isozyme Data

Data from six different enzyme systems (aconitase [ACO], alcohol dehydrogenase [ADH], isocitrate dehydrogenase [IDH], leucine aminopeptidase [LAP], malic dehydrogenase [MDH], and phosphoglucomutase [PGM]) were collected for 204 individuals (accession information listed in Ragone, 1991) (table 10.1). The samples came largely from the Breadfruit Institute collection and comprised 6 *A. camansi*, 3 *A. mariannensis*, and 195 breadfruit

Table 10.1 Genetic Diversity Estimates of Breadfruit and Wild Relatives Based on Isozyme and AFLP Data

Species	Locality	Iso/AFLP (*n*)	Isozyme Data % PES	AFLP Data % UZ	% PL	Shannon Index	G_{st}
	Melanesia	28/34	83.3	71	43	.1455	.2985
Breadfruit	Fiji	9/9	83.3	89	30	.1322	
Breadfruit	Solomons	7/7	66.7	100	22	.1040	
Breadfruit	Vanuatu	6/7	66.7	83	23	.1073	
Breadfruit	Rotuma	6/8	50	83	20	.0932	
Breadfruit	PNG	0/3	NA	NA	9	.0518	
	Micronesia	76/87	83.3	58	81	.2841	.3136
Breadfruit	Chuuk	26/9	83.3	96	24	.1195	
Breadfruit	Palau	6/6	83.3	67	21	.1104	
Breadfruit	Marianas	2/21	0	NA	32	.1617	
Breadfruit	Pohnpei	36/47	66.7	58	66	.2716	
Breadfruit	Yap	3/2	66.7	100	5	.0311	
Breadfruit	Kiribati	1/2	NA	NA	9	.1748	
Breadfruit	Kosrae	1/0	NA	NA	NA	NA	
	Polynesia	89/79	100	24	56	.1428	.3802
	E. Polynesia	60/66	33	5	46	.1122	.4235
Breadfruit	Societies	43/45	33.3	4.7	29	.1208	
Breadfruit	Cooks	7/11	50	29	17	.0912	
Breadfruit	Marquesas	7/9	0	14	17	.0790	
Breadfruit	Hawaii	3/1	0	33	NA	NA	
	W. Polynesia	13/13	83.3	69	35	.1430	NA
Breadfruit	Samoa	13/13	83.3	69	35	.1430	
Breadfruit	Tokelau*	16/27	66.7	62.5	27	.1406	
	Non-Pacific	NA	NA	NA	NA	NA	NA
Breadfruit	Jamaica	0/4	NA	NA	2	.0119	
Breadfruit	Seychelles	0/4	NA	NA	1	.0057	
A. camansi	New Guinea, Philippines, Indonesia	6/30	100	100	39	.1617	
A. mariannensis	Micronesia	3/24	50	100	29	.1059	

Estimates were determined for regions (*shaded in gray*) and island groups. Accession numbers of samples used in isozyme analyses are listed in Ragone (1991); samples used in AFLP analyses are listed in Zerega (2003) and Zerega et al. (2004).

*Tokelau cultivars are all recent introductions of hybrid origin and were not included in the Polynesian region calculations.

Iso = isozyme, n = number of samples, % PES = percentage polymorphic enzyme systems, % UZ = percentage unique zymotypes, % PL = percentage polymorphic AFLP loci, G_{st} = between-population differentiation, NA = not applicable because of small sample size.

cultivars from Fiji, the Solomon Islands, Vanuatu, Rotuma, Chuuk, Palau, the Mariana Islands, Pohnpei, Yap, Kiribati, Kosrae, the Society Islands, the Cook Islands, the Marquesas, Hawaii, and Samoa (table 10.1). Each individual was scored for the presence or absence of bands, and each unique pattern of bands identified for an enzyme system represents a unique isozyme phenotype. The combination of phenotypes for each individual over the six enzyme systems is the zymotype for that individual.

To summarize diversity in the regions of the Pacific, data from breadfruit cultivars were pooled together by island groups and by the regions Melanesia, western Polynesia, eastern Polynesia, and Micronesia. In order to determine levels of diversity in enzyme systems in breadfruit, the percentage of polymorphic enzyme systems was calculated (Menancio and Hymowitz, 1989; Ragone, 1991). Additionally, to assess the isozyme diversity of breadfruit within and between regions, the percentage of unique zymotypes was determined for each island and regional population (number of zymotypes in a population divided by total number of individuals in the population; Ragone, 1991).

A total of 45 different bands were scored across all six enzyme systems (9 for ACO, 12 for ADH, 4 for IDH, 8 for LAP, 3 for MDH, and 9 for PGM). Forty-four different isozyme phenotypes were scored across all six enzyme systems (18 for ACO, 7 for ADH, 2 for IDH, 7 for LAP, 4 for MDH, and 6 for PGM). When phenotypes from all six enzyme systems were combined for each individual, 90 unique zymotypes were identified. Although most zymotypes were narrowly distributed, one was found in 35% of individuals and was predominant among eastern Polynesian triploid cultivars. This will be called the Polynesian zymotype. All cultivars sampled from the Society Islands (except one), Hawaii, Marquesas, the Mariana Islands, and Kosrae had this Polynesian zymotype. It was also found to a lesser extent in the Cook Islands (57%), Pohnpei (31%), Palau (17%), Fiji (11%), Samoa (8%), and Chuuk (3.8%).

Breadfruit's closest relatives, *A. camansi* and *A. mariannensis,* exhibit high levels of diversity: 100% of the individuals sampled have unique zymotypes, and 100% (*A. camansi*) and 50% (*A. mariannensis*) of the enzyme systems investigated are polymorphic (table 10.1). Among breadfruit cultivars, levels of isozyme diversity range from extremely low to as high as or higher than those of the wild relatives. The percentage of polymorphic enzyme systems is equally high in Micronesia, Melanesia, and western Polynesia and lowest in eastern Polynesia. The percentage of unique zymotypes is highest in Melanesia, followed by western Polynesia, Micronesia, and eastern Polynesia (table 10.1). These measures and the overwhelming dominance of a single zymotype in eastern Polynesia indicate that eastern Polynesian breadfruit cultivars are the least genetically diverse and probably originated from a much reduced gene pool. Interestingly, the percentage of unique zymotypes among breadfruit cultivars for each major region is lower than the percentage for most of the island groups in the region, indicating that the same zymotypes often are distributed between more than one island group.

Although the distribution of most of the isozyme phenotypes indicate no clear geographic patterns, MDH had one phenotype (A) common to *A. camansi* and to breadfruit in all of the Pacific islands in the study. Three additional phenotypes (B, C, and D) were restricted to *A. mariannensis* and Micronesian breadfruit (figure 10.3). This pattern is similar to the distribution of *A. camansi* and *A. mariannensis* AFLP markers and further supports the hypothesis that Melanesian and Polynesian breadfruit cultivars are derived from *A. camansi,* whereas Micronesian cultivars appear to be of hybrid origin.

AFLP Data

The AFLP technique has greater resolving power than isozymes because it samples across the entire genome, and in the current study AFLP data were able to differentiate between individuals with identical zymotypes. For genetic diversity estimates of breadfruit and wild relatives based on AFLP data, a total of 289 individuals were analyzed. These comprised the 254 samples described earlier, 4 breadfruit cultivars each from Jamaica and the Seychelles, and 27 cultivars from Tokelau (accessions listed in Zerega, 2003). The Tokelau cultivars are believed to be the result of hybridization between recently introduced *A. mariannensis* and diploid *A. altilis* cultivars (Ragone, 1991, 2001) and therefore were not included in the breadfruit origins discussion. Three AFLP primer pair combinations yielded 175 polymorphic markers across all 289 individuals (68 markers from *Eco*RI-ACA/*Mse*I-CTC, 51 markers from *Eco*RI-ACA/*Mse*I-CAT, and 56 markers from *Eco*RI-AAG/*Mse*I-CTG). To summarize levels of diversity, breadfruit cultivars were pooled together by island groups and regional populations.

To determine the genetic diversity of breadfruit cultivars and wild relatives based on AFLP data, the percentage of polymorphic loci (% PL) and the Shannon index (Shannon and Weaver, 1949; Lewontin, 1972) were calculated using Popgene version 1.31 (Yeh et al., 1999). The Shannon index is a diversity measure that reflects richness and distribution of genotypes in a population. It is calculated for each locus ($\sum i \log^2 i$, where i = the frequency of the presence or absence of the band), and the mean diversity is calculated as the average of index values over individual loci. The standard deviations for the Shannon index are not shown in table 10.1, but in all cases they are higher than the mean index because several loci were monomorphic in all populations and had a Shannon index of zero. Additionally, the G_{st} value was calculated (Hartl and Clark, 1989) to measure the proportion of the total genetic variance present in each subpopulation (e.g., the individual island groups) relative to the total genetic variance in the entire population (e.g., Melanesia, eastern and western Polynesia, and Micronesia). A high G_{st} value implies a high degree of differentiation between populations.

Comparison of Diversity Between Breadfruit and Its Closest Relatives

Levels of genetic diversity, as indicated by the percentage of polymorphic AFLP loci, for both *A. camansi* and *A. mariannensis* are as low as or lower than those for breadfruit cultivars from Polynesia, the least genetically diverse of the major Pacific regions. However, when compared with individual island groups, only Pohnpei has greater genetic diversity than *A. camansi*, whereas Fiji, the Marianas, and Samoa also have higher levels than *A. mariannensis*. The Shannon index for *A. mariannensis* is lower than that for breadfruit cultivars in any of the major Pacific regions, although it is higher than levels of diversity in breadfruit in most of the individual island groups. The Shannon index of genotypes among *A. camansi* individuals is greater than that found for the breadfruit cultivars from any of the major Pacific regions except Micronesia. These measures attest to the high levels of genetic diversity that exist among Pacific breadfruit cultivars throughout the islands of Oceania compared with their progenitor species. However, it must be pointed out that the full range of diversity for *A. camansi* and *A. mariannensis* was not represented because *A. mariannensis* from Palau and *A. camansi* from Irian Jaya (western New Guinea) and the Moluccas were not available. Additional sampling may reveal greater genetic diversity in these two species.

Breadfruit Genetic Diversity in Oceania

The diversity measures for Pacific island breadfruit cultivars indicate that Micronesia harbors the greatest levels of genetic diversity, followed by

Melanesia and then Polynesia. Interestingly, however, Polynesian culti-vars are the most genetically differentiated (reflected by higher G_{st} values) (table 10.1). In other words, compared with other regions, a greater per-centage of the genetic diversity in Polynesia is attributed to diversity within individual island groups. This may be explained by the fact that Polynesian cultivars are predominantly vegetatively propagated, which leads to a reduc-tion in gene flow and lower genetic diversity than in the outcrossing culti-vars in Melanesia and Micronesia. At the same time, vegetative propagation also increases differentiation between reproductively isolated individuals. Therefore, vegetative propagation in Polynesia appears to have contributed to a narrower genetic base than in other Pacific regions, but much of the existing diversity is unique to specific island groups. This may help explain the occurrence of hundreds of cultivar names in many of the island groups in Polynesia (Wester, 1924; Wilder, 1928; Handy et al., 1991; Ragone, 1997). In order to further investigate how genetic diversity is partitioned between breadfruit cultivars, a nested analysis of molecular variance (Excoffier et al., 1992) was conducted using Arlequin software version 2.000 (Schneider et al., 2000). Breadfruit diversity was examined between regions, between island groups within regions, and within island groups. By far the largest percentage of the total variance (74.92%) was accounted for within island groups (table 10.2), indicating that individual islands throughout Oceania represent important repositories of breadfruit genetic diversity.

Based on these results, the genetic diversity of breadfruit appears to depend on mode of reproduction. Melanesian and Micronesian cultivars exhibit the highest levels of genetic diversity based on both isozyme and AFLP data. In Melanesia breadfruit comprises primarily seeded, diploid, outcrossing individuals (figure 10.5b, c), which are propagated by seed. In

Table 10.2 Analysis of Molecular Variance Based on AFLP Markers from Breadfruit

Source of Variation	df	Sum of Squares	Variance Components	% of Total Variance	p Value
Between regions	3	239.938	1.21652	14.20	<.001
Between island groups within regions	13	184.306	0.93300	10.88	<.001
Within island groups	188	1194.503	6.42206	74.92	<.001

Degrees of freedom *(df)* are equal to the number of samples minus one.

Micronesia, seeded, outcrossing diploids also occur, and many Micronesian cultivars are of hybrid origin (Fosberg, 1960; Ragone, 2001; Zerega et al., 2004). Thus sexual reproduction and hybridization are responsible for higher levels of genetic diversity throughout these regions. Polynesian, particularly eastern Polynesian cultivars, exhibit the lowest levels of diversity based on both isozyme and AFLP data. In these regions, diploid few-seeded (western Polynesia) and triploid, seedless (eastern Polynesia) cultivars are overwhelmingly predominant (figure 10.5b, c), and propagation is vegetative. Thus genetic diversity in these areas would result primarily from the occurrence and subsequent human selection of desirable somatic mutations.

Diversity Among Non-Pacific Breadfruit

Cultivars from the islands of Jamaica and the Seychelles have the lowest levels of genetic diversity (table 10.1). This is not surprising because only a limited number of cultivars were ever introduced outside the Pacific, effectively creating a genetic bottleneck (Leakey, 1977; Ragone, 1997). For example, Bligh introduced approximately 600 plants representing only five different breadfruit cultivars to the islands of St. Vincent and Jamaica in 1792. These were subsequently spread throughout the Caribbean (Powell, 1973; Leakey, 1977). A single cultivar, *kele kele,* brought by the French from Tonga in 1796 was the ancestor of all seedless breadfruit trees distributed throughout the French tropical colonies (Leakey, 1977; Rouillard and Gueho, 1985). All the breadfruit trees in West Africa are also believed to have stemmed from a single introduction (Smith et al., 1992). This lack of genetic diversity outside the Pacific makes these regions especially susceptible to disease and emphasizes the importance of conserving the diversity of Pacific island breadfruit.

Human-Mediated Dispersal of Breadfruit

Human Migration and Breadfruit Dispersal in Melanesia and Polynesia

Long-distance breadfruit movement through the Pacific islands had to be human mediated because seeds are short-lived, and many cultivars are seedless. Therefore, evidence about human migrations in the Pacific based on linguistics, archaeology, anthropology, and genetics provides a working hypothesis that can be tested against molecular evidence from breadfruit.

Oceania consists of many culturally and linguistically diverse islands, and their settlement was not necessarily a simple event. This being said, scholars from several diverse disciplines generally agree that Polynesia represents a monophyletic entity and was settled via the north coast of New Guinea and then through island Melanesia within the last 4000 years by the Lapita people, a group known for their distinctive pottery and excellent seafaring skills (Kirch and Hunt, 1988; Spriggs, 1989; Intoh, 1997; Lum and Cann, 1998, 2000; Kirch, 2000; Gibbons, 2001). The Lapita are believed to have originated from somewhere in island Southeast Asia, but the exact location from which these Austronesian-speaking people came and how extensively they integrated with the Melanesians who had already been living in New Guinea and the Solomon Islands for more than 40,000 years are still debated (Diamond, 1988; Terrell, 1988; Lum and Cann, 1998; Richards et al., 1998; Kirch, 2000).

Lebot (1999) has demonstrated that several Pacific island crops (banana, *Musa* spp.; sugarcane, *Saccharum* sp.; yam, *Dioscorea alata;* and taro, *Colocasia esculenta*) probably were domesticated in New Guinea or western Melanesia and that genetic diversity decreases from the west (Melanesia) to the east (Polynesia) among cultivars of both taro and kava (*Piper methysticum*) (Lebot, 1992). This is also true for breadfruit, as demonstrated by the isozyme and AFLP data discussed earlier. If it is assumed that the region of origin is the region with the highest genetic variability, these findings correlate well with an eastward colonization through New Guinea and island Melanesia, and into Polynesia. As people sailed east into Polynesia to settle uninhabited islands, they would have been able to take only a subset of a crop's genetic diversity with them, causing the gene pool to decrease with each successive colonization event. In the case of breadfruit, most Melanesian and Polynesian breadfruit cultivars appear to be derived from *A. camansi,* a native New Guinea species. New Guinea, the Bismarck Archipelago, and the Solomon Islands are considered part of "near" rather than "remote" Oceania (Green, 1991) because they are all geographically close and were settled in the late Pleistocene (ca. 40,000 years ago) before the advent of the Austronesian-speaking Lapita people (ca. 3500–5000 years ago; Kirch, 2000). Consequently, the short-lived seeds, cuttings, or young plants of *A. camansi* may have been transported from their native New Guinea by pre-Lapita, non-Austronesian-speaking humans as far east as the Solomon Islands. Human selection of desirable traits gave rise to the domesticated *A. altilis,* but the continued sexual reproduction of plants would explain the dominance of seeded, diploid cultivars in these islands.

However, when the Lapita people arrived and ventured on longer ocean voyages eastward into the more distant unsettled islands of Melanesia and Polynesia in remote Oceania, a shift to vegetative propagation probably would have been necessary to facilitate survival on such long journeys. In fact, the Lapita people are known for their dependence on vegetatively propagated crops such as bananas, taro, yam, sugarcane, kava, and breadfruit (Barrau, 1963; Lebot, 1992; Kirch, 2000). This shift to vegetative propagation would have made long-distance transportation of breadfruit and other crops possible and increased the chances of few-seeded or seedless cultivars originating (through accumulated somatic mutations and meiotic defects) and persisting (through human selection). For example, in regions where vegetative propagation and sexual reproduction both occurred, diploid gametes arising from nondisjunction in meiosis, possibly caused by somatic mutation defects, could have joined with normal haploid gametes to produce triploid seedless cultivars. Indeed, it is on the periphery of near Oceania (eastern Solomon Islands and Vanuatu) where few-seeded diploid cultivars begin to appear and in western Polynesia where diploid seedless and few-seeded as well as triploid seedless cultivars become more common (Ragone, 1997). Seedless triploid cultivars were then preferentially propagated and dispersed eastward (figure 10.6), transforming breadfruit into a staple starch crop in Polynesia.

Human Migration and Breadfruit Dispersal in Micronesia

Breadfruit cultivars found in Micronesia include triploid *A. camansi*–derived "eastern Polynesian" type breadfruit in addition to hybrid cultivars bearing the genetic imprint of both *A. camansi* and *A. mariannensis*. This raises the questions, "From where was the Polynesian type cultivar introduced, and how and where did hybridization in Micronesia occur?"

Because there is no direct evidence of a Polynesian migration into Micronesia, the presence of Polynesian type breadfruit in Micronesia might be explained by trade following European contact. It has been suggested that the Spanish may have introduced the Polynesian type triploid breadfruit into the Philippines in the 1600s (Jarrett, 1959b). Despite its use as a food plant, there was no mention of it by de Morga (1971) in the Philippines in the early 17th century even though de Morga was acquainted with breadfruit through correspondences with Quiros (Markham, 1904). However, breadfruit was mentioned in Camel's (1704) list of Philippine plants in the early 1700s. Therefore, it is possible that the Spanish distributed the

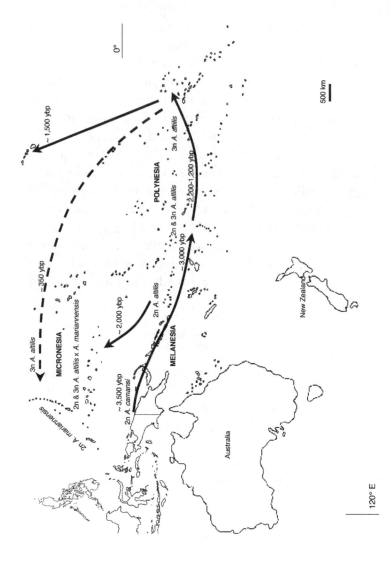

FIGURE 10.6 Map of Oceania with proposed human-mediated breadfruit dispersal routes. Arrows indicate general direction of migrations and are simplified from actual migration routes. Times are estimates and are given in years before present (ybp) (from Kirch, 2000). Species and ploidy levels found within regions are indicated. The dashed arrow represents a post-European contact route.

Polynesian breadfruit into both Micronesia and the Philippines to help provision their colonies.

Regarding the question of breadfruit hybrids in Micronesia, Fosberg (1960) proposed that hybridization and subsequent introgression between introduced triploid sterile Polynesian cultivars and native diploid *A. mariannensis* was occurring in the Mariana Islands. He suggested that this accounted for the great variability in Micronesian cultivars with shared morphological characters of both *A. altilis* and *A. mariannensis*. However, this hypothesis is highly unlikely because triploids very rarely make it through meiosis to successfully produce viable gametes. An alternative hypothesis is that diploid *A. camansi*–derived breadfruit was introduced into the range of *A. mariannensis*, allowing the two species to hybridize. Subsequently, varying degrees of introgression and human selection have led to the diversity of cultivars unique to Micronesia. This hypothesis is supported by another source of evidence that diploid *A. altilis* and *A. mariannensis* can hybridize because recent introductions of the two species in Tokelau have led to fertile hybrids (Ragone, 1991, 2001). The Micronesian hybrids comprise fertile and sterile diploids and sterile triploids (Ragone 2001). These triploids arose from a separate event than the seedless autotriploids in Polynesia and probably result from hybrid diploid gametes (through nondisjunction in meiosis) joining with normal haploid gametes from other diploid hybrids, *A. altilis*, or *A. mariannensis*. An alternative explanation for the presence of *A. camansi* and *A. mariannensis* markers in Micronesian breadfruit is that these two species hybridized with one another. However, the ranges of the two species do not overlap, and there is no evidence that they have ever overlapped (Zerega et al., 2004).

How does our knowledge of human migrations in Micronesia relate to the hypothesis about the origin of hybrid Micronesian breadfruit outlined above? The human settlement of the culturally and linguistically heterogeneous islands of Micronesia is more complex than that of Polynesia. It probably was settled from several directions at different times, and based on evidence from linguistics, archaeology, and genetics, several nonexclusive hypotheses have been proposed. These include migrations from New Guinea into Palau and Yap (Lum and Cann, 2000), independent colonizations of the Mariana Islands and Yap from Southeast Asia (Kirch, 2000; Lum and Cann, 2000), and a direct (or indirect through the Kiribati archipelago) northerly Lapita migration from somewhere between the Bismarck Archipelago and the southeast Solomons–Vanuatu region into central-eastern Micronesia (Caroline Islands [including Chuuk, Kosrae, Pohnpei],

Kiribati, and the Marshall Islands) (Lebot and Lévesque, 1989; Petersen, 1995; Kirch, 2000). Subsequent secondary migrations also occurred among the islands of Micronesia (Kirch, 2000; Lum and Cann, 2000).

It is unlikely that *A. camansi* was introduced into Micronesia from New Guinea because there are no historical accounts of its presence in Micronesia, and it grows on only a few Micronesian islands today as the result of recent introductions (Ragone, 2001). However, a northerly Lapita migration (transporting diploid *A. camansi*–derived breadfruit) from the southeast Solomons–Vanuatu region into central-eastern Micronesia, followed by subsequent human migrations and trading within Micronesia (Kirch, 2000; Lum and Cann, 2000), could have brought diploid *A. camansi*–derived breadfruit into the range of wild *A. mariannensis* (Mariana Islands and Palau), allowing the two species to hybridize (figure 10.6; Zerega et al., 2004). There has been debate about whether a northerly Lapita migration into Micronesia occurred directly into the high islands of the Carolines or indirectly via island hopping through the atolls of the Kiribati archipelago (Petersen, 1995). Because breadfruit cultivars without *A. mariannensis* traits do not grow well in harsh atoll conditions (Ragone, 1988), a human migration successfully transporting breadfruit probably was direct across open water as opposed to going through the low atolls of Kiribati, where purely *A. camansi*–derived cultivars would have fared poorly. Genetic and cultural evidence from kava (*Piper methysticum*), another cultivated Pacific plant, also suggests a direct migration (Lebot and Lévesque, 1989; Petersen, 1995). Such a direct route from Melanesia into Micronesia may have been reciprocal because *A. mariannensis*–diagnostic markers are also present in some breadfruit cultivars in Vanuatu and eastward into Polynesia (figure 10.3). Thus, a small percentage of breadfruit cultivars with *A. mariannensis* markers could have subsequently been dispersed into Polynesia with the eastward Lapita migration.

Conclusions

Two species (*A. camansi* and *A. mariannensis*) and at least two different events (vegetative propagation coupled with human selection in Melanesia and Polynesia and introgressive hybridization in Micronesia) were involved in the origins of breadfruit. Thousands of years of cultivation and selection of breadfruit have led to a wealth of morphological diversity and unique breadfruit cultivars suited to different purposes and environments. Genetic erosion is evident in non-Pacific regions, where only a limited number of cultivars were introduced. Today genetic erosion is also a concern in many areas of the Pacific. Because of urbanization and the ease of obtaining and preparing introduced foods, the importance of traditional foods such as breadfruit has diminished and endemic island cultivars have been lost. This is exacerbated by global climate change and rising sea levels, which threaten the very existence of some Pacific islands and the breadfruit cultivars unique to them. However, there is a growing interest in reducing food imports, more fully using locally grown crops, and encouraging young people to learn and perpetuate traditional cropping systems. As a result, the potential exists for breadfruit to once again become a much more widely grown and used tropical crop. Despite thousands of years of evolution in domestication, breadfruit research and commercial utility are still in their infancy. Additional research to improve the future potential and conservation of breadfruit is under way, including the development of a morphological descriptor list to identify cultivars and in-depth yield studies. Additional projects on pollination biology, development of breadfruit food products with a long shelf life suitable for a commercial market, and the collection of cultivars from underrepresented areas for deposit in both ex situ living gene banks and in situ conservation collections will all contribute to the future use and conservation of breadfruit.

Acknowledgments

The authors would like to thank R. Kiapranis, M. Kostka, V. Novotny, L. Raulerson, W. Raynor, A. Rinehart, and D. Zerega for field assistance, the editors and anonymous reviewers for their comments on the manuscript, the National Tropical Botanical Garden, National Research Institute of Papua New Guinea (PNG), the Forest Research Institute of Papua New Guinea (Lae, PNG), the Binatang Research Center (Madang, PNG), and the Nature Conservancy in Pohnpei, Federated States of Micronesia. This research was made possible through funding from the Lewis B. and Dorothy Cullman Foundation, National Science Foundation grant DEB-0073161, the Garden Club of America, and the Explorers Club.

References

Aalbersberg, W. G. L., C. E. A. Lovelace, K. Madhoji, and S. V. Parkinson. 1988. Davuke, the traditional Fijian method of pit preservation of staple carbohydrate foods. *Ecology of Food and Nutrition* 21: 173–180.

Andrews, L. 1990. Breadfruit varieties in the Windward Islands. *Tropical Fruit Newsletter* 1: 3–4.

Atchley, J. and P. A. Cox. 1985. Breadfruit fermentation in Micronesia. *Economic Botany* 39: 326–335.

Barrau, J. 1963. *Plants and the Migrations of Pacific Peoples.* Bishop Museum Press, Honolulu, HI, USA.

Bligh, W. 1792. *A Voyage to the South Sea, Undertaken by Command of His Majesty, for the Purpose of Conveying the Bread-Fruit Tree to the West Indies, in His Majesty's Ship the* Bounty, *commanded by Lieutenant William Bligh.* George Nicol, London, UK.

Brantjes, N. B. M. 1981. Nectar and the pollination of breadfruit, *Artocarpus altilis* (Moraceae). *Acta Botanica Neerlandica* 30: 345–352.

Camel, J. G. 1704. Appendix. In J. Ray (ed.), *Historia Plantarum,* Vol. 3, 52. S. Smith and B. Walford, London, UK.

Coenen, J. and J. Barrau. 1961. The breadfruit tree in Micronesia. *South Pacific Bulletin* 11: 37–39, 65–67.

de Morga, A. 1971. *Sucesos de las Islas Filipinas.* Hakluyt Society, London, UK.

Diamond, J. M. 1988. Express train to Polynesia. *Nature* 336: 307–308.

Dumont d'Urville, M. J. 1832. Notice sur les îles du Grand Océan et sur l'origine des peuples qui le habitent. *Société de Géographie Bulletin* 17: 1–21.

Excoffier, L., P. Smouse, and J. Quattro. 1992. Analysis of molecular variance inferred from metric distances among DNA haplotypes: Application to human mitochondrial DNA restriction data. *Genetics* 131: 479–491.

Fosberg, F. R. 1960. Introgression in *Artocarpus* in Micronesia. *Brittonia* 12: 101–113.

Gibbons, A. 2001. The peopling of the Pacific. *Science* 291: 1735–1737.

Graham, H. D. and E. Negron de Bravo. 1981. Composition of the breadfruit. *Journal of Food Science* 46: 535–539.

Green, R. C. 1991. Near and remote Oceania: Disestablishing "Melanesia" in culture history. In A. Pawley (ed.), *Man and a Half: Essays in Pacific Anthropology and Ethnobiology in Honour of Ralph Bulmer,* 491–502. Polynesian Society, Auckland, NZ.

Handy, E. S. C., E. G. Handy, and M. K. Pukui. 1991. *Native Planters in Old Hawaii. Their Life, Lore, and Environment.* Bishop Museum, Honolulu, HI, USA.

Hartl, D. L. and A. G. Clark. 1989. *Principles of Population Genetics,* 2nd ed. Sinauer Associates, Sunderland, MA, USA.

Hasan, S. M. Z. and A. R. Razak. 1992. Parthenocarpy in seedless breadfruit (*Arthocapus incircus* (Thumb.) L.). *Acta Horticulturae* 321: 648–652.

Intoh, M. 1997. Human dispersal into Micronesia. *Anthropological Science* 105: 15–28.

Jarrett, F. M. 1959a. Studies in *Artocarpus* and allied genera, I. General considerations. *Journal of the Arnold Arboretum* 40: 1–29.

Jarrett, F. M. 1959b. Studies in *Artocarpus* and allied genera, III. A revision of *Artocarpus* subgenus *Artocarpus. Journal of the Arnold Arboretum* 40: 113–155, 327–368.

Kirch, P. V. 2000. *On the Road of the Winds: An Archaeological History of the Pacific Islands Before European Contact.* University of California Press, Berkeley, CA, USA.

Kirch, P. V. and T. L. Hunt. 1988. Radiocarbon dates from the Mussau Islands and the Lapita colonization of the Southwest Pacific. *Radiocarbon* 30: 161–169.

Leakey, C. L. A. 1977. *Breadfruit Reconnaissance Study in the Caribbean Region.* CIAT/InterAmerican Development Bank, Cali, Colombia.

Lebot, V. 1992. Genetic vulnerability of Oceania's traditional crops. *Experimental Agriculture* 28(3): 309–323.

Lebot, V. 1999. Biomolecular evidence for plant domestication in Sahul. *Genetic Resources and Crop Evolution* 46: 619–628.

Lebot, V., and J. Lévesque. 1989. The origin and distribution of kava. *Allertonia* 5: 223–280.

Lewontin, R. C. 1972. The apportionment of human diversity. *Evolutionary Biology* 6: 381–398.

Lum, J. K. and R. L. Cann. 1998. mtDNA and language support a common origin of Micronesians and Polynesians in island Southeast Asia. *American Journal of Physical Anthropology* 105: 109–119.

Lum, J. K and R. L. Cann. 2000. mtDNA lineage analyses: Origins and migrations of Micronesians and Polynesians. *American Journal of Physical Anthropology* 113: 151–168.

Markham, C. 1904. *The Voyages of Pedro Fernandez de Quiros 1595 to 1606.* Hakluyt Society, London, UK.

Marte, R. 1986. Nontraditional fruit crops in the Windward Islands. *Proceedings Interamerican Society Tropical Horticulture* 30: 15–24.

Menancio, D. I. and T. Hymowitz. 1989. Isozyme variation between diploid and tetraploid cytotypes of *Glycine tabacina* (Labill.) Benth. *Euphytica* 42: 79–87.

Momose, K., A. Hatada, R. Yamaoka, and T. Inoue. 1998. Pollination biology of the genus *Artocarpus*, Moraceae. *Tropics* 7: 165–172.

Nei, M. 1978. Estimation of average heterozygosity and genetic distance from a small number of individuals. *Genetics* 89: 583–590.

Petersen, G. 1995. The complexity of power, the subtlety of kava. *Canberra Anthropology* 18: 34–60.

Powell, D. 1973. *The Voyage of the Plant Nursery, H.M.S. Providence, 1791–1793.* Institute of Jamaica, Kingston, Jamaica.

Purseglove, J. W. 1968. *Artocarpus altilis.* In J. W. (ed.), *Tropical Crops (2), Dicotyledons,* 377–384. Longman, London, UK.

Ragone, D. 1988. *Breadfruit Varieties in the Pacific Atolls.* Integrated Atoll Development Project, UNDP, Suva, Fiji.

Ragone, D. 1991. *Collection, Establishment, and Evaluation of a Germplasm Collection of Pacific Island Breadfruit.* Ph.D. dissertation, University of Hawaii, Honolulu, HI, USA.

Ragone, D. 1997. *Breadfruit, Artocarpus altilis (Parkinson) Fosberg, Promoting the Conservation and Use of Underutilized and Neglected Crops,* 10. International Plant Genetic Resources Institute, Rome, Italy.

Ragone, D. 2001. Chromosome numbers and pollen stainability of three species of Pacific island breadfruit (*Artocarpus*, Moraceae). *American Journal of Botany* 88: 693–696.

Richards, M., S. Oppenheimer, and B. Sykes. 1998. mtDNA suggests Polynesian origins in eastern Indonesia. *American Journal of Human Genetics* 63: 1234–1236.

Roberts-Nkrumah, L. B. 1993. Breadfruit in the Caribbean: A bicentennial review. *Extension Newsletter Department of Agriculture University of the West Indies (Trinidad and Tobago)* 24: 1–3.

Rouillard, G. and J. Gueho. 1985. History of the horticultural, medicinal, and economic plants of Mauritius. *Revue Agricole et Sucrière de l'Ile Maurice* 64: 151–167.

Sakai, S., M. Kato, and H. Nagamasu. 2000. *Artocarpus* (Moraceae)–gall midge pollination mutualism mediated by a male-flower parasitic fungus. *American Journal of Botany* 87: 440–445.

Schneider, S., D. Roessil, and L. Excoffier. 2000. *Arlequin ver. 2.000: A Software for Population Genetics Data Analysis.* Genetics and Biometry Laboratory, University of Geneva, Switzerland.

Shannon, C. E. and W. Weaver. 1949. *The Mathematical Theory of Communication.* University of Illinois Press, Urbana, IL, USA.

Smith, N. J. H., J. T. Williams, D. L. Plucknett, and J. P. Talbot. 1992. *Tropical Forests and Their Crops,* 296–303. Comstock Publishing Associates, Ithaca, NY, USA.

Spriggs, M. 1989. The dating of island Southeast Asia Neolithic: An attempt at chronometric hygiene and linguistic correlation. *Antiquity* 63: 587–613.

Sunarto, A. T. 1981. Fertility test of *Artocarpus altilis*. *Berita Biologi* 2: 118.

Terrell, J. 1988. History as a family tree, history as an entangled bank: Constructing images and interpretations of prehistory in the South Pacific. *Antiquity* 62: 642–657.

Vos, P., R. Hogers, M. Bleeker, M. Rijans, T. Van de Lee, M. Hornes, A. Frijters, J. Pot, J. Peleman, M. Kuiper, and M. Zabeau. 1995. AFLP: A new technique for DNA fingerprinting. *Nucleic Acids Research* 23: 4407–4414.

Wester, P. J. 1924. The seedless breadfruits of the Pacific archipelagos. *Philippine Agricultural Review* 17: 24–39.

Wilder, G. P. 1928. *Breadfruit of Tahiti*. Bishop Museum Bulletin 50. Bishop Museum, Honolulu, HI, USA.

Yeh, F. C., T. Boyle, Y. Rongce, Z. Ye, and J. M. Xiyan. 1999. *Popgene* version 1.31, Microsoft Windows–based freeware for population genetic analysis. www.ualberta.ca/˜fyeh/.

Zerega, N. J. C. 2003. *Molecular Phylogenetic and Genome-Wide Analyses of* Artocarpus *(Moraceae): Implications for the Systematics, Origins, Human-Mediated Dispersal, and Conservation of Breadfruit.* Ph.D. dissertation, New York University, New York, NY, USA.

Zerega, N. J. C., D. Ragone, and T. J. Motley. 2004. Complex origins of breadfruit: Implications for human migrations in Oceania. *American Journal of Botany* 91: 760–766.

Roger Malapa, Jean-Louis Noyer,
Jean-Leu Marchand, and Vincent Lebot

CHAPTER 11

Genetic Relationship Between *Dioscorea alata* L. and *D. nummularia* Lam. as Revealed by AFLP Markers

The greater yam, *Dioscorea alata* L., is the most widely cultivated species of yam in the tropics. It is grown for its starchy tubers that are harvested from 6–9 months after planting. Its origin has been a long-standing enigma of Oceanian ethnobotany and is still a subject of debate (Barrau, 1956; Bourret, 1973; Hahn, 1991; Degras, 1993). This chapter attempts to clarify its taxonomic status and position within section Enantiophyllum using amplified fragment length polymorphism (AFLP) markers. Additionally, a brief review of traditional uses and folk classification in Vanuatu, Melanesia, and cytogenetic research is also presented and considered in light of historical data to address the geographic distribution and dispersal of edible yams in Oceania.

Taxonomic Classification

Edible yams are twining vines that annually develop thickened tubers at the stem bases, which serve as storage organs to carry the plant through a period of dry season dormancy. At the onset of the rainy season, tubers begin to sprout, and new plants are produced for the next growing period. Stems are sometimes armed and twine either to the right or left according to species.

Leaves are entire, palmately lobed, or compound and are arranged alternately or oppositely on the stems.

Most *Dioscorea* species are dioecious and bear male and female flowers on separate plants. Female plants produce paniculate inflorescences with round trilocular capsules that contain two seeds per locule, and male plants produce inflorescences in panicles with small sessile flowers. Seeds are winged. However, flowering is erratic and seeds are seldom produced. Therefore reproduction is ensured mainly by vegetative propagation through aerial bulbils or underground tubers.

According to Degras (1993), the genus *Dioscorea* was originally described by Linnaeus in 1753 when he considered the three taxa *D. alata, D. bulbifera* L., and *D. pentaphylla* L. The classification of the genus under the Dioscoreaceae family and its division into botanical sections was initiated by Uline (1898). Knuth (1924) established the prevailing systematics of the genus *Dioscorea,* placing *D. alata* into section Enantiophyllum. Following this system, section Enantiophyllum includes all the species with a rightward stem twining direction, and all develop entire leaves (Degras, 1986). In section Enantiophyllum at least 15 species are edible, and among them only the greater yam develops wings on stems. Therefore, it was named *D. alata* (from Latin *ala,* "wing") because it was the first winged yam to be included in the Linnean classification system (Burkill, 1948–1954).

In Vanuatu, all species of section Enantiophyllum are edible and include the local species *D. alata, D. nummularia* Lam., *D. transversa* Br., two unidentified taxa "netsar"[1] and "rul," and the introduced *D. cayenensis-rotundata* Lam. and Poir. of African origin. Species of other sections are also cultivated there. They include the local *D. bulbifera, D. esculenta* (Lour.) Burk., *D. pentaphylla,* and the introduced South American *D. trifida* L. (Weightman, 1989). These four species are classified respectively into sections Opsophyton, Combilium, Botryosicyos, and Macrogynodium and differ from species of section Enantiophyllum by their gross morphological characteristics, including species that twine to the left (Burkill, 1948–1954; Ding and Gilbert, 2000). Furthermore, *D. bulbifera* (aerial yam) develops big aerial tubers for which the species is cultivated, and *D. esculenta* (Chinese yam) produces underground tubers that are protected by a crown of spiny roots. Both species produce entire and cordate leaves. In contrast, *D. pentaphylla* and *D. trifida* (cush cush yam) develop compound (five leaflets) and palmate (three to five lobes) leaves, respectively, and *D. trifida* has winged stems. Both species are cultivated for their underground tubers.

Ethnotaxonomy

In Oceania, as elsewhere, yams are cultivated primarily for their starchy tubers, and various cooking preparations have been described (Barrau, 1958). In Vanuatu, the national dish *laplap*[2] is traditionally made from fresh, finely ground tuber flesh mixed with coconut milk. The pudding is then spread and covered with *Heliconia indica* leaves and steamed in earthen ovens. *Laplap* is prepared mainly using tubers of *D. alata, D. nummularia,* and *D. transversa,* whereas the other yam species have to be boiled, baked, or roasted in order to be palatable (Bourrieau, 2000).

The folk classification of plants is polytypic in Vanuatu, as is also the case in New Guinea (Hayes, 1976). In Vanuatu yam growers use an approach based on similarities including morphological, ecological, and chemical criteria. In the north of Malakula Island, for instance, 11 groups of homogeneous morphotypes are distinguished according to their aerial phenotypic traits and their underground tuber morphology and organoleptic properties (Barrau, 1956; table 11.1). Four groups of morphotypes correspond to well-defined botanical species: *D. alata,* "bapa"; *D. esculenta,* "rontak"; *D. bulbifera,* "norenbo"; and *D. pentaphylla,* "imbo." Other groups are related to *D. nummularia* ("buts," "buts rom," "net," "timbek") and *D. transversa* ("maro"), and "rul" and "netsar" represent unidentified species.

The relationships between morphotypes are determined by their distinct ecological adaptations and needs because they can be either spontaneous or cultivated under well-developed tree canopies or in cleared gardens. Such adaptations also reflect their perennial or annual vegetative growth habit to which farmers respond with adapted horticultural practices (Barrau, 1962).

"Buts" (wild forms), "buts rom" (cultivated), and "rul" (cultivated) are found exclusively under living trees. They have several spiny stems and produce tubers in bundles that are harvested throughout the year from senescent stems without uprooting the whole plant. Tubers have no dormancy, and fragments can be replanted immediately. Such perennial types are known in Malakula as "buts" and are commonly called *wild yams* in Vanuatu. "Maro," "net," "netsar," and "timbek" are also perennial forms but are cultivated and trellised on dead trees to optimize sunlight. They produce several spiny stems, but farmers often eliminate the new shoots to conserve only one or two stems when big tubers are needed. Tubers are harvested annually by uprooting the whole plant. "Net" and "timbek" can be harvested 7 months after planting (by cutting down the green vines), and tuber fragments can be replanted immediately, whereas "netsar" and "maro" are harvested mainly

Table 11.1 Traditional Uses and Folk Classification of Yams in Vanuatu

Ecology & dry matter		Morphotype	Lifespan	Species	Tuber characteristics	Alimentary uses*	Ceremonial uses
forest	wild yams	'buts'	perennial	*D. nummularia*	Bundle, root stalk or slender. Cylindrical	R	none
		'buts rom'	perennial	*D. nummularia*	Bundle. Cylindrical	R/G	none
		'rul'	perennial	*Dioscorea* spp.	Bundle, Cylindrical or globose	R/B/G	2nd grade
	strong yams	'net'	perennial	*D. nummularia*	Bundle. Cylindrical	R/G	2nd grade
		'timbek'	perennial	*D. nummularia*	Bundle, anastomosing. Cylindrical/ flattened	R/B/G	2nd grade
		'maro'	perennial	*D. transversa*	Bundle. Cylindrical/ flattened or globose	R/G/B/Bk	2nd grade
		'netsar'	perennial	*Dioscorea* spp.	Bundle, short. Cylindrical	R/G	none
cleared garden	soft yam	'bapa'	annual	*D. alata*	Rarely in bundle. Various shapes	R/G/B/Bk	1st grade
	sweet yam	'rontak'	annual	*D. esculenta*	Bundle. Cylindrical to subglobose	R/B/Bk	none
	n.a.**	'norenbo'	annual	*D. bulbifera*	Aerial. Round	B/Bk	none
	n.a.**	'imbo'	annual	*D. pentaphylla*	Globose or flat	B/Bk	none

*Baked (Bk); boiled (B); grounded (G); roasted (R)
**'norenbo' (*D. bulbifera*) and 'imbo' (*D. pentaphylla*) do not have their common Bislama names although they are considered in the folk classification as being definitely distinct from *soft yam* (*D. alata*), *strong yams* (*D. nummularia, D. transversa, Dioscorea* spp.), *sweet yam* (*D. esculenta*) and *wild yams* (*D. nummularia, Dioscorea* spp.).

upon senescence of the aerial parts of the plants. In Malakula Island, these horticultural types are commonly known as "batun bapa" and *strong yams*.

Both *wild yams* and *strong yams* are robust climbers of high growth vigor and are characterized by the production of several tubers in bundles with flattened, cylindrical, globose, or subglobose shapes (table 11.1). Their tubers have higher dry matter content than the annual species *D. alata, D. esculenta, D. bulbifera,* and *D. pentaphylla* (Bradbury and Holloway, 1988). Thus, perennial types are never planted together with annual cultivars because of their different habitat needs and growth vigor.

At the intraspecific level, farmers identify different varieties called *kaen* (= kind). They usually use a binary labeling system that describes successively the tuber morphology (morphotype) and its flesh characteristics (chemotype and color). This type of classification is mostly used for *D. alata*, "bapa," to distinguish the long tubers (called "romets") from the short ones (called "letslets") and their white or anthocyanin-colored flesh. It is by combining the morphotype and the chemotype that farmers identify and label the different *kaen* they are cultivating. A *kaen* represents a feature that is constant and identifiable among cultivars within species and determines its mode of cooking preparation (table 11.1).

Although *Dioscorea* species are cultivated throughout the Indo-Pacific region, their religious and cultural importance is unique in Melanesia (Lea, 1966; Coursey, 1967; Bourret, 1973; Weightman, 1989; Degras, 1993). Such an ethnocentric attachment is reserved mostly for "bapa" (*D. alata*). In Vanuatu, *D. alata* is the first crop in the cultural rotation cycle of the shifting agriculture system. The yam lifecycle begins with different cultural and magic practices and ends with the celebration of the *new yam* event upon harvest (Weightman, 1989). It also plays an important role in the local subsistence economy, which is based on product exchanges between island inhabitants of the inland and the coastal regions (Bonnemaison, 1996). *D. alata* cultivars with regular long and cylindrical "romets" have a commercial and prestigious value in ceremonial exchanges, chiefs' ordinances, and yam growers' retirements (Weightman, 1989).

"Batun bapa" (cultivars of *D. nummularia* and *D. transversa*) are also used for ceremonial feasts, although they always rank second to "bapa" (*D. alata*). Therefore farmers of Malakula, Malo, and Santo recognize two grades of ceremonial yams: Grade 1 is represented by "bapa" (*D. alata, soft yam*), and grade 2 comprises distinct morphotypes of "batun bapa" (*strong yams*) including "maro" (*D. transversa*), "net" and "timbek" (*D. nummularia*), and "rul" (table 11.1).

Origins of *D. alata*

There is consensus on neither the origin nor the area of domestication of *D. alata* because the species does not occur in the wild but is found only in cultivation. Therefore, for Prain and Burkill (1939), *D. alata* is a true cultigen that has been selected from the two closely related Southeast Asian wild species, *D. hamiltonii* Hook. and *D. persimilis* Prain and Burk., or from their natural hybrids. Both species are characterized by long, deeply buried tubers that superficially resemble some cultivated but inferior varieties of *D. alata*.

Burkill (1951) hypothesized that deep tubers evolved as a protection from wild pigs and that human selection resulted in the short-tubered, compact varieties within *D. alata*. For varieties with upward-curving tubers, which eventually push their way out of the soil, he also implicated human selection because this growth habit makes harvesting easy. This hypothesis has been widely accepted (Alexander and Coursey, 1969; Martin, 1974; Hahn, 1991; Degras; 1993). Recently Mignouna et al. (2002) adopted it to interpret the segregating ratio analysis among sexual progenies of *D. alata* and emphasized an allotetraploid genomic structure of the species involving the two different genomes: PPHH (P = *D. persimilis*, H = *D. halmitonii*).

Barrau (1956) reported that the Southeast Asian species *D. alata, D. bulbifera, D. esculenta,* and *D. pentaphylla* are cultivated from west Melanesia to eastern Polynesia. These regions also include two species that are difficult to distinguish from one another, *D. nummularia* and *D. transversa.* They are not found in continental Asia, and *D. transversa* is reported only in western Oceania, including insular Southeast Asia, northern Australia (Telford, 1986), Vanuatu (Malapa et al., in press), and New Caledonia (Bourret, 1973). *D. transversa* has not yet been found in New Guinea, although Yen (1982) reports that some cultivars of *D. nummularia* from New Guinea do not match the type specimen deposited in the herbarium of the Philippines. Thus it is possible that *D. transversa* also occurs in New Guinea. The morphological confusion between *D. alata, D. nummularia,* and *D. transversa* is also reported in the Philippines (Cruz and Ramirez, 1999), Indonesia (Sastrapradja, 1982), Vanuatu (SPYN, 2001), and New Caledonia (Bourret, 1973). Because these sympatric species belong to section Enantiophyllum and bear striking morphological similarities to *D. alata,* one may assume that they could have contributed to its genetic makeup.

Because the region of greatest variability of *D. alata* is not compatible with the range of distribution of its proposed wild relatives (*D. hamiltonii* and *D. persimilis*), the area of domestication has remained unclear. De Candolle (1886) placed the geographic origin of *D. alata* on the Indo-Malayan peninsula. Prain and Burkill (1939) suggested an area in the northern part of the southeast Asian peninsula, following Vavilov's center of origin of cultivated plants in the Assam–Burma region. These two species occur naturally in the Southeast Asian peninsula, where *D. hamiltonii* occupies the western range from east India to west Burma (Coursey, 1967), and *D. persimilis* thrives in the eastern range from southwest China to Vietnam (Ding and Gilbert, 2000). So far, it has not been demonstrated that the two species overlap. Furthermore, the greatest phenotypic variation of *D. alata* is observed south of the Southeast Asian peninsula, in Indonesia, Malaysia, and Melanesia, where several authors have reported that local cultivars with primitive (irregular tuber shapes and spiny stems) and improved phenotypes (regular shape and shallow tubers) exist (Bourret, 1973; Coursey, 1976; Martin and Rhodes, 1977; Ochse and van den Brink, 1977; Sastrapadja, 1982; Degras, 1993; Lebot et al., 1998; Cruz and Ramirez, 1999; Malapa, 2000).

Despite extensive prospects and inventories, the area of domestication and origin of *D. alata* remains problematic because cultivars have been widely distributed since prehistoric times (Barrau, 1956; Coursey, 1967). In this context, Harlan (1971) suggested that the study of cultivated plants and their origins is difficult and necessitates an interdisciplinary framework including anthropology, archaeology, geography, geology, genetics, and linguistics.

The situation is complex and does not lend itself to a simple answer. Our objectives are to combine molecular, morphological, cytogenetic, and folk classification techniques to address questions that have been difficult to answer in the past. Because the most in-depth study regarding *D. alata* genetic diversity has been conducted outside its area of origin (Martin and Rhodes, 1977), we assume that such an investigation should now be conducted in a geographic region where the species exhibits tremendous cultural importance and morphological variation. In Melanesia, the archipelago of Vanuatu offers both for *D. alata* and related species.

Morphological and Molecular Variation

Since 1999, the South Pacific Yam Network (SPYN, 2001) has established in Vanuatu a collection of 376 accessions: *D. alata* (331), *D. bulbifera* (8), *D. cayenensis-rotundata* (2), *D. esculenta* (15), *D. nummularia* (4), *D. pentaphylla* (6), *D. transversa* (9), and *D. trifida* (1). Malapa (2000) used this germplasm collection to assess the morphological variation between cultivars of *D. alata,* and Malapa et al. (in press) used a core sample of 100 accessions, selected to represent the morphological variation and the different geographic origins of the Vanuatu germplasm collection, for ploidy level assessment and comparative analysis of AFLP fingerprinting patterns.

Here, we use these collections along with two additional ambiguous morphotypes, "netsar" and "rul," for morphological, ploidy level, and molecular analyses using AFLP. We report the morphological variation between the 331 accessions of *D. alata* as expressed in terms of 28 phenotypic trait frequencies (table 11.2). For molecular analysis, a subset of 56 accessions (table 11.3) was used to summarize the results of Malapa et al.

Table 11.2 Frequencies of Accessions Exhibiting
Morphological Traits

Descriptor Category Frequency	(%)
Young stem color Green	9.1
Purplish green	44.6
Brownish green	1.4
Purple	16.2
Yellowish green	27.7
Other	1
Young wing color Green	5.6
Green with purplish edge	61.4
Purple	21.4
Other	11.6
Young leaf color Yellowish	13
Pale green	5.6
Purplish green	42.1
Purple	35.1
Other	4.2
Young leaf vein color Yellowish	11.9
Pale green	9.8
Dark green	54
Purplish green	22.8
Other	1.5
Young leaf petiole color Green with purple base	7.4
Green with purple leaf junction	0.4
Green with purple at both ends	6.7
Purplish green with purple base	8.4
Purplish green with purple at both ends	34
Green	31.5
Purple	9.5
Brownish green	0.3
Other	1.8
Mature stem color Green	11.6
Purplish green	9.8
Brownish green	77.9
Purple	0.7
Mature stem wing color Green	17.9
Green with purple edge	75.4
Purple	6

(*continued*)

Table 11.2 (continued)

Descriptor Category Frequency	(%)
Other	0.7
Mature leaf color Pale green	51
Dark green	48.4
Purple	0.3
Other	0.3
Mature leaf vein color Yellowish	1.4
Green	84.6
Dark green	9.1
Purplish green	4.9
Mature leaf petiole color Green with purple base	12.3
Green with purple leaf junction	0.4
Green with purple at both ends	36.1
Purplish green with purple base	0.3
Purplish green with purple at both ends	3.5
Green	47.4
Skin color at tuber head White	9.8
Yellow	33.3
Orange	1
Pink	22.8
Purple	33.1
Flesh color at central section White	73
Yellowish white	1
Yellow	0.4
Light purple	1
Purple	3.9
Purple with white	7.7
White with purple	12.6
Outer purple, inner yellowish	0.4
Flesh color of lower part White	71.2
Yellowish white	3.9
Yellow	0.7
Light purple	2.8
Purple	6.3
Purple with white	3.5
White with purple	11.2
Outer purple, inner yellowish	0.4
Mature leaf shape Elongate	39.2

(continued)

Table 11.2 *(continued)*

Descriptor Category Frequency	(%)
Ovoid	5.3
Cordiform	53
Curled under	0.7
Cupped	1.8
Distance between lobes Intermediate	63.9
Very distant	36.1
Petiole length 6–9 cm	55.4
≥10 cm	44.6
Number of stems Single	4.3
Few	40.3
Many	55.4
Mature spines on stem base None	82.1
Few	15.1
Many	2.8
Mature stem wing size ≤1 mm	78.2
>1 mm	21.8
Internode length of mature stem <9 cm	7
9–18 cm	87.4
>18 cm	5.6
Leaf size Narrow (<10 cm)	10.1
Medium (10–15 cm)	8
Large (>15 cm)	81.9
Tuber shape Round	3.2
Oval	8.1
Cylindrical	46.6
Flattened	1.4
Triangular	2.5
Irregular	38.2
Number of tubers per plant 1–2.5	74.3
2.5–5	22.1
5–7.5	2.5
≥7.5	1.1
Flowering Absent	94
Present	6
Sex Female	11.8
Male	88.2

(continued)

Table 11.2 (continued)

Descriptor Category Frequency	(%)
Aerial tubers Absent	92.6
Present	7.4
Maturity of tuber <6 months	6.3
6–9 month	93.7
Yield per plant ≤0.5 kg	18.9
0.5–2.5 kg	57.2
2.5–4.5 kg	17.2
4.5–6.5 kg	3.2
≥6.5 kg	3.5

(in press) with additional comparisons to folk classification in order to assess the potential of AFLP for assigning ambiguous morphotypes into homogeneous species and elucidate their genetic relationships and taxonomic position with *D. alata, D. nummularia,* and *D. transversa.* (For materials and methods, see Malapa et al., in press). Two studies were conducted using distinct species, samples, and AFLP primer pair combinations (tables 11.3 and 11.4):

Study I: Four primer pair combinations were used for six species of section Enantiophyllum of Oceanian, Southeast Asian, and African origins. These primer pairs produced 156 bands, of which only 7 were monomorphic.

Study II: Eleven primer pair combinations were used for *D. alata, D. transversa, D. persimilis, D. cayenensis-rotundata, D. pentaphylla, D. bulbifera, D. esculenta,* and *D. trifida.* They produced a total of 493 bands, of which 64% were species-specific.

Intraspecific Variability

Morphology

The results show that phenotypic pigmentation varies greatly from white to brown, depending on organ, growth stage, and cultivar (table 11.2). Most cultivars produce tubers with white flesh (73%). Also, the shape and size of organs vary. Cylindrical (46.6%) and irregular (38.2%) tuber shapes are prevalent, whereas round tubers are uncommon (3.2%). The numbers of tubers per plant and yields per plant are also highly variable. Most accessions

Table 11.3 Accessions Analyzed per Species in Two Successive AFLP Studies

Codes	Names	Institutions	Origins	Species	Sections	Study I	Study II
A102	n.d.	CIRAD	Burkina Faso	D. cayenensis-rotundata	Enantiophyllum	—	x
Ala-1	n.d.	CIRAD	W. French Indies	D. alata	Enantiophyllum	—	x
CTRT30	Yanon kossu	CIRAD	Benin	D. cayenensis-rotundata	Enantiophyllum	—	x
CTRT52	Togo 46	CIRAD	Togo	D. esculenta	Combilium	—	x
CTRT75	Cuba 6	CIRAD	Cuba	D. trifida	Macrogynodium	—	x
Daby	n.d.	CIRAD	n.d.	D. abyssinica	Enantiophyllum	x	—
Dper	Cu mai	VASI	Vietnam	D. persimilis	Enantiophyllum	x	x
T-Guy	MP 2	CIRAD	Guyana	D. trifida	Macrogynodium	—	x
Togo43	n.d.	CIRAD	Togo	D. bulbifera	Opsophyton	—	x
VN072	Cu mo	VASI	Vietnam	D. alata	Enantiophyllum	x	—
VN073	Cu mo	VASI	Vietnam	D. alata	Enantiophyllum	x	—
VN099	Cu mo	VASI	Vietnam	D. alata	Enantiophyllum	x	—
VN121	Cu mo	VASI	Vietnam	D. alata	Enantiophyllum	x	—
VU009	Dam kabi	VARTC	Vanuatu	D. alata	Enantiophyllum	x	—
VU012	Blarghlin	VARTC	Vanuatu	D. alata	Enantiophyllum	x	—
VU015	n.d.	VARTC	Vanuatu	D. alata	Enantiophyllum	x	—
VU376	Wailou white	VARTC	Vanuatu	D. cayenensis-rotundata	Enantiophyllum	x	—

VU408	Manioc	VARTC	Vanuatu	*D. alata*	Enantiophyllum	x	—
VU434	Pili	VARTC	Vanuatu	*D. alata*	Enantiophyllum	x	x
VU444	Tamate ajuju	VARTC	Vanuatu	*D. alata*	Enantiophyllum	x	—
VU497	Maliok	VARTC	Vanuatu	*D. alata*	Enantiophyllum	x	x
VU519	Salomon	VARTC	Vanuatu	*D. alata*	Enantiophyllum	x	—
VU528	Sinoua	VARTC	Vanuatu	*D. alata*	Enantiophyllum	x	—
VU556	Valise	VARTC	Vanuatu	*D. alata*	Enantiophyllum	x	—
VU564	Makila	VARTC	Vanuatu	*D. alata*	Enantiophyllum	x	—
VU571	Letslets rorosiv	VARTC	Vanuatu	*D. alata*	Enantiophyllum	x	—
VU579	Letslets bokis	VARTC	Vanuatu	*D. alata*	Enantiophyllum	x	—
VU590	n.d.	VARTC	Vanuatu	*D. alata*	Enantiophyllum	x	—
VU606	n.d.	VARTC	Vanuatu	*D. transversa*	Enantiophyllum	x	—
VU616	Rolbu	VARTC	Vanuatu	*D. nummularia*	Enantiophyllum	x	x
VU618	Braswaea	VARTC	Vanuatu	*D. transversa*	Enantiophyllum	x	—
VU621	Taktak bungen	VARTC	Vanuatu	*D. transversa*	Enantiophyllum	—	x
VU631	Musadega	VARTC	Vanuatu	*D. esculenta*	Combilium	—	x
VU640	Vagavaga	VARTC	Vanuatu	*D. esculenta*	Combilium	—	x
VU662	Strong yam	VARTC	Vanuatu	*D. transversa*	Enantiophyllum	x	x
VU687	Kahut	VARTC	Vanuatu	*D. transversa*	Enantiophyllum	x	—
VU692	Taniru	VARTC	Vanuatu	*D. transversa*	Enantiophyllum	x	—

(continued)

Table 11.3 (continued)

Codes	Names	Institutions	Origins	Species	Sections	Study I	Study II
VU702	Katipanaum	VARTC	Vanuatu	*Dioscorea* spp.	Enantiophyllum	x	—
VU705	Ros apin	VARTC	Vanuatu	*D. alata*	Enantiophyllum	—	x
VU706	Namio	VARTC	Vanuatu	*D. alata*	Enantiophyllum	x	—
VU711	n.d.	VARTC	Vanuatu	*D. transversa*	Enantiophyllum	x	—
VU712	Wingosu	VARTC	Vanuatu	*D. alata*	Enantiophyllum	x	—
VU713	n.d.	VARTC	Vanuatu	*D. nummularia*	Enantiophyllum	x	—
VU715	n.d.	VARTC	Vanuatu	*Dioscorea* spp.	Enantiophyllum	x	—
VU717	Nupumori	VARTC	Vanuatu	*D. alata*	Enantiophyllum	x	—
VU735	Noplon	VARTC	Vanuatu	*D. alata*	Enantiophyllum	x	—
VU737	n.d.	VARTC	Vanuatu	*D. pentaphylla*	Botryosicyos	—	x
VU745	Tam matua	VARTC	Vanuatu	*D. alata*	Enantiophyllum	x	—
VU746	Nioutec	VARTC	Vanuatu	*D. alata*	Enantiophyllum	x	—
VU747	Nowewa	VARTC	Vanuatu	*D. nummularia*	Enantiophyllum	x	—
VU749	Waïlou yellow	VARTC	Vanuatu	*D. cayenensis-rotundata*	Enantiophyllum	x	—
VU754	Noulelcae	VARTC	Vanuatu	*D. alata*	Enantiophyllum	x	—
VU757	Narouvanua	VARTC	Vanuatu	*D. alata*	Enantiophyllum	x	—
VU758	Inomotjamja	VARTC	Vanuatu	*D. pentaphylla*	Botryosicyos	—	x
VU759	Konore	VARTC	Vanuatu	*D. transversa*	Enantiophyllum	x	—
VU760	Noureangdan	VARTC	Vanuatu	*D. alata*	Enantiophyllum	x	—

n.d. = not determined, CIRAD = Centre International de Recherches Agronomiques pour le Développement (France), VASI = Vietnam Agricultural Science Institute, VARTC = Vanuatu Agricultural Research and Technical Centre.

Table 11.4 Number of Polymorphic Bands Revealed per Primer Pair and per Study

Primer Pair	Numbers of Polymorphic Bands	
	Study I	Study II
E-AAC/M-CTA	—	39
E-AAC/M-CAG	40	53
E-ACA/M-CAT	—	52
E-ACC/M-CTA	—	55
E-ACC/M-CAT	54	46
E-ACA/M-CAA	—	40
E-ACG/M-CTA	—	43
E-AAC/M-CAT	31	32
E-ACT/M-CTA	—	40
E-ACT/M-CTC	—	53
E-ACA/M-CAC	—	40
E-ACA/M-CAG	31	—
Total	156	493
Mean	39	44.8
SD	10.86	7.5

produce 1–2.5 tubers per plant (74.3%) and yield 0.5–2.5 kg per plant. Very few accessions produce aerial bulbils (7.4%) or flowers (6%), and most of the flowering plants are male. Phenetic analysis of the morphological data revealed a vast continuum of variations; that is, no clear groupings were revealed, and no obvious structure of the variation was observed.

This result supports previous findings of phenotypic variation using morphological descriptors of aerial and underground organs (Bourret, 1973; Martin and Rhodes, 1977; Sastrapadja, 1982; Lebot et al., 1998; Cruz and Ramirez, 1999). Martin and Rhodes (1977:10) stated that although great phenotypic variability has been observed among their worldwide collection of 235 accessions, the classification of cultivars based on 28 morphological characters failed to reproduce any strict division on both morphological and geographic grounds because cultivars of distinct origins clustered together within a group of similar morphotypes and vice versa, like the "anastomosing branches of a tropical banyan tree." Thus the use of morphological traits for classifying cultivars seems unreliable within *D. alata* because they are extremely variable, and no investigator has had the opportunity to see more than a small fraction of the existing variability.

Ploidy Levels

Cytogenetic markers have been used to analyze ploidy levels in *Dioscorea* spp. According to Smith (1937), Southeast Asian species have a haploid chromosome set of x = 10, and most of the results found in the literature indicate the existence of 2n = 4x, 6x, and 8x for *D. alata,* 2n = 4x, 6x, 8x, and 10x for *D. bulbifera,* 2n = 4x, 6x, 9x, and 10x for *D. esculenta,* 2n = 4x, 8x for *D. pentaphylla,* and 2n = 8x for *D. transversa* (Miege, 1952; Coursey, 1967; Essad, 1984; Abraham and Nair, 1991; Degras, 1993; Gamiette et al., 1999; Egesi et al., 2002; Malapa et al., in press). Thus the absence of diploid (2n = 2x = 20) forms among cultivars and related wild species suggests that polyploidy is a common state among edible yam species and cultivars (Miege, 1952).

Malapa et al. (in press) analyzed the ploidy levels of 53 accessions of *D. alata* using both root tip counts and flow cytometry. Tetraploids (29 accessions), hexaploids (5 accessions), and octoploids (19 accessions) were identified, but no diploids were found. The existence of three levels of ploidy supports previous studies and confirms that polyploidy is common among *D. alata* cultivars. Tetraploids and octoploids are widely distributed throughout Vanuatu, whereas hexaploids were collected mainly from the southern part of the archipelago.

Comparisons between morphotypes and cytotypes revealed that tetraploids, hexaploids, and octoploids tend to assemble very distinct morphotypes. Tetraploids have narrow leaves, whereas hexaploids and octoploids have thick, dark green, waxy leaves with a cordate base. All cytotypes include flowering plants, and female plants are absent among hexaploids.

Comparisons between genetic variation and cytotypes did not reveal any grouping pattern according to specific bands or the total number of bands per accession. Taken together, these results indicate that hexaploid and octoploid cultivars probably resulted from autopolyploidization involving tetraploid cultivars.

DNA Fingerprinting

Recently, neutral molecular markers have been applied for fingerprinting and genetic diversity analyses among *Dioscorea* spp. Random amplified polymorphism DNAs (Asemota et al., 1996; Ramser et al., 1996), AFLPs (Mignouna, 1998), and isozymes (Lebot et al., 1998) proved to be highly repeatable. The latter authors studied the genetic relationships between 269 accessions of *D. alata* of the Pacific, Asia, Africa, and the Caribbean. The four polymorphic enzyme systems used revealed that genetic heterogeneity exists among clonal populations, but no correlation was found between genetic groups, geographic origins, or phenotypic traits. Their results suggest that the species has been widely distributed and that genetic recombinations have occurred in the past. Although today it is difficult to observe fruiting plants in cultivation, it appears that before these genotypes were brought into cultivation and clonally propagated, they were outcrossing.

Our AFLP results also reveal highly polymorphic patterns within *D. alata.* These patterns are highly discriminant and allow the unique fingerprinting of cultivars because most of them clustered separately (figure 11.1 and table 11.3). These results indicate that our sample is composed of distinct clonal populations as well as duplicates (accessions 497 and 556).

Within *D. transversa,* pairwise genetic distance analysis indicates that accessions of *D. transversa* have a narrow genetic basis (93% similarity) and could be differentiated into two clonal populations, including cv. "maro" (606 and 618) and cv. *langlang* (662, 687, 692, 711, and 759) (figure 11.1). The former cultivar produces globose to round tubers, and the latter produces cylindrical to flattened tubers. Farmers also distinguish them from one another based on tuber morphology, although their aerial morphotypes are similar (table 11.1). Thus, AFLPs reveal that *strong yams* include genetically heterogeneous species and cultivars expressing different morphotypes, including "maro" and "netsar."

Within *D. nummularia,* AFLPs also distinguish two groups among the *wild yams* and separated the morphotype "buts" (616 and 713) from the cultivated "buts rom" (747) (figure 11.1 and table 11.1). These morphotypes clustered separately and have a wider genetic basis than *D. alata* and *D. transversa.*

Interspecific Variability

DNA Fingerprinting

In study I, cluster analysis of the distance matrix reveals five major groups (figure 11.1). Cluster 1 includes accessions of *D. alata.* Cluster 2 includes *D. transversa* and the morphotypes "netsar" (702) and "rul" (715) (*Dioscorea* spp.), whereas the *wild yams* species *D. nummularia* (616, 713, and 747) is found in cluster 3. Finally, in clusters 4 and 5 are found the Southeast Asian *D. persimilis* and the West African *D. abyssinica* and *D. cayenensis-rotundata* (376 and 749), respectively.

Interspecific clustering patterns also divide the perennial morphotypes (table 11.1) into two groups, the first one including species *D. transversa* and

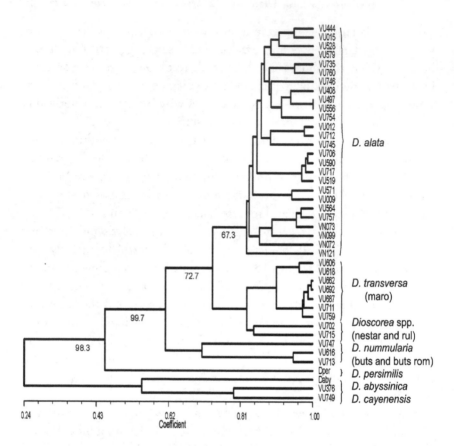

FIGURE 11.1 UPGMA representation, based on four AFLP primer pair combinations, of the Dice similarities between 42 clones representing 5 *Dioscorea* species of the Enantiophyllum section. Bootstrap values are indicated under branches (study I).

Table 11.5 Specific AFLP Bands for ***D. alata***, ***D. nummularia***, and ***D. transversa*** (Study I)

Primer Pairs	Total Bands	Polymorphic Bands	Specific Bands		
			D. alata	*D. transversa*	*D. nummularia*
E+AAC/M+CAG	47	40	4	10	15
E+AAC/M+CAT	43	31	8	9	6
E+ACA/M+CAG	43	31	6	7	4
E+ACC/M+CAT	58	54	3	7	15
Mean	47.75	39	5.25	8.25	10
SD	7.09	10.86	2.22	1.50	5.83

Dioscorea spp. and the second one including *D. nummularia* (figure 11.1). The first group is genetically much closer to *D. alata* than the second group, as indicated by the higher number of common bands. Furthermore, comparison of species-specific bands within clusters indicates that the *strong yams* "netsar" and "rul" have fewer bands as compared with *D. alata, D. transversa,* and *D. nummularia* (table 11.5). These results suggest that the latter three species are distinct from "netsar" and "rul" but share a common genetic background with them.

In study II, interspecific pairwise similarity averages indicated that *D. alata* is most closely related to *D. transversa* (table 11.6). These two species are also closely related to *D. persimilis* and *D. cayenensis-rotundata,* conforming to their taxonomic position into section Enantiophyllum and their geographic origins in that *D. alata, persimilis,* and *D. transversa* are Asian–Oceanian, whereas *D. cayenensis-rotundata* is African. These relationships can also be extended to the other sections and species including *D. bulbifera* (sect. Opsophyton), *D. esculenta* (sect. Combilium), *D. pentaphylla* (sect. Botryosicyos), and *D. trifida* (sect. Macrogynodium). Pairwise genetic similarities between the overall species indicate that African and Asian species are closer to each other and that the South American *D. trifida* is distant from the rest of the group (table 11.6). These results are also supported by unweighted pair group method with arithmetic mean (UPGMA) and bootstrap analysis (87–100%), as indicated for the major branches of the dendrogram (figure 11.2).

The global tree topologies (figures 11.1 and 11.2), supported by bootstrap values, reveal interspecific relationships that are consistent between study I and study II. They indicate that *D. alata* is closer to

Table 11.6 Averages of Genetic Distances and Similarities (%) between 8 *Dioscorea* Species (Study II)

	Dala	Dtra	Dper	Dcay	Dpen	Dbul	Desc	Dtri
Dala	87.7	—	—	—	—	—	—	—
Dtra	53.7	94.1	—	—	—	—	—	—
Dper	25.7	31.4	100	—	—	—	—	—
Dcay	14.7	14.7	13	78.3	—	—	—	—
Dpen	8.8	10.9	7	2.6	84.7	—	—	—
Dbul	8.4	6.8	10.1	5.8	2.2	100	—	—
Desc	5.7	5.2	5	3.9	6.2	3	94.2	—
Dtri	5.5	5.1	4.9	0.3	1.1	3.3	1.3	74.5

Dala = *D. alata,* Dtra = *D. transversa,* Dper = *D. persimilis,* Dcay = *D. cayenensis-rotundata,* Dpen = *D. pentaphylla,* Dbul = *D. bulbifera,* Desc = *D. esculenta,* Dtri = *D. trifida.*

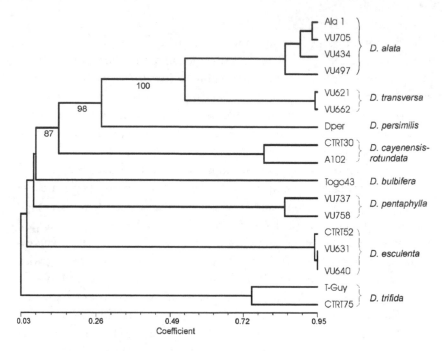

FIGURE 11.2 UPGMA representation, based on 11 ᴀꜰʟᴘ primer pair combinations, of the Dice similarities between 17 clones representing 8 *Dioscorea* species. Bootstrap values are indicated under lines (study II).

D. transversa and *D. nummularia* than to *D. persimilis.* Furthermore, the section Enantiophyllum representatives from Africa are genetically distant from their Southeast Asian–Oceanian counterparts and form a distinct gene pool, as revealed by the species-specific ᴀꜰʟᴘ fingerprinting patterns. Also, American *Dioscorea* are genetically very distant from African and Asian *Dioscorea* species (figure 11.2 and table 11.6).

Thus these results do not contradict an early genetic divergence of New World species stemming from isolation at the end of the Cretaceous period when South America split from Gondwanaland, about 70 million years ʙᴘ (Coursey, 1967). The modern American *Dioscorea* are distinct from Old World species, and no section is common to both the Old World and the New World,[3] suggesting that American *Dioscorea* species evolved independently from Old World species at an early time that involved proto-*Dioscorea* ancestors. In contrast, Old World species evolved together in both Gondwana and Laurasia. The separation of the Asiatic and the African *Dioscorea* probably was much more recent because two sections (Enantiophyllum and Opsophyton) are represented by very similar species on the two continents, and one species

(*D. bulbifera*) is common to both. At present, the two groups are separated by the deserts of the Middle East that probably were formed in the late Miocene or early Pliocene ages, about 10 million years BP (Coursey, 1967).

Conclusions

The analyses of molecular and phenotypic data together with archaeological, ethnobotanical, and linguistic evidence reveal that the diversification of *D. alata* occurred in Melanesia, where the species has a strong sociocultural importance. Thus both primitive (long irregular tubers and spiny stems) and improved (compact tubers) varieties, as described by Martin and Rhodes (1977), also exist in Vanuatu, as well as the numerous intermediate morphotypes that are recognized in local classifications. They have different uses and exhibit important phenotypic variations, as expressed by their different colors and shapes of both aerial and underground organs (tables 11.1 and 11.2).

Both molecular and cytogenetic markers reveal that the great phenotypic variability found within *D. alata* also has a genetic and genomic basis. Thus AFLPs reveal high polymorphism within *D. alata,* and cultivars can be uniquely fingerprinted at the molecular level. Intraspecific level analyses of AFLP banding patterns indicate that *D. alata* is genetically heterogeneous despite its vegetative propagation. Malapa et al. (in press) also confirm the genomic heterogeneity of the species using chromosome counts and flow cytometry analyses of ploidy levels. Their results are congruent with previous studies indicating the existence of tetraploids, hexaploids, and octoploids (Abraham and Nair, 1991; Hamon et al., 1992; Gamiette et al., 1999). This genetic and genomic heterogeneity suggests that sexual recombination exists, as already revealed with isozymes (Lebot et al., 1998), and that higher ploidy levels may have arisen from autopolyploidization involving tetraploids.

Another major finding of this study is the genetic relatedness between species of section Enantiophyllum of Oceania. Malapa et al. (in press) suggest that the perennial yams (*strong yams* and *wild yams*) of Vanuatu form a species complex that could not be limited exclusively to *D. nummularia,* as generally reported (Barrau, 1956; Weightman, 1989; Allen, 2001). They show that the "maro" morphotype belongs to the Oceanian *D. transversa* and that this species is closely related to *D. alata* and *D. nummularia.*

The present study using AFLPs also confirms this for the ambiguous morphotypes "netsar" and "rul." It indicates that these two morphotypes belong to a taxon that is genetically distinct from *D. alata, D. transversa,* and *D. nummularia.* Furthermore, AFLPs reveal that the morphotype "rul" is genetically closer to the morphotype "netsar" and "maro" than to "buts" and "buts rom," confirming its classification into the traditional groups of *wild yams*. Thus, the clustering pattern based on AFLP data supports the folk classification of section Enantiophyllum species based on morphological characters and horticultural needs but not the general pattern of classification of *Dioscorea* spp. as based on dry matter content.

These results suggest that the phenotypic similarities observed between *D. alata* and "netsar," between *D. transversa* and "rul," and between *D. nummularia* and "rul" are also confirmed at the molecular level because "maro," "netsar," and "rul are closely related to *D. alata* and *D. nummularia* and cluster at an intermediate position between these two species. Taken together, these findings suggest that section Enantiophyllum in Vanuatu includes at least four distinct species belonging to a common gene pool that should be centered within their natural range of geographic distribution (figure 11.3). However, AFLP markers could not unravel the genetic basis of their relationships because of their dominant nature.

Phenetic analyses of molecular data reveal that section Enantiophyllum is a well-supported group. Pairwise genetic distances between species of section Enantiophyllum indicate that the cultivated (*D. cayenensis-rotundata*) and wild (*D. abyssinica*) African species are closely related to each other but are genetically distant from the Southeast Asian–Oceanian species *D. persimilis, D. alata, D. transversa,* and *D. nummularia*. These results probably indicate the existence of two divergent gene pools within African and Asian Enantiophyllum. Such genetic divergence between both continents has already been revealed within section Opsophyton, using the unique species *D. bulbifera*, which is common to Africa and Asia (Terauchi et al., 1991). Therefore these results suggest that cultivated species of section Enantiophyllum of Africa, Asia, and Oceania could have been domesticated from local wild resources. This conclusion is also supported by phylogenetic analyses based on cpDNA and restriction fragment length polymorphism, which revealed that cultivars of *D. cayenensis-rotundata* had been domesticated in Africa (Terauchi et al., 1992) and that cultivars of *D. bulbifera* (Terauchi et al., 1991) were independently domesticated from the local wild relatives located in Africa, Asia, Australia, and New Guinea.

Because AFLPs reveal that *D. alata* is more closely related to Oceanian species than to the wild Asian *D. persimilis,* we suggest that species related to *D. alata* have evolved independently from Asian species to form a divergent gene pool within Oceania. Furthermore, the genetic relationships between *D. alata, D. nummularia,* and *D. transversa* and the existence of ambiguous

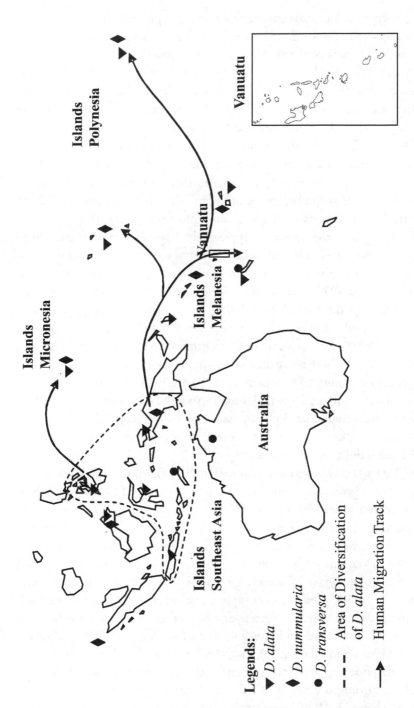

Legends:

▶ *D. alata*

◆ *D. nummularia*

● *D. transversa*

- - - Area of Diversification
      of *D. alata*

→ Human Migration Track

FIGURE 11.3 Origin and human dispersal of *D. alata, D. nummularia,* and *D. transversa* cultivars in Oceania.

morphotypes in Vanuatu indicate that these taxa probably have a wild common ancestor in insular Southeast Asia and New Guinea, where the diversification of *D. alata* occurred (Martin and Rhodes, 1977).

Finally, these results have obvious bearings on the origin of *D. alata* cultivars of Vanuatu. Archeological evidence indicates that the early settlers of that archipelago came from insular Southeast Asia in about 3000 BP (Spriggs, 1997). They traveled southeast across the coast of New Guinea and into insular Melanesia that had already been settled by Papuan hunter–gatherers during the Pleistocene era, about 40,000 BP (Bellwood, 1985). In New Guinea they moved from the northeast to the southeast coast of that continental island before reaching the Melanesian arc (circa 3500 BP) and the Polynesian islands (circa 2500–1500 BP) (Spriggs, 1997).

This migration theory is also supported by linguistic affinities and ethnobotanical and anthropological evidence of root crop–based Melanesian civilizations for which yams satisfy both spiritual and physical needs (Coursey, 1967; Bourret, 1973; Tryon, 1985; Weightman, 1989). Such affinities are observed from the Maprik (Lea, 1966) to the East Sepik regions (www.art-pacific.com) of New Guinea and from the Bismarck Archipelago (Degras, 1993) south to Vanuatu (Weightman, 1989) and New Caledonia (Bourret, 1973). In these civilizations, where subsistence economy is based on product exchanges, *D. alata* cultivars have a high commercial value and are exchanged for products including introduced crops and animals. This sociocultural importance probably has enforced the geographic dispersal of cultivars into Oceania through the use of planting materials such as tubers and aerial bulbils (Barrau, 1958).

AFLP markers showed that although the distribution of *D. alata* is pantropical, it is genetically related to *D. nummularia* and *D. transversa*, which are restricted to western Oceania. They also revealed that *D. alata* could not have been domesticated directly from *D. persimilis* because of the existence of species-specific electromorphs (i.e., bands that exist in one species but not in the others). *D. alata* is also genetically related to unidentified Vanuatu endemic taxa of section Enantiophyllum. These findings are also supported by the present phenotypic variation report and previous studies that indicated that the greatest diversity of *D. alata* is located between insular Southeast Asia and the Solomons (de Candolle, 1886; Coursey, 1976; Martin and Rhodes, 1977; Ochse and van den Brink, 1977). The molecular findings are in agreement with other genetic and cytogenetic results regarding the origin of other widespread Oceanian crops such as banana (Lanaud, 1999), breadfruit (Zerega et al., 2004), and sugarcane (see chapter 3, this volume). They indicate that domestication and diversification occurred independently in different regions from Southeast Asia to Polynesia via New Guinea (Lebot, 1999).

The interdisciplinary findings suggest that *D. alata, D. nummularia,* and *D. transversa* belong to a natural flora of common origin that is located at the crossroads of the human migration track into Oceania. This flora has been exploited since the late Pleistocene era (circa 40,000 BP) by the early settlers of insular Melanesia who probably conducted the clonal selection of their cultivars from local resources.

Notes

1. In this chapter, two categories of languages are used to illustrate the folk classification and the traditional uses of yams in Vanuatu. Names in quotation marks are borrowed from the nomenclature system of the local Wala-Rano language of north Malakula Island. Names in italics are borrowed from the common Bislama language. This language is derived from English and is spoken throughout the archipelago, where more than a hundred local languages exist (Bonnemaison, 1996).

2. *Laplap* is the Bislama name for "soso ur." In this chapter, Bislama names (when they exist) are used in preference to Wala-Rano names because they are commonly used throughout Vanuatu and cited in the literature as well (Weightman, 1989).

3. "The only exception is the section *Stenophora*, which occurs in North America as well as in Europe and Asia. It is suggested that representatives of this section migrated into North America across the Bering Straits land-bridge formed in the climatically mild Miocene age" (Coursey, 1967:230).

References

Abraham, K. and G. P. Nair. 1991. Polyploidy and sterility in relation to sex in *Dioscorea alata* L. (Dioscoreaceae). *Genetica* 83: 93–97.

Alexander, J. and D. G. Coursey. 1969. The origins of yam cultivation. In P. J. Ucko and G. H. Dimbleby (eds.), *The Domestication and Exploitation of Plants and Animals*, 405–425. Duckworth, London, UK.

Allen, M. G. 2001. *Change and Continuity: Land Use and Agriculture on Malo Island, Vanuatu*. M.S. thesis, Australian National University, Sydney, Australia.

Asemota, H. N., J. Ramser, C. Lopéz-Peralta, K. Weising, and G. Kahl. 1996. Genetic variation and cultivar identification of Jamaican yam germplasm by random amplified polymorphic DNA analysis. *Euphytica* 92: 341–351.

Barrau, J. 1956. Les ignames alimentaires des îles du Pacifique Sud. *Journal d'Agriculture Tropicale et de Botanique Appliquée* 3: 385–401.

Barrau, J. 1958. Subsistence agriculture in Melanesia. *Bernice Bishop Museum Bulletin* 219: 1–112.

Barrau, J. 1962. *Les Plantes Alimentaires de l'Océanie: Origine, Distribution et Usages*. Doctoral thesis, Faculté des Sciences, Marseilles, France.

Bellwood, P. S. 1985. *Prehistory of the Indo-Malaysian Archipelago*. Academic Press, Sydney, Australia.

Bonnemaison, J. 1996. *Gens de Pirogue et Gens de la Terre*. ORSTOM, Paris, France.

Bourret, D. 1973. *Etude Ethnobotanique des Dioscoreacées Alimentaires. Ignames de Nouvelle Calédonie*. Thesis, Faculté des Sciences, Université de Paris, Paris, France.

Bourrieau, M. 2000. *Valorisation des Racines et Tubercules Tropicaux Pour l'Alimentation Humaine en Océanie: Le Cas du Laplap au Vanuatu.* Master's thesis, ENSAI/SIARC, Montpellier, France.

Bradbury, J. and W. D. Holloway. 1988. *Chemistry of Tropical Root Crops: Significance for Nutrition and Agriculture in the Pacific.* ACIAR Monograph No. 6, Canberra, Australia.

Burkill, I. H. 1948–1954. Dioscoreaceae. In C. G. G. J. Van Steenis (ed.), *Flora Malesiana,* Vol. 4, Part I, 293–335. Noordhoff-Kolff N. V., Jakarta, Indonesia.

Burkill, I. H. 1951. The rise and decline of the greater yam in the service of man. *Advancement of Science* 7(28): 443–448.

de Candolle, A. 1886. *Origin of Cultivated Plants,* 2nd ed. Reprinted by Hafner, New York, NY, USA.

Coursey, D. G. 1967. *Yams.* Longmans, London, UK.

Coursey, D. G. 1976. Yams. In N. W. Simmonds (ed.), *Evolution of Crop Plants,* 70–74. Longmans, London, UK.

Cruz, V. M. V. and D. A. Ramirez. 1999. Variation in morphological characters in a collection of yams (*Dioscorea* spp.). *Philippines Journal of Crop Sciences* 24: 27–35.

Degras, L. 1986. *L'igname, techniques agricoles et productions tropicales.* Maisonneive & Larose, Paris, France.

Degras, L. 1993. *The Yam, a Tropical Root Crop.* Macmillan, London, UK.

Ding, Z. Z. and M. G. Gilbert. 2000. Dioscoreaceae. In Z. Y. Wu and P. H. Raven (eds.), *Flora of China,* Vol. 24. Science Press, Beijing, and Missouri Botanical Garden Press, St. Louis, MO, USA.

Egesi, C. N., M. Pillay, R. Asiedu, and J. K. Egunjobi. 2002. Ploidy analysis in water yam, *Dioscorea alata* L. *Euphytica* 128: 225–230.

Essad, S. 1984. Variation géographique des nombres chromosomiques de base et polyploïdie dans le genre *Dioscorea* à propos du dénombrement des espèces *transversa* Br., *pilosiuscula* Bert. et *trifida* L. *Agronomie* 7: 611–617.

Gamiette, F., F. Bakry, and G. Ano. 1999. Ploidy determination of some yam species (*Dioscorea* spp.) by flow cytometry and conventional chromosomes counting. *Genetic Resources and Crop Evolution* 46:19–27.

Hahn, S. K. 1991. Yams. In N. W. Simmonds (ed.), *Evolution of Crop Plants,* 112–120. Longmans, London, UK.

Hamon, P., J. P. Brizard, J. Zoundjihékpon, C. Duperray, and A. Borgel. 1992. Etude des index d'ADN de huit espèces d'ignames (*Dioscorea* spp.) par cytométrie en flux. *Canadian Journal of Botany* 70: 996–1000.

Harlan, J. R. 1971. Agriculture origins: Centers and noncenters. *Science* 74: 468–474.

Hayes, T. E. 1976. Plant classification and nomenclature in Ndumba, Papua New Guinea Highlands. *American Ethnologist* 3: 253–270.

Knuth, R. 1924. Dioscoreaceae. In A. Engler (ed.), *Das Pflanzenreich* IV, 43(87): 1–387. Engelmann, Leipzig, Germany.

Lanaud, C. 1999. Use of molecular markers to increase understanding of plant domestication and to improve the management of genetic resources: Some examples with tropical species. *Plant Genetic Resources Newsletter* 119: 26–31.

Lea, D. A. M. 1966. Yam growing in the Maprik area. *Papua New Guinea Agriculture Journal* 18: 5–15.

Lebot, V. 1999. Biomolecular evidence for plant domestication in Sahul. *Genetic Resources and Crop Evolution* 46: 619–628.

Lebot, V., B. Trilles, J. L. Noyer, and J. Modesto. 1998. Genetic relationships between *Dioscorea alata* L. cultivars. *Genetic Resources and Crop Evolution* 45: 499–508.

Malapa, R. 2000. *Etude de la Diversité Génétique des Cultivars de* D. alata *L. du Vanuatu par les Marqueurs Morpho-Agronomiques et* AFLP. DEA de Génétique, Adaptation et Productions Végétales, ENSA de Rennes, Rennes, France.

Malapa, R., G. Arnau, J. L. Noyer, and V. Lebot. In press. Genetic diversity of the greater yam (*Dioscorea alata* L.) and relatedness to *D. nummularia* Lam. and *D. transversa* Br. as revealed with AFLP markers. *Genetic Resources and Crop Evolution.*

Martin, F. W. 1974. Tropical yams and their potential. Part 3. *Dioscorea alata. USDA Agriculture Handbook* 495: 40.

Martin, F. W. and A. M. Rhodes. 1977. Intra-specific classification of *Dioscorea alata. Tropical Agriculture* (Trinidad) 54: 1–13.

Miege, J. 1952. *Contribution à l'Étude Systématique des* Dioscorea *Ouest Africains.* Thesis, Université de Paris, France.

Mignouna, H. D. 1998. Analysis of genetic diversity in Guinean yams (*Dioscorea* spp.) using AFLP fingerprinting. *Tropical Agriculture* (Trinidad) 75: 224–229.

Mignouna, H. D., R. A. Mank, T. H. N. Ellis, N. van den Bosch, R. Asiedu, M. M. Abang, and J. Peleman. 2002. A genetic linkage map of water yam (*Dioscorea alata* L.) based on AFLP markers and QTL analysis for anthracnose resistance. *Theoretical and Applied Genetics* 105: 726–735.

Ochse, J. J. and R. C. B. van den Brink. 1977. Dioscoreaceae. In A. Asher and B. V. Co (eds.), *Vegetables of the Dutch East Indies,* 229–247. A. Asher & Co., Amsterdam, The Netherlands.

Prain, D. and I. H. Burkill. 1939. An account of the genus *Dioscorea.* 1. Species which turn to the right. *Annals of the Royal Botanical Garden, Calcutta* 14: 211–528.

Ramser, J., C. Lopéz-Peralta, R. Wetzel, K. Weising, and G. Kahl. 1996. Genomic variation and relationships in aerial yam (*Dioscorea bulbifera* L.) detected by random amplified polymorphic DNA. *Genome* 39:17–25.

Sastrapadja, S. 1982. *Dioscorea alata*: Its variation and importance in Java, Indonesia. In J. Miege and S. N. Lyonga (eds.), *Yams. Ignames,* 45–49. Clarendon Press, Oxford, UK.

Smith, B. W. 1937. Notes on cytology and distribution of the Dioscoreaceae. *Torrey Botanical Club Bulletin* 64: 189–197.

Spriggs, M. 1997. The island Melanesia. In P. Bellwood. and I. Glover (eds.), *The People of South-East Asia and the Pacific,* 1–313. Blackwell, Oxford, UK.

SPYN. 2001. *South Pacific Yam Network Annual Report.* CIRAD, Montpellier, France.

Telford, I. R. H. 1986. Dioscoreaceae. *Flora of Australia* 46: 196–202.

Terauchi, R., V. A. Chikaleke, G. Thottappily, and S. K. Hahn. 1992. Origin and phylogeny of Guinea yams as revealed by RFLP analysis of chloroplast DNA and nuclear ribosomal DNA. *Theoretical and Applied Genetics* 83: 743–751.

Terauchi, R., T. Terachi, and K. Tsunewaki. 1991. Intraspecific variation of chloroplast DNA in *Dioscorea bulbifera* L. *Theoretical and Applied Genetics* 81: 461–470.

Tryon, D. T. 1985. The peopling of the Pacific. *Journal of Pacific History* 19: 147–158.

Uline, E. B. 1898. Eine monographie der Dioscoreollon. *Botanische Jahrbücher für Systematik Pflanzengeschichte und Pflanzengeographie* 25: 126–165.

Weightman, B. 1989. *Agriculture in Vanuatu. A Historical Review.* Grosvenor Press, Portsmouth, UK.

Yen, D. E. 1982. The history of cultivated plants. In R. J. May and N. Hank (eds.), *Melanesia: Beyond Diversity,* 281–295. Australian National University, Canberra, Australia.

Zerega, N. J. C., D. Ragone, and T. J. Motley. 2004. Complex origins of breadfruit: Implications for human migrations in Oceania. *American Journal of Botany* 91: 760–766.

PART 4

VARIATION OF PLANTS
UNDER SELECTION

Agrodiversity and Germplasm Conservation

Barbara A. Schaal, Kenneth M. Olsen,
and Luiz J. C. B. Carvalho

CHAPTER 12

Evolution, Domestication, and Agrobiodiversity in the Tropical Crop Cassava

Cassava (*Manihot esculenta*), Euphorbiaceae, is the sixth most important crop globally (Mann, 1997). It is the primary staple crop for more than 500 million people worldwide, serving mostly the poor in tropical developing countries (Best and Henry, 1992). It is the major source of calories in sub-Saharan Africa, where it is grown primarily for its starchy roots, although it can serve as a leaf crop as well (Cock, 1985). Cassava is an inexpensive source of starch and is currently being developed for industrial uses as well as a source of animal feed, primarily in Asia. Nonetheless, most of the world's cassava is consumed by subsistence farmers in Africa and Latin America. In the past cassava was considered an orphan crop. Because the majority of cassava consumers live in poverty with little access to cash, efforts at cassava crop improvement have lacked the economic stimulus that commerce provides. Consequently, much of the basic biology of the crop and its closely related species has gone understudied until recently, in stark contrast to the extensive work on cash crops such as corn, wheat, rice, and soybeans.

Cassava has great potential to increase the food security of people in the developing world. Average yields for cassava in Africa are 8 ton/ha, but potential yields are 80 ton/ha (*FAO News*, November 5, 2002). Unlike most other crops, plants continue to deposit carbohydrates into the storage roots as long as the plant is actively growing. In general, cassava tolerates moderate levels of drought, and it grows well in the nutrient-poor soils of the tropics. Moreover, the crop is easy to plant and grows especially well in marginal agricultural areas of the world. New plants are started from planting sticks; pieces of stem are stuck into the ground, and 6–9 months later the farmer returns to harvest some or all of the storage roots.

Although cassava has many benefits as a crop of the tropics, it has several limitations that challenge the human populations that subsist on it. Historically, the crop has been cultivated mainly to produce cassava flour and starch. Selection for these traits has resulted in the storage root of most cultivars having high caloric content and low vitamin, protein, and mineral levels. People who subsist solely on cassava are subject to nutritional deficiency diseases such as kwashiorkor and night blindness. Other challenges include root susceptibility to postharvest deterioration. The roots of cassava plants are nonperenniating, and spoilage occurs quickly after harvest through physiological deterioration caused initially by oxidation followed by colonization and growth of various saprophytic microorganisms. In addition, cassava plants are susceptible to various pathogens, including a bacterial blight and the African cassava mosaic virus, a devastating pathogen found in Africa that can reduce yields by as much as 70%. The challenge for crop breeders is to increase nutrition and pathogen resistance while reducing postharvest deterioration. Finally, the value of cassava as a cash crop is low. Many other crops provide inexpensive sources of starch, and the quality of cassava starch is low, making it less desirable as a source of raw material for the food industry. The lack of a cash crop for many farmers of the developing world has significant effects on the well-being of families, who cannot buy such things as medicines and books without cash. Any modification of cassava to enhance its value as a market crop would have a direct and positive effect on the lives of poor families. As we discuss in this chapter, landraces of cassava that sequester carbohydrates other than amylose have great potential as cash crops.

Traditionally, many plant breeders have turned to landraces, wild ancestors, and closely related species as a source of traits for future crop improvement. For example, some of the rice varieties developed during the Green Revolution contained pathogen-resistant genes from rice's wild progenitor, *Oryza rufipogon*. Our own work with cassava has centered on the native and agricultural biodiversity of *Manihot*. We have investigated the wild ancestor of cassava, determined the site of domestication, and studied the relationship of the cassava to its wild relatives, and one of us (L.J.C.B.C.) has discovered and characterized the biodiversity of landraces in the Amazon Basin. These studies identify and describe potential reservoirs of germplasm for the improvement of cassava. In addition, cassava serves as a model system for understanding evolution in a recently arisen plant genus and for understanding domestication in a perennial, clonally reproducing crop.

Systematics of *Manihot*: Cassava's Close Relatives

Manihot (Euphorbiaceae) is a genus of an estimated 98 neotropical species (Rogers and Appan, 1973). Plants range in habit from herbs to small trees. There are two centers of diversity for the genus: one in Brazil, with some 80 species, and another in Mexico, with approximately 17 species. Four major centers of species diversity are found in Brazil. The region of greatest diversity is in the Central Plateau, with 58 of 80 species. Eight species are found each in the northeast and southeast of Brazil, with six species localized in the Amazon region (for distribution maps, see www.cenargen. embrapa.br). The taxonomy of *Manihot* is enigmatic. Early work by Rogers (1965) and Rogers and Appan (1973) noted the overlap in morphology between species, the phenotypic plasticity of characters, the lack of chromosome variation, and the limited number of taxonomically informative traits. Rogers and Appan and, more recently, Allem (Allem, 1987, 1994, 1999) have struggled with a morphologically based taxonomy. The paucity of reliable taxonomic traits has made studies on the origin of cassava difficult because based on morphology alone, several species of *Manihot* from Mexico, Central America, and South America are potential ancestors, with no single species being morphologically so similar to cassava to be unambiguously assigned the wild ancestor.

The lack of clear morphological affinities of cassava with any single wild *Manihot* species led to hypotheses that cassava may be a hybrid derivative, a hybrid between two to several species. This "compilospecies" origin is most closely associated with the species complex in Mexico, but a similar hybrid origin has also been suggested among South American species (Rogers, 1963, 1965; Rogers and Appan, 1973; Ugent et al., 1986; Sauer, 1993). As an alternative to a multiple-species origin, Costa Allem in the early 1990s suggested that wild *Manihot* populations occurring in Brazil were so similar to domesticated cassava that they were part of the same species (Allem, 1987, 1994). These wild populations, *M. esculenta* ssp. *flabellifolia,* differ from domesticated cassava almost entirely in traits that appear to be associated with domestication such as shortened internodes, thickened stems, swollen leaf scars, a more erect stature, and increased size of storage roots.

To address the issue of cassava's origins, we initially used a traditional molecular phylogenetic approach to understanding species relationships (Hillis et al., 1996). We obtained collections of cassava species from Mesoamerica and from South America, including species that had been previously identified as potential progenitors of cassava such as *M. aesculafolia* from Mexico, *M. cartagineneis* from Colombia, and a suite of species that Allem considered the secondary gene pool for cassava, *M. pilosa, M. triphylla,* and *M. esculenta* ssp. *flabellifolia.* We used several DNA sequences to reconstruct the phylogeny of this group of species, including chloroplast DNA, and nuclear regions such as the internal transcribed spacer region of ribosomal DNA, linamerase, and aspartate transaminase. Typical results are shown in figure 12.1. First and most notable was the overall sequence similarity between species within the genus. *Manihot* is thought to be a recently arisen genus, based on the morphological similarity of species and the lack of chromosome differentiation. These molecular results indicating low sequence divergence (1–2% in some cases) are consistent with a hypothesis of recent origin. In many cases the relationship between species could not be resolved, as indicated by polytomies on the tree. However, in all cases the phylogenies consisted of two clades, supported by bootstrap

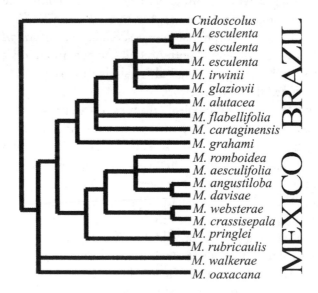

FIGURE 12.1 Phylogeny reconstruction in *Manihot:* maximum parsimony tree using internal transcribed spacer. Only the node separating the Mexican and Brazilian species is well supported (bootstrap >90%).

values greater than 90%; one clade represents the South American species, and the other represents species from Mesoamerica. These results are similar to Bertram's (1993) findings from chloroplast DNA restriction site data that also indicated South American and Mesoamerican lineages within the genus. Finally, in every case, cassava is nested within the South American clade.

Another result noted in our molecular sequencing studies was the almost identical sequence similarity between cassava and *M. esculenta* ssp. *flabellifolia*. Although species relationships could not be statistically resolved within the South American clade, even with multiple loci, in every case cassava showed greater sequence similarity to *M. esculenta* ssp. *flabellifolia* than to other *Manihot* species. The close genetic similarity of this subspecies to cassava has been noted in a number of amplified fragment length polymorphisms (AFLPs) and other DNA marker studies (Fregene et al., 1994; Roa et al., 1997). These molecular data and Allem's morphological data (Allem, 1994), which first indicated *M. esculenta* ssp. *flabellifolia* as a potential ancestor, make a strong case for examining the relationship between populations of this native taxon and domesticated cassava by high-resolution population genetic analyses.

Cassava's Wild Ancestor

M. esculenta ssp. *flabellifolia* occurs in the transition zone between the southern Amazon forest and the drier Cerrado region of Brazil and Peru. Populations are found in gallery forests, the mesic forest patches often associated with river drainages. *M. esculenta* ssp. *flabellifolia* is a clambering understory vine-like shrub. A total of 27 populations of *M. esculenta* ssp. *flabellifolia* from its entire known range in Brazil were collected for genetic analysis (Olsen, 2000). In addition, several populations of *M. pruinosa* were also collected to test the traditional hybridization hypotheses. *M. pruinosa* is a member of the secondary gene pool of cassava and is the closest relative of cassava that over-laps in distribution, within the eastern part of *M. esculenta* ssp. *flabellifolia*'s geographic range (Allem, 1999). Domesticated cassava was represented by a collection of 20 landraces of cassava, the "world core collection" maintained by the Centro International de Agricultura Tropical in Cali, Colombia.

Populations were scored for two distinct types of genetic markers (Olsen and Schaal, 1999, 2001). We used different markers because both the precision and type of evolutionary inferences that can be drawn vary depending on marker choice. First, DNA sequence variation of the gene glyceraldehyde

3-phosphate dehydrogenase (*G3pdh*) was used for a historical, phylogeographic analysis. An analysis of sequence variation can provide information on the mutational relationships between variants (haplotypes), which are related by a haplotype network. In turn, a haplotype network can be used to infer historical, evolutionary processes and provide information on the geographic sorting of lineages. Second, we used a suite of five microsatellite loci, with a total of 73 alleles, to analyze genetic variation and population differentiation. Microsatellites provide high levels of variability, and alleles are codominant, which allows detailed population genetic analysis. The comparative analysis with these two markers sought to document that *M. esculenta* ssp. *flabellifolia* was indeed the wild progenitor of cassava and to determine the geographic site of domestication.

Figure 12.2 shows the results of the phylogeographic study of the *G3pdh* locus. Variants of *G3pdh* (haplotypes) are shown on the haplotype tree. The tree orders haplotypes based on their mutational relationships, with each line connecting haplotypes representing a single nucleotide substitution. Haplotype trees can be constructed either by hand or by a computer program, using parsimony to order the haplotypes. Several conclusions are apparent from figure 12.2. First, cassava contains much less haplotype diversity than does the wild taxon *M. esculenta* ssp. *flabellifolia*. Only 6 of the 23 haplotype variants of *M. esculenta* are found in domesticated cassava. The haplotypes of cassava are a subset of those found in *M. esculenta* ssp. *flabellifolia*, with the exception of a single haplotype not detected in the wild populations. These data are consistent with a hypothesis of cassava being derived from *M. esculenta* ssp. *flabellifolia*. Second, there appears to be no evidence of hybridization with *M. pruinosa*. The haplotypes of *M. pruinosa*, often shared with *flabellifolia*, are not observed in cassava. The absence of *M. pruinosa* haplotypes in domesticated cassava weakens the case for hybridization being a dominant process in cassava domestication. Finally, the geographic location of domestication can be inferred by the distribution of alleles contained both in cassava and in wild *M. esculenta* ssp. *flabellifolia* populations. Populations of *flabellifolia* that contain alleles also found in cassava are geographically limited to the southern and western parts of *flabellifolia*'s range, the transition zone between the humid Amazon forest and the dry Cerrado (figure 12.3).

One of the criticisms of evolutionary studies based on single gene sequences is that the sequence may reveal only a gene tree. That is, the relationships between haplotypes of a gene are reflected, not necessarily the relationships between populations. The organismal tree, the phylogeny of the species or

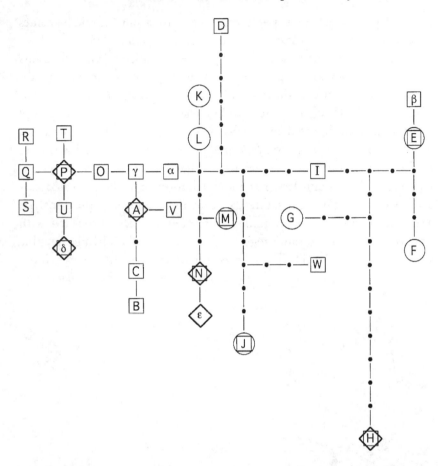

FIGURE 12.2 Haplotype network for *G3pdh*. Letters represent observed haplotypes. Lines represent single mutations, and dots are haplotypes not detected in the sample. Squares represent haplotypes of *M. esculenta* ssp. *flabellifolia*; diamonds represent haplotypes found in domesticated cassava. Circles are haplotypes found in *M. pruinosa*.

populations, may be different from the relationships between alleles at a locus. In phylogenetic studies multiple genes, often from different genomes such as the nuclear, chloroplast, or mitochondrion, are used to address this concern. Population-level studies often use high-resolution markers to assess relationships. Microsatellites and some other markers, such as AFLPs, represent a broader segment of the genome and thus provide good distance measures (Gepts, 1993). (Distance measures have their own assumptions and limitations, particularly for inferring historical relationships.) A disadvantage of microsatellites is the difficulty of developing a set of markers for a specific plant species; we were fortunate to have a suite of microsatellites already developed for cassava (Chavarriaga-Aguirre et al., 1998).

In our microsatellite study, the analyses were consistent with the results of the phylogeographic study. First, cassava appeared to contain a subset of the variation contained in wild *M. esculenta* ssp. *flabellifolia* populations. Only 15 of the observed 73 alleles found in *M. esculenta* ssp. *flabellifolia* were detected in cassava, again suggesting that cassava is a derivative of *M. esculenta* ssp. *flabellifolia*. As an aside, if *flabellifolia* were feral cassava populations, as has been suggested by some, these populations should contain less, not more variation than domesticated cassava. Second, the microsatellite data and the phylogeographic analysis are concordant: Cassava contains microsatellite alleles associated with populations from the same region of Brazil as in the phylogeographic study (figure 12.3). Thus we have concluded from these two independent data sets that cassava was domesticated from *M. esculenta* ssp.

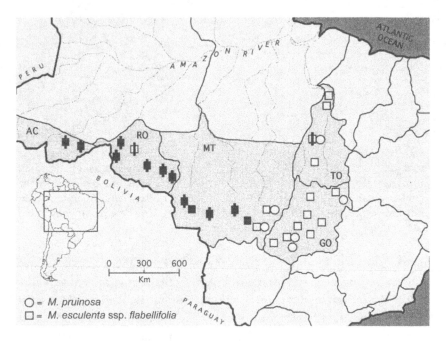

FIGURE 12.3 Wild *Manihot* populations and their genetic relationship with cultivated cassava. Black shading indicates populations containing *G3pdh* haplotypes detected in cassava samples. Vertical bars indicate populations clustered with cassava accessions in a distance analysis of microsatellite allele frequencies. AC = Acre, GO = Goiás, MT = Mato Grosso, RT = Rondônia, TO = Tocantins.

flabellifolia in the southern region of the transition zone between the lower Amazon forest and the Cerrado. Interestingly, this region is thought to be the location of an agricultural complex that includes domestication of the peanut, chili pepper, and jack bean (Pearsall, 1992; Piperno and Pearsall, 1998). Archaeological sites from this region suggest very early agricultural settlements, at about 4800 years BP (Miller, 1992). Finally, although neither the *G3pdh* data nor the microsatellite data indicate that substantial interspecific hybridization has taken place during the origin of cassava, the study cannot totally exclude hybridization, either before or, more importantly, after domestication. There are numerous reports of natural hybridization among *Manihot* species, based primarily on morphological evidence. Morphology of *Manihot* species is notoriously plastic, making such studies difficult to evaluate. Additional genetic studies would be extremely useful for understanding the role of hybridization in the evolution of the genus.

Morphology and Domestication

Determining the wild ancestor of cassava allows us to examine changes associated with domestication. Morphological distinctions between *M. esculenta* ssp. *flabellifolia* and modern cassava cultivars identify traits that have been altered during domestication. *M. esculenta* ssp. *flabellifolia* is a clambering vinelike shrub with a rudimentary storage root. The species is highly plastic; if the surrounding forest is removed, regrowth forms an erect shrub with a hard storage root that contains small amounts of starch and high fiber content. This erect habit is much more similar to that of domesticated cassava than the vinelike form. A second change associated with domestication is the development of a fleshy storage root high in starch, certainly the most striking difference between the ancestor and crop. A third major change is related to vegetative propagation of the crop. *M. esculenta* ssp. *flabellifolia* flowers freely, whereas in modern cassava cultivars flowering may be limited, often to small number of flowers and partial fruit set. This reduction in flowering of the modern varieties could result from the long history of vegetative propagation of the crop; many plants show a trade-off between sexual and vegetative reproduction. This later change in particular has implications for modern cassava germplasm collections around the world. For instance, most of the germplasm collections of cassava are based on local cultivars, and few accessions are derived from crosses of conventional breeding programs, in part because of flowering limitations that in turn restrict the ability to cross specific cultivars. Thus finding and identifying new landraces of cassava that have both good flowering ability and useful and variable traits are critical for enhancing cassava as a staple crop.

Agrobiodiversity in the Amazon

Because cassava was domesticated in the lower Amazon region, one expects that this would be a geographic area where humans have had a long traditional association with cassava. Many visitors to the region have noted the diverse uses of cassava among villages. Cassava meal is used in sauces, flour is baked into flatbreads, the leaves are ground and cooked, and several fermented drinks are produced from the storage root. This diversity of uses is quite different from the use of cassava cultivated in much of the rest of the world, where it is grown as a source of starch for flour. Moreover, in the Amazon cassava is grown mostly in small, intercropped fields or in backyard gardens that often contain several distinct landraces of cassava. Again this is in contrast to the monoculture of cassava observed for improved varieties grown in many parts of the world. Given that both the uses of cassava and the mode of cultivation are more diverse in the Amazon region (Carvalho and Schaal, 2001), there may be underlying variability in key agronomic characters for cassava that could be useful in addressing the challenges cassava faces as a crop. The discovery of the site of domestication allows agricultural biologists to focus their areas of study and collection.

One of us (L.J.C.B.C.) has made several field trips to the Brazilian Amazon to learn of new uses and varieties of cassava and to collect diverse storage root variants. Smallholder farmers, isolated rural communities, local markets, and regions with different systems of cassava cultivation were visited in the states of Mato Grosso, Rondonia, Amazon, Para, Marajo Island, and Amapa. The landraces in this region showed an astounding diversity in unusual storage root traits related to root shape, color, and structure as well as carbohydrate content and type (figure 12.4). A field test for starch based on iodine allows one to identify starch in a cross-section of the storage root and to identify the type of starch and the pattern of starch distribution. Cross-sections of various landraces of cassava clearly show diversity in both the presence or absence of starch and the pattern of starch distribution (figure 12.4). Biochemical studies of the carbohydrates of these landraces revealed a new type of cassava, sugary cassava, which contained large amounts of free sugar (primarily glucose). These landraces also contained amylose-free starch and glycogen-like starch (phytoglycogen) (Carvalho et al., 2004). Many of these landraces

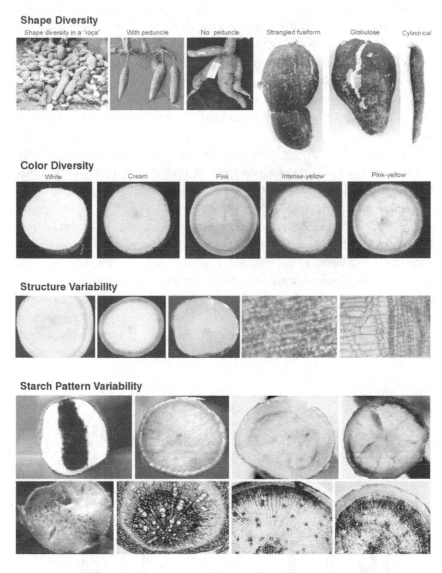

Shape Diversity

Shape diversity in a "roça" With peduncle No peduncle Strangled fusiform Globulose Cylindrical

Color Diversity

White Cream Pink Intense-yellow Pink-yellow

Structure Variability

Starch Pattern Variability

FIGURE 12.4 Shape, color, structure, and starch pattern variation between landraces of cassava. Cassava varies both in overall root color and shape and in the deposition of starch. Colors range from white to pink and intense yellow. Cassava roots also vary in the pattern of secondary xylem and parenchyma cells in the root. The dark regions in the lower 8 photographs are cells stained with iodine to detect starch. The presence of starch and the pattern of deposition vary between landraces.

with unique carbohydrates have been domesticated for specific local uses. For example, the landrace that accumulates phytoglycogen is used as a food for very young children. Phytoglycogen is a highly branched molecule with short linear polymerized glucose, which makes it soluble in cold water and much easier to digest than common starch, amylopectin. The sugary cassava, with high amounts of free sugar, is used to prepare a glucose syrup for local dessert dishes and a sweet smoked cassava cake. The sugary cassava is also used to produce a fermented alcohol drink used during the community's religious ceremonies.

Landraces also vary in pigments found in the storage root. Color variants of cassava often are observed in several germplasm collections around the world (India; Moorthy et al., 1990; Brazil: Ortega-Flores, 1991; Guimaraes and Barros, 1971; Marinho et al., 1996; and Colombia: Iglesias et al., 1997). The variants found in the Amazon are unusual in their diversity of colors, their carotenoid content, and their tissue-specific patterns of pigment distribution across the root. Figure 12.4 shows the range of color variants from standard white cassava to intense yellow, cream, and pink cassava. These color variants are closely associated with the type of carotenoid present. Biochemical analysis (figure 12.5a) indicates accumulation of a number of different carotenoid forms, including β-carotene, lycopene, and lutein in amount higher than previously reported. This diversity in the

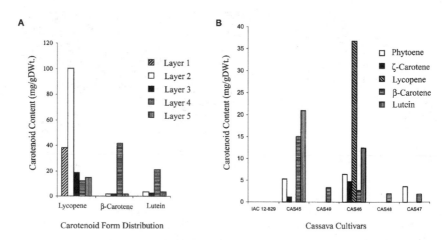

FIGURE 12.5 Diversity in carotenoid forms and content in pigmented cassava in the Amazon. **(a)** Different carotenoid forms and content in the storage root of local landraces. **(b)** Distribution pattern of carotenoid content across storage root in local landraces.

amount and type of carotenoids underscores the high potential of cassava for development of an improved staple food combining macronutrients (starch) with micronutrients (β-carotene, provitamin A). Importantly, the accumulation of β-carotene in the intense yellow cassava is accompanied by a higher protein content in the storage root. The intense yellow cassava has 40% more aqueous extractable protein content than does standard white cassava (figure 12.4), representing an opportunity for cassava to also supply protein as a staple food crop. The color diversity has been used locally in the Amazon in several food preparations. For example, the cassava juice called *tucupi*, made from the intense yellow cassava rich in β-carotene, is used in a soup. The fresh yellow roots are boiled and served in many dishes. Colored cassava varieties also are pickled as a way to preserve the roots.

How do these carbohydrate and pigment biochemical variants arise? Analysis of enzymes in the starch synthesis pathway and protein blot analysis indicate that in several of the novel carbohydrate varieties, specific enzymes of the starch biosynthetic pathway are no longer active, and their corresponding proteins are no longer present in a crude protein extract. Thus it appears that mutations that have reduced or eliminated key enzyme functions have in turn altered the flow of carbohydrates through the starch metabolic pathway. The absence of enzyme activity results in the accumulation of novel carbohydrates. Gene expression analysis shows that the sugary cassava that accumulates phytoglycogen no longer expresses the gene coding for starch branching enzyme I (Carvalho et al., 2004). In the case of pigmented cassava, we speculate that the accumulation of novel pigments is also the result of either natural mutations of key enzymes in the pigment synthesis pathway or mutations in the sequestering protein of the chromoplast in a particular cassava variety. These alternative explanations are being explored. It is quite likely that once these mutations initially occurred within a landrace, they are selected by the native peoples to enhance the concentration of either type of novel compound.

Conclusions

The results of what were initially solely academic studies have proven useful in an applied sense. First, the identification of *M. esculenta* ssp. *flabellifolia* as the wild progenitor of cassava allows one to target both the species and specific populations of *Manihot* that potentially have the most important germplasm for cassava improvement. These species and populations may contain useful genes for important traits of agronomic interest. Second, populations of the wild ancestor as well as primitive landraces and modern mass-selected landraces could improve our understanding of the morphological and genetic processes associated with domestication. Third, focusing landrace collection on the Amazon region has identified several extremely important landraces that can be used to address some of the challenges that confront human populations subsisting on cassava. In addition, these landraces also offer opportunities for better understanding the biological processes that lead to useful biochemical variants. Because the carbohydrate diversity of cassava landraces is the result of specific enzymes losing function, could gene knockouts for other metabolic pathways lead to other useful biochemical variants? The high-sugar cassava and new carbohydrate variants could serve as a cash crop for poor farmers. The β-carotene variant could be very useful for preventing night blindness caused by vitamin A deficiency, a scourge for many poor populations in the tropical developing world.

Many more landraces and the biodiversity of native populations remain to be characterized. Unfortunately, both native populations of *M. esculenta* ssp. *flabellifolia* and traditional landraces are threatened as more land is cleared for modern agriculture. Conserving the agrobiodiversity of cassava should be of high priority, given the value and potential benefit that can be derived from cassava's germplasm resources.

References

Allem, A. C. 1987. *Manihot esculenta* is a native of the Neotropics. *Plant Genetic Resources Newsletter* 71: 22–24.

Allem, A. C. 1994. The origin of *Manihot esculenta* Crantz (Euphorbiaceae). *Genetic Resources and Crop Evolution* 41: 133–150.

Allem, A. C. 1999. The closest wild relatives of cassava (*Manihot esculenta* Crantz). *Euphytica* 107: 123–133.

Bertram, R. B. 1993. *Application of Molecular Techniques to Genetic Resources of Cassava (*Manihot esculenta Crantz–Euphorbiaceae): Interspecific Evolutionary Relationships and Intraspecific Characterization*. Ph.D. dissertation, University of Maryland at College Park, MD, USA.

Best, R. and G. Henry. 1992. Cassava: Towards the year 2000. In W. M. Roca and A. M. Thro (eds.), *Report of the First Meeting of the International Network for Cassava Genetic Resources,* 3–11. Centro Internacional de Agricultura Tropical, Cali, Colombia.

Carvalho, L. J. C. B. and B. A. Schaal. 2001. Assessing genetic diversity in cassava (*Manihot esculenta* Grantz) germplasm collection in Brazil using PCR-based markers. *Euphytica* 120: 133–142.

Carvalho, L. J. C. B., C. R. B. de Souza, J. C. M. Cascardo, and C. B. Junior. 2004. Identification and characterization of a novel cassava (*Manihot esculenta* Crantz) clone with high free sugar content and novel starch. *Plant Molecular Biology* 56: 643–659.

Chavarriaga-Aguirre, P., M. M. Maya, M. W. Bonierbale, S. Kresovich, M. A. Fregene, J. Tohme, and G. Kochert. 1998. Microsatellites in cassava (*Manihot esculenta* Crantz): Discovery, inheritance and variability. *Theoretical and Applied Genetics* 97: 493–501.

Cock, J. H. 1985. *Cassava: New Potential for a Neglected Crop.* Westfield, London, UK.

FAO News. 2002. Partnership formed to improve cassava, staple food of 600 million people. November 5, Food and Agriculture Organization of the United Nations. www.fao.org/english/newsroom/news/2002/10541-en.html.

Fregene, M. A., J. Vargas, J. Ikea, F. Angel, J. Tohme, R. A. Asiedu, M. O. Akoroda, and W. M. Roca. 1994. Variability of chloroplast DNA and nuclear ribosomal DNA in cassava (*Manihot esculenta* Crantz) and its wild relatives. *Theoretical and Applied Genetics* 89: 719–727.

Gepts, P. 1993. The use of molecular and biochemical markers in crop evolution studies. In M. K. Hecht, R. J. MacIntyre, and M. T. Clegg (eds.), *Evolutionary Biology*, Vol. 27, 51–94. Plenum, New York, NY, USA.

Guimaraes, M. L. and M. S. C. Barros. 1971. Sobre a ocorrência de caroteno em variedades de mandioca amarela. *Boletim Técnico da Divisão de Tecnologia Agricultura e Alimentção* 4: 1–4.

Hillis, D. M., B. K. Maple, A. Larson, S. K. Davis, and E. A. Zimmer. 1996. Nucleic acids IV: Sequencing and cloning. In D. M. Hillis, C. Moritz, and B. K. Maple (eds.), *Molecular Systematics*, 321–381. Sinauer, Sutherland, MA, USA.

Iglesias, C., J. Mayer, L. Chavez, and F. Calle. 1997. Genetic potential and stability of carotene content in cassava roots. *Euphytica* 94: 367–373.

Mann, C. 1997. Reseeding the Green Revolution. *Science* 277: 1038–1043.

Marinho, H. A., J. J. B. N. Xavier, R. M. Miranda, and J. S. Castro. 1996. Estudos sobre carotenóides com atividade de pro-vitamina "A" em cultivares de mandioca (*Manihot esculenta* Crantz) em ecossistema de terra firme de Manaus, Amazonas, Brasil. *Acta Amazônica* 1: 127–136.

Miller, E. T. 1992. *Archaeology in the Hydroelectric Projects of Eletronorte: Preliminary Results.* Centrais Eletricas do Norte do Brasil S.A., Brasília, Brazil.

Moorthy, S. N., J. S. Jos, R. B. Nair, and M. T. Sreekumari. 1990. Variability of β-carotene content in cassava germplasm. *Food Chemistry* 36: 233–236.

Olsen, K. M. 2000. *Evolution in a Recently Arisen Species Complex: Phylogeography of Manihot esculenta Crantz (Euphorbiaceae).* Ph.D. dissertation, Washington University, St. Louis, MO, USA.

Olsen, K. M. and B. A. Schaal. 1999. Evidence on the origin of cassava: Phylogeography of *Manihot esculenta. Proceedings of the National Academy of Sciences (USA)* 96: 5586–5591.

Olsen, K. M. and B. A. Schaal. 2001. Microsatellite variation in cassava and its wild relatives: Further evidence for a southern Amazonian origin of domestication. *American Journal of Botany* 88: 131–142.

Ortega-Flores, C. I. 1991. *Carotenoides com atividade pro-vitamica A e teores de cianeto em diferentes cultivares de mandioca no Estado de São Paulo.* M.S. thesis, Universidade de São Paulo, Brazil.

Pearsall, D. M. 1992. The origins of plant cultivation in South America. In C. W. Cowan, P. J. Watson, and N. L. Benco (eds.), *The Origins of Agriculture: An International Perspective*, 173–205. Smithsonian Institution Press, Washington, DC, USA.

Piperno, D. R. and D. M. Pearsall. 1998. *The Origins of Agriculture in the Lowland Neotropics.* Academic Press, New York, NY, USA.

Roa, A. C., M. M. Maya, M. C. Duque, J. Tohme, A. C. Allem, and M. W. Bonierbale. 1997. AFLP analysis of relationships among cassava and other *Manihot* species. *Theoretical and Applied Genetics* 95: 741–750.

Rogers, D.J. 1963. Studies of *Manihot esculenta* Crantz and related species. *Bulletin of the Torrey Botanical Club* 90: 43–54.

Rogers, D.J. 1965. Some botanical and ethnological considerations of *Manihot esculenta*. *Economic Botany* 19: 369–377.

Rogers, D.J. and S.G. Appan. 1973. *Manihot and Manihotoides (Euphorbiaceae): A Computer Assisted Study*. Hafner Press, New York, NY, USA.

Sauer, J.D. 1993. *Historical Geography of Crop Plants: A Select Roster*. CRC, Boca Raton, FL, USA.

Ugent, D., S. Pozorski, and T. Pozorski. 1986. Archaeological manioc (*Manihot*) from Coastal Peru. *Economic Botany* 40: 78–102.

David M. Spooner and
Wilbert L. A. Hetterscheid

CHAPTER 13

Origins, Evolution, and Group Classification of Cultivated Potatoes

Potato is the world's most productive vegetable and provides a major source of nutrition and income to many societies. The story of the potato begins with wild potato species that look very similar to the cultivated potato today. Wild potatoes are widely distributed in the Americas from the southwestern United States to southern Chile, but the first cultivated potatoes probably were selected from populations in the central Andes of Peru and Bolivia sometime between 6000 and 10,000 years ago. These wild species and thousands of indigenous primitive cultivated landrace populations persist throughout the Andes, with a second set of landrace populations in Chiloé Island, the adjacent islands of the Chonos Archipelago, and mainland areas of lowland southern Chile. These Chilean populations probably arose from Andean populations that underwent hybridization with the wild species *Solanum tarijense,* found in southern Bolivia or northern Argentina. The first record of potato out of South America is from the Canary Islands in 1562, and the potato rapidly became cultivated in Europe and then worldwide. Selection and breeding transformed the potato into a set of modern cultivars with more uniform colors and shapes and with improved agronomic qualities such as greater disease resistance and yield. Current opinion invokes the earliest European introductions from Andean landraces, with the introduction of Chilean landraces only after late blight disease killed many potato populations in Europe in the 1840s. We suggest early introductions of cultivated potatoes from both the Andes and Chile, with the Chilean landraces becoming the predominant modern breeding stock long before the 1840s. There is also a controversy about the classification of potato as Linnean species treated under the International Code of Botanical Nomenclature (ICBN) or as Groups under the International Code of Nomenclature of Cultivated Plants (ICNCP). We support a recent Group classification of the landrace populations and here propose the first Group classification of the modern cultivars, placing all under the single name (denomination class) of *Solanum tuberosum*.

Cultivated Potato in the Context of Tuber-Bearing Species in *Solanum* Section *Petota*

The cultivated potato and its tuber-bearing wild relatives, (*Solanum* L. sect. *Petota* Dumort.) are monophyletic (Spooner et al., 1993) and are distributed from the southwestern United States to central Argentina and adjacent Chile (Hijmans and Spooner, 2001). Indigenous primitive cultivated (landrace) potatoes are grown throughout middle to high (about 3000–3500 m) elevations in the Andes from western Venezuela to northern Argentina, and then in south-central Chile, concentrated in the Chonos Archipelago. Landrace populations in Mexico and Central America are recent, post-Columbian introductions (Hawkes, 1967; Ugent, 1968; Glendinning, 1983). Potatoes can be divided into three artificial groups based entirely on use: wild species, cultivated indigenous landrace populations growing in the Andes and southern Chile, and modern cultivars initially developed in Europe in the 1500s and later spread worldwide. The landrace populations are highly diverse, with a great variety of shapes and skin and tuber colors not often seen in modern varieties (figure 13.1). There are fewer than 200 wild species (Spooner and Hijmans, 2001).

Ploidy levels in *S. tuberosum* L. and in section *Petota* range from diploid ($2n = 2x = 24$), to triploid ($2n = 3x = 36$), to tetraploid ($2n = 4x = 48$), to pentaploid ($2n = 5x = 60$); the wild species also have hexaploids ($2n = 6x = 72$). This chapter focuses on the origin and taxonomy of *S. tuberosum*, beginning with its selection from wild Andean species in the *S. brevicaule* complex, to the origin of Andean and Chilean landraces, to first introductions of Andean and Chilean landraces to Europe, to the current breeding efforts of modern cultivars.

Hawkes (1990) provided the last attempt to formally classify wild potatoes and recognized 21 series, which included tuber-bearing and non–tuber-bearing species. Studies by Spooner and Sytsma (1992), Spooner et al. (1993), and Spooner and Castillo (1997) showed that the non–tuber-bearing species do not belong to section *Petota* and that not all Hawkes's series are monophyletic.

Origin of Cultivated Potatoes from the *S. brevicaule* Complex

We lack well-resolved multigene phylogenies to divide section *Petota* into formal taxonomic groups, but one phenetic group, the *S. brevicaule* complex, has long attracted the attention of biologists because of its similarity to cultivated potatoes (Correll, 1962; Ugent, 1970; Grun, 1990; Van den Berg et al., 1998; Miller and Spooner, 1999). Some members of this group, endemic to central Peru, Bolivia, and northern Argentina, probably were ancestors of the landraces. The species in the complex share the pinnately dissected leaves, round fruits, and rotate to rotate-pentagonal corollas of cultivated potato and are largely sexually compatible with each other and with cultivated potato (Hawkes, 1958; Hawkes and Hjerting 1969, 1989; Ochoa, 1990, 1999; Huamán and Spooner, 2002). The complex includes diploids, tetraploids, pentaploids, and hexaploids. Most are weedy plants, sometimes occurring in or near cultivated potato fields, from about 2500–3500 m. It is so hard to identify species in the group that experienced potato taxonomists Hawkes and Hjerting (1989) and Ochoa (1990) provide different identifications to identical collection numbers of the *S. brevicaule* complex in 38% of the cases (Spooner et al., 1994). Many species grow as weeds in or adjacent to cultivated potato fields and form crop–weed complexes (Ugent, 1970). Morphological data (Van den Berg et al., 1998) and single- to low-copy nuclear restriction fragment length polymorphism data (Miller and Spooner, 1999) failed to clearly differentiate wild species in the complex from each other or from most landraces, and the most liberal taxonomic interpretation of these studies was to recognize only three wild taxa: the Peruvian populations of the *S. brevicaule* complex, the Bolivian and Argentinean populations of the *S. brevicaule* complex, and *S. oplocense*. However, even these three groups could be distinguished only by computer-assisted use of widely overlapping character states, not by species-specific characters (a polythetic morphological species concept). Accordingly, it is difficult to designate species-specific progenitors of the landraces, as Hawkes (1990) has done by designating *S. leptophyes* Bitter and *S. sparsipilum* (Bitter) Juz. and Bukasov as progenitors of the cultivated diploid *S. stenotomum*.

FIGURE 13.1 Representative landraces **(A)** from the Andes (from Graves, 2001) and **(B)** from Chile (courtesy of Andres Contreras, Universidad Austral de Chile) and **(C)** modern cultivars (USDA Agricultural Research Magazine image gallery, www.ars. usda.gov/is/graphics/photos/).

Ploidy Level and Gene Flow Within and Between Cultivated and Wild Species

Bukasov (1939) was the first to count chromosomes of the cultivated potatoes and discovered diploids, triploids, tetraploids, and pentaploids. Ploidy level soon became a major character to distinguish one cultivated species from another. Cultivated potato fields in the Andes contain mixtures of landraces at all ploidy levels (Ochoa, 1958; Jackson et al., 1980; Brush et al., 1981; Johns and Keen, 1986; Quiros et al., 1990, 1992; Zimmerer, 1991), which often co-occur and hybridize with wild potato species (Ugent, 1970; Grun, 1990; Rabinowitz et al., 1990). Watanabe and Peloquin (1989, 1991) showed both diploid and unreduced gametes to be common in the wild and cultivated species, potentially allowing gene transfer between different ploidy levels. The boundary between cultivated and wild often is vague, and some putative wild species may be escapes from cultivation (Spooner et al. 1999).

Treatment of Cultivated Potatoes as Linnean Taxa

Cultivated potatoes have been classified as species under the ICBN (Greuter et al., 2000). The widely used species classification of Hawkes (1990) recognizes seven cultivated species (and subspecies): *S. ajanhuiri, S. chaucha, S. curtilobum, S. juzepczukii, S. phureja* ssp. *phureja, S. phureja* ssp. *hygrothermicum, S. phureja* ssp. *estradae, S. stenotomum* ssp. *stenotomum, S. stenotomum* ssp. *goniocalyx, S. tuberosum* ssp. *andigenum* (as *andigena*), and *S. tuberosum* ssp. *tuberosum*. In contrast, Ochoa (1990, 1999) recognizes 9 species and 141 infraspecific taxa (subspecies, varieties, and forms, including his unlisted autonyms) for the Bolivian cultivated species alone, and Russian potato taxonomists Bukasov (1971) and Lechnovich (1971) recognize 21 cultivated species, including separate species status for *S. tuberosum* ssp. *andigenum* and ssp. *tuberosum* (as *S. tuberosum*) (Huamán and Spooner, 2002).

Treatment of Cultivated Potatoes as Groups

Dodds (1962) suggested that there was poor morphological support for most cultivated species, and he recognized only *S. ×curtilobum, S. ×juzepczukii,* and *S. tuberosum,* with five Groups recognized in the latter. The classifications of Dodds (1962) and Hawkes (1990) are regularly used today, creating confusion among users. Groups are classification categories used by the ICNCP (Brickell et al., 2004) to group cultivated plants with traits that are of use to agriculturists. The term *Group* replaces *Cultivar Group* of the prior ICNCP (Trehane et al., 1995). The ICNCP associates cultivated plant names with denomination classes. A denomination class is a nomenclatural device found in the ICNCP, not the ICBN. It is defined (ICNCP Article 5) as a taxon, or a designated subdivision of a taxon, or a particular Group, within which cultivar epithets must be unique. The botanical genus is the denomination class by default. However, *S. tuberosum* is the denomination class recognized by the International Union for the Protection of New Varieties of Plants (UPOV) as a tool in naming potato cultivars in countries that signed the UPOV treaty and as such possess the mechanism of breeders' rights protection.

Huamán and Spooner (2002) studied the morphological distinction of the potato landraces with numerical phenetics and showed a gradation of support for the cultivated species of Hawkes (1990). For example, the best support was shown for *S. ajanhuiri, S. chaucha, S. curtilobum, S. juzepczukii,* and *S. tuberosum* ssp. *tuberosum,* but there was little or no support for the other six taxa. However, most characters, except tuber dormancy for *S. phureja* ssp. *phureja* and relative position of pedicel articulation for *S. ajanhuiri, S. curtilobum,* and *S. juzepczukii,* overlap extensively with those of other species. In other words, the only morphological support is provided by a complex of characters, all of which are shared with other taxa (polythetic support). Huamán and Spooner (2002) group all landrace populations of cultivated potatoes into the single denomination class, *S. tuberosum,* with eight Groups: Ajanhuiri Group, Andigenum Group, Chaucha Group, Chilotanum Group (*S. tuberosum* ssp. *tuberosum* from Chile), Curtilobum Group, Juzepczukii Group, Phureja Group, and Stenotomum Group.

This gradation of support (groups defined only by shared characters) makes a taxonomic decision of cultivated potatoes under the ICBN or ICNCP difficult. An argument could be made for *S. ajanhuiri, S. curtilobum, S. juzepczukii,* and *S. tuberosum* ssp. *tuberosum* to be recognized as species and the other taxa as Groups under a separate cultivated species *S. andigenum.* Support for the separate species treatment of *S. tuberosum* ssp. *tuberosum* is provided by Raker and Spooner (2002), who demonstrated that most of the landrace populations of the Chilotanum Group (from Chile) can be distinguished with microsatellite data from most populations of the Andigenum Group (from the Andes), and molecular support probably will be provided for the Ajanhuiri, Curtilobum, and Juzepczukii groups because of their independent hybrid origins involving other wild species. Despite these ambiguities, Huamán and Spooner (2002) classify all cultivated landraces under the single denomination class *S. tuberosum* because of the following lack of monophyly, taxonomic difficulties, and classification philosophy:

- Polythetic morphological support predominates (Huamán and Spooner, 2002).
- Origins are reticulate (Hawkes, 1990; Huamán et al., 1982; Schmiediche et al., 1982; Cribb and Hawkes, 1986).
- Multiple origins are possible (Hosaka, 1995).
- There are evolutionarily dynamic populations with continuing hybridization of crops to weeds (Ugent, 1970).
- Some accessions of wild and cultivated species are so similar that classification as cultivated or wild often rests on whether they are collected in the wild or in a cultivated field (Spooner et al., 1999).
- ICNCP classification philosophy is more logical for cultivated species.

Chilean and Andean Hypotheses of the First Introductions of Potato to Europe

Juzepczuk and Bukasov (1929) proposed that Chilean potato landraces originated from indigenous primitive Chilean tetraploid wild species and that the first European modern cultivars were introductions of Chilean landraces. They argued that the Chilean landraces were already adapted to the long days of Europe (Andean landraces form tubers under short days) and have a leaf morphology more similar to that of European landraces than Andean landraces.

In contrast, Salaman (1949), Salaman and Hawkes (1949), Hosaka and Hanneman (1988b), Grun (1990), Hawkes (1990), and Hawkes and Francisco-Ortega (1993) collectively suggested the following:

- *S. tuberosum* ssp. *tuberosum* in Chile arose from ssp. *andigenum* from the Andes, either directly or through a cross with an unidentified wild species. Grun (1979, 1990) found that the cytoplasmic types of Chilean landraces of *S. tuberosum* and modern potatoes were identical. However, he identified nine cytoplasmic factors that separate ssp. *andigenum* from ssp. *tuberosum* that cause sterility in the presence of specific chromosomal genes, abnormal anthers and pollen, anthers fused to styles, and female sterility. These factors are expressed only when ssp. *tuberosum* is used as a female, and when it is used as a male the crosses are fertile; that is, there are reciprocal crossing differences that affect sterility. Hawkes (1990) identified the putative wild progenitors of Chilean landraces proposed by Juzepczuk and Bukasov (1929) to be nothing more than other landraces, not wild species.

- The first modern potatoes were introduced from the Andes to Europe as *S. tuberosum* ssp. *andigenum*. The first record of potato in Europe is from the Canary Islands in 1562 (Hawkes and Francisco-Ortega, 1993) and the second record from Seville, Spain, in 1570 (Hawkes and Francisco-Ortega, 1992).

- *S. tuberosum* ssp. *andigenum* in Europe rapidly evolved into a wider leaf morphotype with long-day adaptation, a parallel event to long-day selection in Chile, and these evolved forms should be classified as ssp. *tuberosum*, just like the Chilean landraces.

- The fungal disease late blight (*Phytophthora infestans* (Mont.) De Bary) in Europe killed most *tuberosum*-evolved *andigenum* clones in the 1840s, but modern potato was rapidly mass selected and bred for blight resistance with ssp. *tuberosum*, purchased in Panama (as cultivar Rough Purple Chile) but believed to have come from Chile (Plaisted and Hoopes, 1989; Grun, 1990).

Chloroplast and Mitochondrial DNA Evidence

Chloroplast DNA (cpDNA) restriction site data have been used to investigate the wild species progenitors of the putative first cultivated potato *S. stenotomum* (a diploid) and subsequent origins of the other cultivated potatoes. Hosaka and Hanneman (1988a) and Hosaka (1995) documented five chloroplast genotypes (A, C, S, T, and W) in the Andean diploid and tetraploid landraces and in their putative progenitors in the *S. brevicaule* complex. The Chilean landraces had three of these genotypes (A, T, and W) but with a predominant T type cpDNA, characterized by a 241-bp deletion (Kawagoe and Kikuta, 1991), which is rare in the Andes. Hosaka (2002) showed that the only other wild potato species possessing T-type cpDNA were *S. berthaultii, S. neorossii,* and *S. tarijense* from Bolivia and Argentina. However, he also showed that that there were other chloroplast DNA restriction site markers shared only by some populations of *S. tarijense* and Chilean landraces of potato (Hosaka, 2003). He therefore concluded that these populations of *S. tarijense* were maternal parents to Chilean potato, perhaps after hybridization with Andean diploid or tetraploid landraces.

Both chloroplasts and mitochondria are extranuclear (cytoplasmic) organelles that contain their own DNA, but only mitochondria are known to condition the reciprocal crossing differences of male sterility that are evidenced in crosses between Andean and Chilean potato. Lössl et al. (1999) detected five major mitochondrial DNA (mtDNA) types in potato that they designated with Greek symbols α, β, γ, δ, and ε. Interestingly, β-type mtDNA is associated with T-type cpDNA. Lössl et al. (1999) found β-type mtDNA in Chilean landraces and *S. berthaultii* that has T-type cpDNA (*S. tarijense* was not examined). This suggests that Hosaka's cpDNA types are good markers to infer origins of Chilean landraces but that the mtDNA is the actual causal agent conditioning cytoplasmic male sterility.

Our Challenge to the Andean Introduction Hypothesis

Most publications since Salaman (1949) and Salaman and Hawkes (1949) accept the Andean introduction hypothesis without question, and most suggest that Chilean landraces were an important cultivated germplasm source only after the late blight epidemics of the 1840s. Evidence supporting the Andigenum Group as the first European introductions includes the following:

- Early herbarium specimens of potato in Europe had the narrow-leaved phenotype thought to distinguish the Andigenum Group from the Chilotanum Group = *S. tuberosum* ssp. *tuberosum* in Chile (Salaman and Hawkes, 1949).
- The earliest records of cultivated potatoes from the Canary Islands (in 1567; Hawkes and Francisco-Ortega, 1993) and from Seville, Spain (in 1573; Salaman, 1949; Hawkes and Francisco-Ortega, 1992), apparently were harvested late in the year (November and December), suggesting that they were the short day–adapted Andigenum Group. Remnants of these early introductions of Andigenum Group and triploid clones of Andean Chaucha Group persist on the Canary Islands, with putatively more recent introductions of the Chilotanum Group (Gil González, 1997; Casañas et al., 2002).
- The trip from Chile to Europe took longer than from Peru (or Colombia) to Europe, and tubers from Chile would have less of chance to survive this long voyage.
- Artificial selection of Andigenum Group collectively produced some Chilotanum Group–like clones ("neo-*tuberosum*") having greater flowering, shorter stolons, greater yield, earlier tuberization, reduction of cytosterility, and greater late blight resistance (Simmonds, 1966; Glendinning, 1975; Huarte and Plaisted, 1984; Vilaro et al., 1989) that showed the possibility for rapid selection.

We challenge the sole Andean introduction hypothesis and suggest that early introductions to Europe were from the Andes *and* from Chile, and the Chilean introductions became the prominent type well before the 1840s. Our arguments follow:

- Huamán and Spooner (2002; character 13 of figure 3) quantified overlap of leaf shapes between Andigenum Group and Chilotanum Group landraces. Identification is problematic of a limited set (18) of early European introduction potato herbarium specimens to Andean or Chilean origins based on leaf shape alone (Salaman and Hawkes, 1949).
- The historical evidence, including late cultivation of potatoes in Spain and the Canary Islands (Salaman, 1949; Hawkes and Francisco-Ortega, 1992, 1993) combined with extant putatively remnant populations in the Canary Islands (Gil González, 1997; Casañas et al., 2002), makes a strong case for early introductions of the Andigenum Group there. But historical records of early introductions are at best sparse and indefinite (Salaman, 1949; Glendinning, 1983). There probably were multiple introductions of all landrace groups from both the Andes and Chile after the value of potato became known, but they simply were not recorded.
- The argument that Chilean tubers would not have survived the long trip from the Andes to Europe (Hawkes, 1967) ignores the simple possibility of transport of true seeds, potted plants, or even well-preserved tubers. Potatoes certainly were an item of ship's stores from Chile, and there are records as early as 1587 of potatoes crated for shipment to Europe (Glendinning, 1983).
- Juzepczuk and Bukasov's (1929) argument that Chilean landraces were preadapted to the long days of Europe are compelling, and early introductions from Chile would be selected rapidly over Andean clones. Although neo-*tuberosum* clones show the possibility to select for long-day adaptation from Andigenum clones (Simmonds, 1966; Glendinning, 1975; Huarte and Plaisted, 1984; Vilaro et al., 1989), Chilean introductions would not require such intentional selection.
- More than 99% of extant advanced potatoes have T-type DNA typical of most Chilean germplasm (Hosaka, 1993, 1995; Powell et al., 1993; Provan et al., 1999). This includes a clone released before 1836 (cultivar "Yam"; Powell et al., 1993). The Andean introduction proponents explain these facts by an elimination of Andigenum Group clones

after the late blight epidemics and breeding with Chilotanum Group clones. This explanation overlooks the cytoplasmic male sterility of the Chilotanum Group because many crosses as females (but not males) are sterile (Grun, 1979, 1990), and only a cross with Chilotanum Group as female would confer the T-type cpDNA. It also overlooks the fact that Chilotanum Group clones are not known for late blight resistance.

In summary, we consider it likely that both Andigenum Group and Chilotanum Group clones were part of multiple early introductions of potato to Europe and that Chilotanum Group clones quickly became the predominant modern cultivars in Europe, as their derivatives are today worldwide.

Group Classification Under the ICNCP

The most recent edition of the ICNCP (Brickell et al., 2004) lists currently accepted categories to classify and name cultivated plants. Hetterscheid (1994), Hetterscheid and Brandenburg (1995a, 1995b), Hetterscheid et al. (1996), and Spooner et al. (2003) argue for a modernization of the classification and nomenclature of cultivated plants. The use of Linnean categories to classify cultivated plants presents problems because their artificial selection often involves processes very different from the natural evolution of wild plants. These processes often include human-directed multiple origins, extensive interspecific and sometimes intergeneric hybridization, and rapid selection of traits (such as gigantism, lack of dispersability, increased variability of the selected organ, elimination of physical and chemical defenses, change of habit, habitat, and breeding mechanisms) that often obscure origins (Hawkes, 1983; Harlan, 1992). In addition to these biological complications, pedigree records often are lost or intentionally kept secret to guard the proprietary nature of these industrial products. Undoubtedly, parallels occur between artificial and natural selection, such as hybrid origins in wild plants. The difference can be viewed as the scale of intensity between wild plant origins and human-directed selection, with maintenance of cultivated plants that typically cannot survive in nature. These human-selected products require classification codes that are quite different by both necessity and design.

The divergence between the classification objectives for wild and cultivated plants has always been obscured by the use of one common language arising from the taxonomy of wild plants, with the term *taxon* being the main source of confusion (Hetterscheid et al., 1996; Spooner et al., 2003).

Hetterscheid and Brandenburg (1995a, 1995b) introduced an alternative term, *culton* (user-defined groups), to replace *taxon,* a term today used mostly for phylogenetically related organisms.

The name *S. tuberosum* ssp. *tuberosum* may be one of the best examples of the differences between the ICBN and the ICNCP because of its unnatural divisions into cultivated "species." Modern potato cultivars have resulted from crosses between other cultivars and wild species. Fully 16 wild species are documented in pedigrees of different cultivars (Ross, 1986; Plaisted and Hoopes, 1989). Although the pedigrees of many modern cultivars are known, some of them are lost or have always been proprietary (Świeżyński et al., 1997). *S. tuberosum* sensu stricto (as distinct from the other cultivated species) is not a species in the modern concept of related individuals as used by modern evolutionary biologists. The evolutionary dynamics of cultivated plants are not the same as those of wild plants because domestication involves human-driven, special-purpose, artificial selection. The latter leads to a very different diversity of organisms ("industrial products") than what we call biodiversity for wild plants.

Past attempts to classify cultivated plants into the ICBN-based hierarchical systems have problems. The complex and diverse origins of potato are typical of many crop cultivars. An ICBN-based taxonomy of cultivated plants stimulates an inflated number of taxonomic ranks. Ongoing breeding of new cultivars continuously challenge the utility of these ranks, and the classifications become cumbersome. ICBN-based classifications of cultivated plants are plagued by complex typification, diagnosing, and nomenclatural discussions disputing relationships. Such classifications fail to serve the practical needs of users of cultivated plants where cultivar protection, marketing, and useful divisions of plants demand nomenclatural stability.

Name inflation caused by ICBN-based classifications of cultivated plants has become extreme. Fully 55 subspecific ranks for cultivated plants existed (Jirasek, 1961). Jirasek (1966) proposed the following 12 ranks below the species, listed in decreasing order: *specioid, subspecioid, cultiplex, subcultiplex, convarietas, subconvarietas, provarietas, subprovarietas, conculta, subconculta, cultivar,* and *subcultivar.* In such a system, every rank must follow the nomenclatural rules of ICBN; this results in an extreme vulnerability of such cumbersome names to frequent name changes. As impractical as this classification philosophy may seem, even today it is used by many taxonomists of crop plants. Recent classifications of *Brassica oleracea* (cabbage) illustrate this point. Although a much lower number of categories are used, they are still all embedded in nested classification systems for cultivar classification. Even ICBN-based ranks such as *subspecies, varietas,* and *forma* are misused to encompass groupings of plants of purely cultivated origin.

For example, the following is a complex and nested system of classification for vegetable kohlrabi in Mansfeld's World Database of Agricultural and Horticultural Crops (Mansfeld, 1986) that is a mixture of ICBN and ICNCP nomenclature: *Brassica oleracea* ssp. *oleracea* convar. *acephala* var. *gongylodes.*

To avoid this cumbersome and complex way of classifying cultivated plants, we propose that cultivated plants be classified solely by the one code that properly and exclusively deals with this subject, the ICNCP. The basis of this code lies with the very nature of the concept of the cultivar (Brickell et al., 2004, article 2). A few types of cultivars are as follows:

- Clones (several types)
- Graft chimeras
- Assemblages of plants grown from seed
- Inbred lines
- Multilines (assemblages of inbred lines)
- F_1 hybrids
- Hybrids of various complexity
- Genetically modified plants

To date the only ICBN-based systematic categories for cultivated plant classification are the cultivar and the Group (Brickell et al., 2004; Greuter et al., 2000, article 3). Names of culta belonging to either category may be associated loosely with ICBN-based taxa for reference based on a suggested phylogenetic background but must be treated with restraint (see *Brassica oleracea*). The combination of genus name and cultivar epithet suffices to uniquely identify a cultivar, and the latter may subsequently be put in a Group.

The Group

In order to minimize instability resulting from name changes in a hierarchical Linnean-based system, the Group is an appropriate device to eliminate Latin in a name below the generic level. It provides a means of creating classifications purely based on user criteria, ignoring Linnean systems based on relationships that often disregard criteria essential to practical user-driven classifications. The generic name seems to be the one globally used, Latin part of a crop name, but new insights into relationships can change even the genus name. For example, recently an attempt was made to reclassify the garden strawberry from *Fragaria* to *Potentilla* (Mabberley, 2002), but a subsequent DNA study of the Rosoideae (Eriksson et al., 2003), established the monophyly of *Fragaria*. Spooner et al. (1993) reclassified tomatoes from *Lycopersicon* to *Solanum* based on chloroplast DNA restriction site and morphological data, and subsequent molecular studies unequivocally support the nesting of tomato in *Solanum* (e.g., Olmstead and Palmer, 1997; Bohs and Olmstead, 1999).

Several successful attempts at Group classifications replacing the more cumbersome Linnean hierarchy and nomenclature have been made (e.g., Hetterscheid and van den Berg, 1996 [*Aster* L.]; Hoffman, 1996 [*Philadelphus* L.]; Hetterscheid et al., 1999 [onions]; Lange et al., 1999 [beet]; Huamán and Spooner, 2002 [potato landraces]). Van den Berg (1999) discusses the advantages of modern Group classifications over older, more cumbersome ones.

However, stability in names for Groups is not permanent, and they can change based on evolving needs. Contrary to such changes in Linnean classifications, the wishes of the user group at large is the decisive factor that leads to a new classifications rather than intricacies of the Botanical Code or decisions of individual taxonomists. One user group may be best served by a Group classification based on pest resistance, another by ornamental value. Accordingly, several coexisting special-purpose classifications are possible (Spooner et al., 2003). Pitfalls of Group names are that they carry no information on crop origins, and coexisting Group classifications could create confusion.

Names of Groups

Article 7 of the ICNCP (Brickell et al., 2004) lays down the fundamentals for naming Groups. It states that any word or words in a modern language, or even a Latin name, may form a Group name, provided it stabilizes historical reference. Such descriptive names as "Early Red Group" or "Sweet Yellow Group" are possible as Group names. Also, a group may be named after a widely known cultivar in the group to improve recognition. For example, one could imagine a "Bintje Group," based on a well-known Dutch potato cultivar "Bintje." This system also creates the possibility of using translations of Group names into other languages. Thus, a term such as Early Red Group would become *Frühe Rote Gruppe* in German. When a Group name is used in the full name of a cultivar, it reads like *Solanum tuberosum* (Early Red Group) "Mother's Finest."

Which Name for the Potato?

Vilmorin (1881, 1886, 1902), Kohler (1909, 1910), Milward (1912), and Stuart (1915) proposed early group classifications of cultivated potato; Stuart (1915) provides details on these classifications. In sum, the classifications were based on color, shape, and size of the tubers; tuber eye depth; color of the potato sprouts in the dark; color of the flowers; and vine type. None of the early group classifications (to 1915) were widely adopted or persist.

The following more recent publications have informally grouped potatoes: the Potato Association of America's North American Potato Variety Inventory (www.ume.maine.edu/PAA/PVI.htm, 498 cultivars from the United States), the Potato Association of America variety images and descriptions (www.ume.maine.edu/PAA/var.htm, 49 cultivars from the United States), the European Cultivated Potato Database (www.europotato.org/, 4000 cultivars from Europe), Świeżyński et al. (1997, 1998; 2000 and 130 cultivars, respectively, from the United States and Europe), and Hamester and Hils (2003; 3200 cultivars worldwide). We surveyed these publications and the potato Web sites listed in the next paragraph for characters currently in use to divide modern cultivars.

Tuber skin color and shape were the most common characters that grouped potatoes. For instance, the U.S. National Potato Board (www.potatohelp.com/potato101/varieties.asp) groups potatoes as russets (tan to brown-skinned tubers with netted skin), round whites, long whites, long reds, yellow flesh, and blue and purple flesh. Potato skin and flesh characters are not parallel descriptors, but potatoes commonly are divided into these classes using these two traits. Maturity is commonly used in descriptor lists, such as the Potato Association of America's variety images and descriptions. The British Potato Council Variety Handbook (www.potato.org.uk/seedSearch.asp?sec=446&con=458) divides potatoes by tuber size, skin color, flesh color, eye depth, tuber shape (short oval, oval, long oval, round), skin texture (smooth, rough, russet), and corolla traits (color, number, size, peduncle length). Similarly, Schneider and Douches (1997) divide potatoes into tuber skin color and shape classes in order to provide an additional discriminator, in combination with molecular marker data, for cultivar fingerprinting. All cultivar descriptions (e.g., the North American Potato Variety Inventory or the British Potato Council Variety Handbook) class potatoes, irrespective of morphology, into early and late varieties. One type of classification grades potatoes within market classes by tuber quality as it relates to compliance with specific tolerances for tuber sizes, defects, diseases, and other factors; different countries provide different names for these quality classes. However, these are transient quality factors that vary by year and locality and are not suitable as potential Groups.

Hamester and Hils (2003) provide worldwide coverage and a wide range of morphological and disease characterization data, making this publication the most useful resource for quantifying distribution of traits in cultivated potato. We therefore use it for our analysis of use categories that we present here. We analyzed all 20 disease scores from this publication as a single proportion and combined the five general use categories of processing, French fries, chips, dried products, and starch into a processing category that we compared with table use. Most of the records are from Europe (79%), with lesser numbers from North America (8%), Central and South America (6%), Asia (4%), Africa (2%), and Oceania (1%). The dates of release begin at 1760 ("Red Icelandic") and 1836 ("Fortyfold"), with 953 cultivars released from 1990–2002. In our analysis figure 13.2 shows distributions of maturity, tuber shape, skin color, eye depth, flesh color, and disease resistance for all 3530 cultivars. Space constraints preclude displaying use data, but they are as follows: table stock, 1707 cultivars; processing, 779; either processing or table, 1567; and both processing and table, 459.

Clone-specific disease resistance data have use for breeders or growers, but the multitude of disease variants (20) and unknown traits (figure 13.2) make their use for Group classifications unmanageable. Similarly, the use data are of interest to growers and processors, but there are so many mixed use categories as to be impractical for classifications. Flesh color and eye depth are rarely used in classifications, except for blue-fleshed potatoes for specialty markets. The most commonly used potato cultivar classification traits are tuber skin color, skin texture (although this is not part of the Hamester and Hils database), tuber shape, and maturity. Fifteen of the 27 variants of tuber skin color are very rare, with percentages of less than 1%, with the predominant types as yellow (55%), red (12%), white (8%), light yellow, buff (8%), light yellow–white (3%), and light red, pink (3%). Skin texture is divided into two categories of russets and smooth-skinned potatoes.

Figure 13.2 is the first graphic presentation of the variation in worldwide modern potato cultivars. It demonstrates that any Group classification based on very simple categories will be subject to interpretation of intergrading categories of many similar traits. For example, it may be

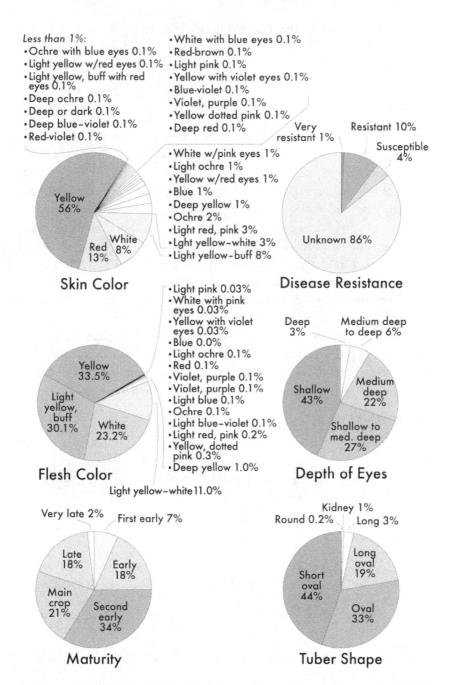

FIGURE 13.2 Distribution of selected modern potato cultivar traits on a worldwide basis as determined from data in Hamester and Hils (2003).

difficult to consistently distinguish skin colors of deep yellow, yellow, light yellow–buff, and light yellow–white (figure 13.2), and similar challenges would arise with tuber shape.

Conclusions

We propose to maintain the name *S. tuberosum* as the umbrella name (as a denomination class) for all potato Groups. Huamán and Spooner (2002) use this approach to classify potato landraces into Groups, thereby discarding all specific and infraspecific ICBN-based names and the catch-all subspecies name "ssp. *tuberosum*" for modern cultivated potatoes. The purpose of this chapter is to present a story of modern cultivated potato in the context of taxonomy and historical data to show that the name *S. tuberosum* sensu stricto is not a species in the proper sense of the word. Rather, this name has been applied to a diverse set of modern clones, of complex hybrid origins, involving other cultivars and wild species. We argue that it is better to classify modern cultivated potato into Groups that reflect actual use by breeders, growers, and processors. We present the first graphic presentation of tuber traits from Hamester and Hils (2003) that may be used to form a formal Group classification for user groups and potato scientists.

Acknowledgments

We thank the International Potato Center for permission to use figure 13.1a, Professor Andres Contreras to use figure 13.1b, Sarah Stephenson and two anonymous reviewers for comments on an earlier draft of our manuscript, and Andrew Rouse for artwork.

References

Bohs, L. and R. G. Olmstead. 1999. *Solanum* phylogeny inferred from chloroplast DNA sequence phylogeny. In M. Nee, D. E. Symon, R. N. Lester, and J. P. Jessop (eds.), *Solanaceae IV, Advances in Biology and Utilization,* 97–110. Royal Botanic Gardens, Kew, UK.

Brickell, C. D., B. R. Baum, W. L. A. Hetterscheid, A. C. Leslie, J. McNeill, P. Trehane, F. Vrugtman, and J. H. Wiersma. 2004. International code of nomenclature for cultivated plants, ed. 7. *Regnum Vegetabile* 144: 1–123.

Brush, S. B., H. J. Carney, and Z. Huamán. 1981. Dynamics of an Andean potato agriculture. *Economic Botany* 35: 70–88.

Bukasov, S. M. 1939. The origin of potato species. *Physis (Buenos Aires)* 18: 41–46.

Bukasov, S. M. 1971. Cultivated potato species. In S. M. Bukasov (ed.), *Flora of Cultivated Plants,* Vol. IX, 5–40. Kolos, Leningrad, Russia.

Casañas, R., M. González, E. Rodríguez, A. Marrero, and C. Díaz. 2002. Chemometric studies of chemical compounds in five cultivars of potatoes from Tenerife. *Journal of Agricultural Food Chemistry* 50: 2076–2082.

Correll, D. S. 1962. The potato and its wild relatives. *Contributions from Texas Research Foundation, Botanical Studies* 4: 1–606.

Cribb, P. J. and J. G. Hawkes. 1986. Experimental evidence for the origin of *Solanum tuberosum* subspecies *andigena*. In W. G. D'Arcy (ed.), *Solanaceae: Biology and Systematics*, 383–404. Columbia University Press, New York, NY, USA.

Dodds, K. S. 1962. Classification of cultivated potatoes. In D. S. Correll (ed.), The potato and its wild relatives. *Contributions from Texas Research Foundation, Botanical Studies* 4: 517–539.

Eriksson, T., M. S. Hibbs, A. D. Yoder, C. F. Delwiche, and M. J. Donoghue. 2003. The phylogeny of the Rosoideae (Rosaceae) based on sequences of the internal transcribed spacers (ITS) of nuclear ribosomal DNA and the TNRL/F region of chloroplast DNA. *International Journal of Plant Science* 164: 197–211.

Gil González, J. 1997. *El Cultivo Tradicional de la Papa en la Isla de la Tenerife*. Associación Granate, University of La Laguna, Tenerife, Spain.

Glendinning, D. R. 1975. Neo-*tuberosum:* New potato breeding material. 2. A comparison of neo-*tuberosum* with selected Andigena and with Tuberosum. *Potato Research* 18: 343–350.

Glendinning, D. R. 1983. Potato introductions and breeding up to the early 20th century. *New Phytologist* 94: 479–505.

Graves, C. (ed.). 2001. *Potato: Treasure of the Andes*. International Potato Center, Lima, Peru.

Greuter, W., J. McNeill, F. R. Barrie, H. M. Burdett, V. Demoulin, T. S. Filgueiras, D. H. Nicolson, P. C. Silva, J. E. Skog, P. Trehane, N. J. Turland, and D. L. Hawksworth (eds.). 2000. International Code of Botanical Nomenclature (St. Louis Code). *Regnum Vegetabile* 138: 1–474.

Grun, P. 1979. Evolution of cultivated potato: A cytoplasmic analysis. In J. G. Hawkes, R. N. Lester, and A. D. Skelding (eds.), The Biology and Taxonomy of the Solanaceae. *Linnean Society of London Symposium Series* 7: 655–665.

Grun, P. 1990. The evolution of cultivated potatoes. *Economic Botany* 44 (3 suppl.): 39–55.

Hamester, W. and U. Hils. 2003. *World Catalogue of Potato Varieties*. Agrimedia GmbH, Bergen/Dumme, Germany.

Harlan, J. R. 1992. *Crops and Man*, 2nd ed. American Society and Agronomy, and Crop Science Society of America, Madison, WI, USA.

Hawkes, J. G. 1958. Kartoffel. I. In H. Kappert and W. Rudorf (eds.), Taxonomy, cytology and crossability. *Handbuch der Pflanzenzüchtung*, 2nd ed., Vol. 3, 1–43. Paul Parey, Berlin, Germany.

Hawkes, J. G. 1967. The history of the potato. Masters Memorial Lecture, 1966. *Journal of the Royal Horticultural Society* 92: 207–224, 249–262, 288–302, 364–365.

Hawkes, J. G. 1983. *The Diversity of Crop Plants*. Harvard University Press, Cambridge, MA, USA.

Hawkes, J. G. 1990. *The Potato: Evolution, Biodiversity, and Genetic Resources*. Belhaven Press, London, UK.

Hawkes, J. G. and J. Francisco-Ortega. 1992. The potato in Spain during the late 16th century. *Economic Botany* 46: 86–97.

Hawkes, J. G. and J. Francisco-Ortega. 1993. The early history of the potato in Europe. *Euphytica* 70: 1–7.

Hawkes, J. G. and J. P. Hjerting. 1969. *The Potatoes of Argentina, Brazil, Paraguay and Uruguay: A Biosystematic Study*. Oxford University Press, Oxford, UK.

Hawkes, J. G. and J. P. Hjerting. 1989. *The Potatoes of Bolivia: Their Breeding Value and Evolutionary Relationships*. Oxford University Press, Oxford, UK.

Hetterscheid, W. L. A. 1994. The culton concept: Recent developments in the systematics of cultivated plants. *Acta Botanica Neerlandica* 43: 78 (Abstract).

Hetterscheid, W. L. A. and W. A. Brandenburg. 1995a. The culton concept: Setting the stage for an unambiguous taxonomy of cultivated plants. *Acta Horticulturae* 413: 29–34.

Hetterscheid, W. L. A. and W. A. Brandenburg. 1995b. Culton versus taxon: Conceptual issues in cultivated plant systematics. *Taxon* 44: 161–175.

Hetterscheid, W. L. A. and R. G. van den Berg. 1996. Cultonomy of Aster. *Acta Botanica Neerlandica* 45: 173–181.

Hetterscheid, W. L. A., R. G. van den Berg, and W. A. Brandenburg. 1996. An annotated history of the principles of cultivated plant classification. *Acta Botanica Neerlandica* 45: 123–134.

Hetterscheid, W. L. A., C. van Ettekoven, R. G. van den Berg, and W. A. Brandenburg. 1999. Cultonomy in statutory registration exemplified by *Allium* L. crops. *Plant Varieties and Seeds* 12: 149–160.

Hijmans, R. and D. M. Spooner. 2001. Geographic distribution of wild potato species. *American Journal of Botany* 88: 2101–2112.

Hoffman, M. H. A. 1996. Cultivar classification of *Philadelphus* L. (Hydrangeaceae). *Acta Botanica Neerlandica* 45(2): 199–209.

Hosaka, K. 1993. Similar introduction and incorporation of potato chloroplast DNA into Japan and Europe. *Japanese Journal of Genetics* 68: 55–61.

Hosaka, K. 1995. Successive domestication and evolution of the Andean potatoes as revealed by chloroplast DNA restriction endonuclease analysis. *Theoretical and Applied Genetics* 90: 356–363.

Hosaka, K. 2002. Distribution of the 241 bp deletion of chloroplast DNA in wild potato species. *American Journal of Potato Research* 79: 119–123.

Hosaka, K. 2003. T-type chloroplast DNA in *Solanum tuberosum* L. ssp. *tuberosum* was conferred from some populations of *S. tarijense* Hawkes. *American Journal of Potato Research* 80: 21–32.

Hosaka, K. and R. E. Hanneman, Jr. 1988a. Origin of chloroplast DNA diversity in the Andean potatoes. *Theoretical and Applied Genetics* 76: 333–340.

Hosaka, K. and R. E. Hanneman, Jr. 1988b. The origin of the cultivated tetraploid potato based on chloroplast DNA. *Theoretical and Applied Genetics* 76: 172–176.

Huamán, Z., J. G. Hawkes, and P. R. Rowe. 1982. A biosystematic study of the origin of the diploid potato, *Solanum ajanhuiri. Euphytica* 31: 665–675.

Huamán, Z. and D. M. Spooner. 2002. Reclassification of landrace populations of cultivated potatoes (*Solanum* sect. *Petota*). *American Journal of Botany* 89: 947–965.

Huarte, M. A. and R. L. Plaisted. 1984. Selection for *tuberosum* likeness in the vines and in the tubers in a population of neotuberosum. *American Potato Journal* 61: 461–473.

Jackson, M. T., J. G. Hawkes, and P. R. Rowe. 1980. An ethnobotanical field study of primitive potato varieties in Peru. *Euphytica* 29: 107–113.

Jirasek, V. 1961. Evolution of proposals of taxonomical categories for the classification of cultivated plants. *Taxon* 10: 34–45.

Jirasek, V. 1966. The systematics of cultivated plants and their taxonomic categories. *Preslia* 38: 267–284 (in Czech).

Johns, T. and S. L. Keen. 1986. Ongoing evolution of the potato on the altiplano of western Bolivia. *Economic Botany* 40: 409–424.

Juzepczuk, S. W. and S. M. Bukasov. 1929. A contribution to the question of the origin of the potato. *Proceedings of the U.S.S.R. Congress Genetics, Plant, and Animal Breeding* 3: 592–611 (in Russian).

Kawagoe, Y. and Y. Kikuta. 1991. Chloroplast DNA evolution in potato (*Solanum tuberosum* L.). *Theoretical and Applied Genetics* 81: 13–20.

Kohler, A. R. 1909. Potato experiments and studies at University Farm. *Minnesota Agricultural Experiment Station Bulletin* 114: 311–319.

Kohler, A. R. 1910. Potato experiments and studies at University Farm in 1909. *Minnesota Agricultural Experiment Station Bulletin* 118: 90–100.

Lange, W., W. A. Brandenburg, and T. S. M. de Bock. 1999. Taxonomy and cultonomy of beet (*Beta vulgaris* L.). *Botanical Journal of the Linnean Society* 130: 81–96.

Lechnovich, V. S. 1971. Cultivated potato species. In S. M. Bukasov (ed.), *Flora of Cultivated Plants* Vol. IX: *Potato,* 41–302. Publishing House Kolos, Leningrad, Russia.

Lössl, A., N. Adler, R. Horn, U. Frei, and G. Wenzel. 1999. Chondriome-type characterization of potato: mt α, β, γ, δ, ε and novel plastid–mitochondrial configurations in somatic hybrids. *Theoretical and Applied Genetics* 98: 1–10.

Mabberley, D. 2002. *Potentilla* and *Fragaria* (Rosaceae) reunited. *Telopea* 9: 793–801.

Mansfeld, R. 1986. *Verzeichnis landwirtschaftlicher und gaertnerischer Kulturpflanze,* Vol. 1. Springer Verlag, Berlin, Germany. (On the Internet as "Mansfeld's World Database of Agricultural and Horticultural Crops" at mansfeld.ipk-gatersleben.de/mansfeld/).

Miller, J. T. and D. M. Spooner. 1999. Collapse of species boundaries in the wild potato *Solanum brevicaule* complex (Solanaceae, *S.* sect. *Petota*): Molecular data. *Plant Systematics and Evolution* 214: 103–130.

Milward, J. G. 1912. Commercial varieties of potatoes for Wisconsin. *Wisconsin Agricultural Experiment Station Bulletin* 225: 7.

Ochoa, C. M. 1958. Expedición colectora de papas cultivadas a la cuenca del lago Titicaca. I. Determinación sistemática y número cromosómico del material colectado. Ministerio de Agricultura, Programa Cooperativo de Experimentación Agropecuaria (PCEA). *Investigación en Papa* 1: 1–18.

Ochoa, C. M. 1990. *The Potatoes of South America: Bolivia.* Cambridge University Press, Cambridge, UK.

Ochoa, C. M. 1999. *Las Papas de Sudamerica: Peru (Parte 1).* International Potato Center (CIP), Lima, Peru.

Olmstead, R. G. and J. D. Palmer. 1997. Implications for the phylogeny, classification, and biogeography of *Solanum* from cpDNA restriction site variation. *Systematic Botany* 22: 19–29.

Plaisted, R. L. and R. W. Hoopes. 1989. The past record and future prospects for the use of exotic germplasm. *American Potato Journal* 66: 603–627.

Powell, W., E. Baird, N. Duncan, and R. Waugh. 1993. Chloroplast DNA variability in old and recently introduced potato cultivars. *Annals of Applied Botany* 123: 403–410.

Provan, J., W. Powell, H. Dewar, G. Bryan, G. C. Machray, and R. Waugh. 1999. An extreme cytoplasmic bottleneck in the modern European cultivated potato (*Solanum tuberosum*) is not reflected in decreased levels of nuclear diversity. *Proceedings of the Royal Academy of London B* 266: 633–639.

Quiros, C. F., S. B. Brush, D. S. Douches, K. S. Zimmerer, and G. Huestis. 1990. Biochemical and folk assessment of variability of Andean cultivated potatoes. *Economic Botany* 44: 254–266.

Quiros, C. F., R. Ortega, L. van Raamsdonk, M. Herrera-Montoya, P. Cisneros, E. Schmidt, and S. B. Brush. 1992. Increase of potato genetic resources in their center of diversity: The role of natural outcrossing and selection by the Andean farmer. *Genetic Resources and Crop Evolution* 39: 107–113.

Rabinowitz, D., C.R. Linder, R. Ortega, D. Begazo, H. Murguia, D.S. Douches, and C.F. Quiros. 1990. High levels of interspecific hybridization between *Solanum sparsipilum* and *S. stenotomum* in experimental plots in the Andes. *American Potato Journal* 67: 73–81.

Raker, C. and D.M. Spooner. 2002. The Chilean tetraploid cultivated potato, *Solanum tuberosum*, is distinct from the Andean populations; microsatellite data. *Crop Science* 42: 1451–1458.

Ross, H. 1986. *Potato Breeding: Problems and Perspectives*. Paul Parey, Berlin, Germany.

Salaman, R.N. 1949. *The History and Social Influence of the Potato*. Cambridge University Press, Cambridge, UK.

Salaman, R.N. and J.G. Hawkes. 1949. The character of the early European potato. *Proceedings of the Linnean Society, London* 161: 71–84.

Schmiediche, P.E., J.G. Hawkes, and C.M. Ochoa. 1982. The breeding of the cultivated potato species *Solanum × juzepczukii* and *S. × curtilobum*. II. The resynthesis of *S. × juzepczukii* and *S. × curtilobum*. *Euphytica* 31: 395–707.

Schneider, K. and D.S. Douches. 1997. Assessment of PCR-based simple sequence repeats to fingerprint North American potato cultivars. *American Potato Journal* 74: 149–160.

Simmonds, N.W. 1966. Studies on the tetraploid potatoes. II. Factors in the evolution of the Tuberosum Group. *Journal of the Linnean Society, Botany* 59: 43–56.

Spooner, D.M., G.J. Anderson, and R.K. Jansen. 1993. Chloroplast DNA evidence for the interrelationships of tomatoes, potatoes, and pepinos (Solanaceae). *American Journal of Botany* 80: 676–688.

Spooner, D.M. and R. Castillo. 1997. Reexamination of series relationships of South American wild potatoes (Solanaceae: *Solanum* sect. *Petota*): Evidence from chloroplast DNA restriction site variation. *American Journal of Botany* 84: 671–685.

Spooner, D.M., W.L.A. Hetterscheid, R.G. van den Berg, and W.A. Brandenburg. 2003. Plant nomenclature and taxonomy: An horticultural and agronomic perspective. *Horticultural Reviews* 28: 1–60.

Spooner, D.M. and R.J. Hijmans. 2001. Potato systematics and germplasm collecting, 1989–2000. *American Journal of Potato Research* 78: 237–268.

Spooner, D.M., A. Salas-L., Z. Huamán, and R.J. Hijmans. 1999. Wild potato collecting expedition in southern Peru (departments of Apurímac, Arequipa, Cusco, Moquegua, Puno, Tacna) in 1998: Taxonomy and new genetic resources. *American Journal of Potato Research* 76: 103–119.

Spooner, D.M. and K.J. Sytsma. 1992. Reexamination of series relationships of Mexican and Central American wild potatoes (*Solanum* sect. *Petota*): Evidence from chloroplast DNA restriction site variation. *Systematic Botany* 17: 432–448.

Spooner, D.M., R.G. van den Berg, W. García, and M.L. Ugarte. 1994. Bolivia potato collecting expeditions 1993, 1994: Taxonomy and new germplasm resources. *Euphytica* 79: 137–148.

Stuart, W. 1915. Group classification and varietal descriptions of some American potatoes. *United States Department of Agriculture Bulletin* 176: 1–56.

Świeżyński, K.M., K.G. Haynes, R.C.B. Hutten, M.T. Sieczka, P. Watts, and E. Zimnoch. 1997. Pedigree of European and North American potato varieties. *Plant Breeding and Seed Science* 41: 1 (Suppl). Plant Breeding and Acclimatization Institute, Radzików, Poland.

Świeżyński, K.M., M.T. Sieczka, I. Stypa, and E. Zimnoch-Guzowska. 1998. Characteristics of major potato varieties from Europe and North America. *Plant Breeding and Seed Science* 42: 2 (Suppl.). Plant Breeding and Acclimatization Institute, Radzików, Poland.

Trehane, P., C.D. Bricknell, B.R. Baum, W.L.A. Hetterscheid, A.C. Leslie, J. McNeill, S.A. Spongberg, and F. Vrugtman, 1995. International code of nomenclature of cultivated plants *Regnum Vegetabile* 133: 1–175.

Ugent, D. 1968. The potato in Mexico: Geography and primitive culture. *Economic Botany* 22: 109–123.

Ugent, D. 1970. The potato: What is the origin of this important crop plant, and how did it first become domesticated? *Science* 170: 1161–1166.

Van den Berg, R.G. 1999. Cultivar-group classification. In S. Andrews, A.C. Leslie, and C. Alexander (eds.), *Taxonomy of Cultivated Plants: Third International Symposium,* 135–143. Royal Botanic Gardens, Kew, UK.

Van den Berg, R.G., J.T. Miller, M.L. Ugarte, J.P. Kardolus, J. Nienhuis, and D.M. Spooner. 1998. Collapse of morphological species in the wild potato *Solanum brevicaule* complex (Solanaceae: sect. *Petota*). *American Journal of Botany* 85: 92–109.

Vilaro, F.L., R.L. Plaisted, and R.W. Hoopes. 1989. Comparison of cytoplasmic male sterilities in progenies of Tuberosum × Andigena and Tuberosum × Neo-Tuberosum crosses. *American Potato Journal* 66: 13–24.

Vilmorin, C.P.H.L. de. 1881, 1886, and 1902. *Essai d'un Catalogue Méthodique et Synynomique des Principales Variétés de Pommes de Terre,* eds. 1–3. Chez Vilmorin-Andrieux & Cie., Paris, France.

Watanabe, K. and S.J. Peloquin. 1989. Occurrence of 2n pollen and *ps* gene frequencies in cultivated groups and their related wild species in the tuber-bearing Solanums. *Theoretical and Applied Genetics* 78: 329–336.

Watanabe, K. and S.J. Peloquin. 1991. The occurrence and frequency of 2n pollen in 2x, 4x, and 6x wild, tuber-bearing *Solanum* species from Mexico, and Central and South America. *Theoretical and Applied Genetics* 82: 621–626.

Zimmerer, K. 1991. The regional biogeography of native potato cultivars in highland Peru. *Journal of Biogeography* 18: 165–178.

Eve Emshwiller

CHAPTER 14

Evolution and Conservation of Clonally Propagated Crops
Insights from AFLP Data and Folk Taxonomy of the Andean Tuber Oca (*Oxalis tuberosa*)

Vegetatively propagated crops play an enormous role in feeding the world. They include crops that are important worldwide, such as sugarcane, potato, cassava, sweet potato, banana, and plantain, as well as crops of local or regional importance, such as true yam, edible aroids, and several minor Andean roots and tubers. Many of these crops are grown primarily for subsistence, under traditional, nonindustrialized farming systems, which still represent much of world agriculture. Thus they serve as an important safety net against starvation. These agroecosystems retain great diversity of potential use for future breeding efforts (Elias and McKey, 2000), yet studies of the dynamics of genetic diversity in these systems are few. We lack information about how evolutionary factors, such as selection and gene flow, differ between clonally propagated and seed-propagated crops. To understand the evolution and conservation needs of any crop, we need to learn about several aspects: the crop's origins and what wild species are closely related to it, how human influence has affected its evolution under domestication, how its diversity is distributed, and the factors that affect whether that diversity is maintained or lost. As the first effort in a research program aimed at understanding the dynamics of genetic diversity of cultigens and their wild relatives and the continuing human role in their evolution, this chapter discusses research on the Andean tuber crop oca (*Oxalis tuberosa* Molina).

Oca is one of dozens of crops domesticated millennia ago in the Andean region (Pearsall, 1992). These domesticates were the basis for the Inca empire and earlier Andean civilizations, and they still feed millions of inhabitants of this region. Although the potato has spread around the world, most of the other crops are still poorly known outside the Andean region and have received much less research attention than the potato. Tuber-bearing plants from four unrelated plant families were domesticated as food crops in the Andes: oca (Oxalidaceae), potatoes (*Solanum* spp.; Solanaceae), "ulluco" (*Ullucus tuberosus*; Basellaceae), and "mashua" (*Tropaeolum tuberosum;* Tropaeolaceae). Oca and the other minor tubers have an essential role for food security in rural communities of the Central Andean highlands, where they are consumed daily by many households for several months of each year. The tuber crops are cultivated in the highest agricultural zones, from 2800 to 4100 m in elevation, where cultivating diverse crop species reduces the risk of crop failure caused by drought, frost, or hail in the harsh, unpredictable climate. Because they are not subject to the same pest and disease problems as potatoes, the minor tubers are also important in the Andean crop rotation systems that help control plant pathogens.

Oca is considered second to potatoes among these minor tuber crops in the diet and farming system of millions of Quechua and Aymara peasant farmers in Ecuador, Peru, and Bolivia, and its potential in other parts of the world is demonstrated by its recent commercialization in New Zealand (National Research Council, 1989). Oca is primarily a starchy staple in Andean communities of subsistence farmers, providing some variety from potatoes in a largely tuber-based diet, but it is also rich is vitamins.

Oca Diversity

Oca tubers look like elongated potato tubers (figure 14.1), with their eyes (lateral buds) embedded in prominent transverse ridges, which may be colored differently from the rest of the tuber in some cultivars (cultivated varieties). Although it remains capable of sexual reproduction (Vallenas Ramirez, 1997; Trognitz et al., 1998), oca is propagated exclusively vegetatively in traditional agriculture. Nonetheless, it still maintains phenotypic diversity. Pigmentation of the tuber is particularly variable, with colors ranging from nearly white to nearly black, with shades of pink, red, purple, yellow, and orange, with various patterns of distribution of colors on both the exterior

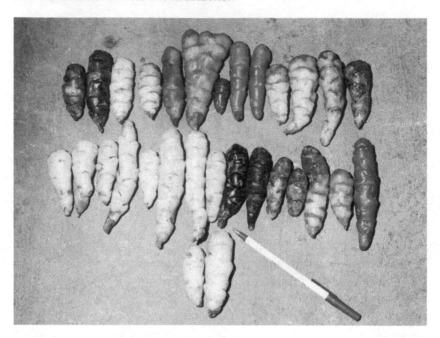

FIGURE 14.1   Array of oca tubers cultivated by a single household in the community of Viacha. The two pale yellow tubers at the bottom are of the sour cultivar *p'osqo*, used exclusively for *khaya*. The others are all *wayk'u* cultivars, including *yana ushpa*, *puka ushpa*, *yuraq kishwar*, *kusipata* (one fasciated), *machasqa*, *puka panti*, *hanq'o q'ello* (yellower and redder variants), *misitu* (of varied tuber shapes), *q'ellu panti*, *q'ellu k'aytu*, and unidentified yellow tubers (see table 14.1 for colors).

and interior of the tuber (IPGRI/CIP, 2001). The high levels of morphological and physiological variation within oca contrast with the low genetic variation in some molecular markers. Low variation has been found in allozymes (del Río, 1990), tuber proteins (Stegemann et al., 1988; Shah et al., 1993), and random amplified polymorphic DNA (A. Donayre, pers. comm., 2000; G. Piedra, pers. comm., 2000). At the same time, variation between cultivars has been described in morphological traits (Castillo Peña, 1974; Arbizu et al., 1997), insect resistance (Apaza Apaza, 1980), phenology (León Salas, 1972; Alarcón Avendaño, 1976), and composition of protein, starch, and dry matter content (Rivero Gonzáles, 1973; Peña Paredes, 1978; Bustinza López, 1979). The numerous vernacular cultivar names reflect this diversity (Rea and Morales, 1980; Arbizu and Robles, 1986; Seminario and Rimarachín, 1995; Terrazas, 1996; Guamán, 1997; Ramírez 2002).

Origins of Polyploidy in Octoploid Oca

Like the potato and many other domesticated plants, oca is polyploid, in this case octoploid (with eight sets of chromosomes). Thus one aspect of understanding oca's evolution involves determining its origin of polyploidy and its phylogenetic relationships with wild species. Specifically, we need to determine not only from what wild species oca was domesticated but also what species contributed genomes to the polyploid crop. Cultivated oca has been found to be octoploid in most studies (de Azkue and Martínez, 1990; Medina Hinostroza, 1994; Valladolid et al., 1994; Emshwiller, 2002b), although there are conflicting reports. The genus *Oxalis* comprises 500–800 species, most of them in South America and southern Africa, making the search for the origins of polyploidy in oca a challenge. Cytological studies revealed that oca was part of the *O. tuberosa* alliance, a group of morphologically similar species that share the same base chromosome number, x = 8 (de Azkue and Martínez, 1990). Other *Oxalis* species have base chromosome numbers from 5 to 12, with 7 most common. Current data suggest that the alliance includes more than the dozen species originally studied by de Azkue and Martínez (1990), probably several dozen species from throughout the central and northern Andes (Emshwiller, 2002a). Molecular studies investigating the origins of oca used DNA sequence data from two loci, the internal transcribed spacer (ITS) of nuclear ribosomal DNA and the chloroplast-expressed (but nuclear-encoded) isozyme of glutamine synthetase (ncpGS). The ITS data confirmed the monophyly of the *O. tuberosa* alliance and the origins of oca from within this group, but ITS had insufficient variation to identify oca's progenitors (Emshwiller and Doyle, 1998). An intron-containing region of ncpGS, however, provided more informative variation than ITS (Emshwiller, 2002a; Emshwiller and Doyle, 2002). Three different sequence classes of ncpGS within an individual plant were separated by molecular cloning for use in phylogenetic analyses. Fixed heterozygosity and separate placement of the sequence classes on the ncpGS gene tree suggested that these three classes represent homeologous loci and that oca is of hybrid origin (allopolyploid) and probably autoallopolyploid (at least one genome is present in more than two copies).

Data from ncpGS identified two wild tuber-bearing taxa, *O. picchensis* of southern Peru and a yet-unnamed species from Bolivia, as progenitor candidates that may have hybridized to form cultivated oca (Emshwiller and Doyle, 2002). Flow cytometry data indicated that *O. picchensis* is tetraploid (Emshwiller, 2002b), and although the ploidy level of the Bolivian taxon is unknown, it is probably also polyploid, based on its fixed heterozygosity for ncpGS sequence classes among the sampled plants. Other sources of data are needed to test this working hypothesis and resolve unanswered questions about the origins of polyploidy in oca. That these origins might be complex is suggested by variation in ncpGS sequences from different plants, especially the absence of the *O. picchensis*–like sequence from one of the nine individual *O. tuberosa* plants sampled. Alternative hypotheses to explain this absence include multiple origins of polyploidy, varying ploidy levels in cultivated oca, introgression of the *O. picchensis*–like sequence through wild-crop gene flow, or loss of this sequence class through chromosomal rearrangements after polyploidization (see reviews in Soltis and Soltis, 1999; Wendel, 2000; Liu and Wendel, 2003). In addition, another wild tuber-bearing taxon from northwestern Argentina, *O. chicligastensis*, appears to be another possible candidate as genome donor for oca, based on both morphology and DNA sequence data (unpublished data). Thus, despite recent progress in the identification of good candidates as the genome donors of polyploid oca, several alternative hypotheses are congruent with the current data. Future studies are planned to use amplified fragment length polymorphism (AFLP) as an independent source of data for examining these working hypotheses.

Ethnotaxonomy and Clonal Crops

The evolution of crops is affected by the management of folk cultivars in traditional agricultural systems, especially in the way in which humans act as agents of selection and dispersal. Thus ethnographic studies combined with genetic studies of crop diversity using molecular markers can elucidate the human influence on crop evolution. Conservation of crop genetic diversity often is said to be linked to knowledge and use; loss of knowledge goes hand in hand with loss of diversity (IPGRI, 2001). Therefore, if we are to understand crop evolution in traditional agriculture and plan for in situ conservation, it is vital to study folk taxonomy. Understanding how crop diversity is named and classified by farmers is key to "how this diversity is perceived and valued by farmers" (Elias et al., 2001a:156) and thus to "understanding behavioral patterns that affect crop evolution" (Quiros et al., 1990:256). Folk nomenclature has been studied in clonal crops such as potato (LaBarre, 1947; Jackson et al., 1980; Brush et al., 1981; Zimmerer, 1991b; Brush and Taylor, 1992), cassava (Boster, 1984, 1985, 1986; Salick et al., 1997; Elias et al., 2000a, 2000b, 2001a, 2001b), sweet potato (Prain et al., 1995; Nazarea, 1998; Prain and Campilan, 1999), and ensete (Shigeta, 1996), and research is ongoing in these and other crops (PLEC, 2001). Studies have compared folk nomenclature with molecular markers (DNA or allozyme) in potato (Quiros et al., 1990; Zimmerer and Douches, 1991; Brush et al., 1995; Zimmerer, 1998) and cassava (Elias et al., 2000a, 2001a, 2001b). However, the generalizability of these results is unknown, and there is a need to expand on these studies and provide comparison with other crops.

Ethnobotanical and Ethnotaxonomic Studies in Pisac District, Southern Peru

To identify factors that affect whether oca genetic diversity is being lost or maintained in traditional Andean agriculture, I conducted an ethnobotanical survey in three indigenous peasant communities (Viacha, Amaru, and Sacaca) in Pisac District, Cusco Department, in southern Peru in 1997 (Emshwiller, 1998). Semistructured interviews in Spanish and Quechua focused on the knowledge and management of the crop by traditional Andean farmers. Information was elicited about how traditional cultivars of oca are named, classified, recognized, acquired, selected, and managed. Some questions focused on how much of a family's harvest went for sale, seed, and home consumption; methods of storage, preparation, and cooking; whether some cultivars were disappearing; pest and disease management; and how propagation material is exchanged between families and between communities.

To study the folk nomenclature and taxonomy of oca variation I asked about the names and characteristics of the cultivars and their preferred uses. Farmers distinguish the culinary traits of tubers, describing them as sweet or sour and their texture as floury, watery, or firm. Similarly to the situation observed by Boster (1984) for cassava, farmers were knowledgeable about these culinary characteristics but did not distinguish between the cultivars in terms of agronomic traits or ecological needs.

This study revealed that oca, like potato, is classified into use categories (sensu Zimmerer, 1991a). Oca tubers are either cooked fresh or preserved in dried form. Sweet cultivars, called *wayk'u* (boiling) oca, usually are exposed to sunlight for a few days to sweeten them and then either boiled whole or roasted in *watia* (temporary earth ovens made of clods of soil). In contrast, sour cultivars are preserved by processing into dried oca tubers called *khaya* (figure 14.1, table 14.1). *Khaya* is prepared by exposing tubers

Table 14.1 Folk Cultivars of Pisac Communities Viacha, Amaru, and Sacaca

Use Category	Folk Cultivars	Subcultivars	Exterior Color	Comments
Khaya	P'osqo		Pale yellow	Very sour, used exclusively for khaya
Wayku (sometimes grouped with khaya)	Kusipata		Magenta pink	Firm texture
Wayku	Puka panti		Magenta pink	
Wayku	Misitu, higos	Misitu	Orangish with brown streaks	Claviform
		Yana misitu	Nearly black	Claviform
		Q'ello misitu	Yellow with darker streaks	Ovoid
		Higos misitu	Orangish with brown streaks	Ovoid
		Tulla misitu	Orangish with brown streaks	Long cylindrical
		K'aspi misitu	Orangish with brown streaks	Long cylindrical
Wayku	Ushpa	Yuraq ushpa	Pinkish white	Floury texture
		Puka ushpa	Mottled red	
		Yana ushpa	Purple-black	
Wayku	Hanq'o q'ello, waqankillay		Yellow at base grading to red apex	Clusters in "yellow group"
Wayku	Q'ello panti, señorita		Pale yellow	Clusters in "yellow group"
Wayku	Q'ello kaytu		Yellow with red eyes	Clusters in "yellow group"
Wayku	Yuraq kishuar		White with pale pink eyes	
Wayku	Puka chiliku		White with pale pink blotches	Chiliku is Quechua pronunciation of the Spanish chaleco = vest
Wayku	Puka p'osqo		Red	Sour but grown with wayku
Wayku	Machasqa		Shiny red	
Wayku	Damaso		Orangish red	

Very roughly, more common cultivars are listed toward the top, less common below. Some unsampled cultivars found in these communities are not listed.

to several alternating days of hot sun and nights of frost until they are completely dry, similarly to the process of making *chuño* from Andean potatoes. The drying period usually is preceded by nearly a month of soaking in a pool of water, which presumably reduces oxalic acid content of these sour cultivars. The use categories *wayk'u* and *khaya* not only seem to differ in oxalic acid composition, but anatomical differences between them have been observed in both modern and archaeological material (Martins, 1976). Cultivars of different use categories are grown in separate fields, whereas cultivars in the same use category usually are grown in mixed plantings, as is also reported for Andean potatoes (Jackson et al., 1980; Brush et al., 1981; Zimmerer, 1991a, 1991b; Brush and Taylor, 1992).

Within the *wayk'u* and *khaya* use categories are individual folk cultivars that are distinguished and named primarily on the basis of tuber color, shape, and texture. In a few cases a name designates a group that is morphologically heterogeneous, and in these cases some farmers distinguish between these subtypes with different names. Here I call these complex cultivars, as contrasted with the simple cultivars that include a single morphotype. One example of a complex cultivar is *misitu,* named for the streaked pattern of secondary pigment (figure 14.2). *Misitu* tubers have a range of colors (brown to black streaks over a base that varies from yellow to orange to brown) and also vary in tuber shape from broad-ovoid to long-claviform (IPGRI/CIP, 2001). Only a few knowledgeable farmers distinguished different kinds of *misitu* with separate names. Another morphologically heterogeneous cultivar was *ushpa,* whose name means "ashes," in reference to this cultivar's preferred floury texture and its blotchy pigmentation pattern. These tubers occurred in a wide range of colors, from nearly white, to shades of red, to nearly black. Farmers might call them all simply *ushpa* or might add a modifier to describe the color.

The possibility that acculturation may be leading to loss of traditional knowledge of oca was suggested by a surprising inconsistency in the use of oca cultivar names. Some inconsistency in the use of names is reported in cassava (Elias et al., 2001a) and sweet potato (Nazarea, 1998), and Quiros et al. (1990:259) reported a "wide range of skill and knowledge [of potato cultivars] among farmers." Even so, I found a higher than expected level of inconsistency in the use of vernacular names of oca cultivars (Emshwiller, 1998). Some cases of the use of different names for the same morphotype did not indicate unreliability but rather cases of synonymy that were recognized as such by the farmers (as also found by Quiros et al., 1990). I observed cases of the use of different names for the same morphotype

FIGURE 14.2 Two *misitu* tubers, showing the streaked pigmentation pattern. The upper tuber is fasciated.

among the three communities (cultivars called *misitu* and *q'ello panti* in Viacha were known as *higos* and *señorita,* respectively, in Sacaca), and farmers noted other instances of synonymy themselves (*hanq'o q'ello* and *waqankillay*). However, knowledge about oca cultivars varied between and within villages, and other cases of the use of different names or the application of the same name to clearly different tubers seemed to reflect this variation. As found in Boster's (1986) study of cassava cultivars, names were applied more consistently to the more common cultivars than to less common cultivars.

The results of this ethnotaxonomic study of oca folk cultivars in Pisac, the assessment of reliability in the use of their names, and the larger ethnobotanical study of factors affecting oca's genetic diversity in Pisac will be published later in more detail. Here I describe a comparison of the genotypes of oca as distinguished by AFLP with the morphotypes and the folk taxonomy of oca variation in the communities of Amaru, Sacaca, and Viacha. The objectives are to determine whether there is a correspondence between use categories and differences in AFLP profiles; whether the "simple" cultivar names refer to a single or to multiple clonal genotypes, and conversely, whether single genotypes bear several names; and how AFLP markers correspond with the morphologically heterogeneous complex cultivars that are distinguished by some but not all farmers. In the latter case of complex cultivars that include subcultivars, some alternative hypotheses include that these subcultivar groups are similar but distinguishable genotypes, dissimilar genotypes that have converged on similar morphological traits, or indistinguishable by AFLP (i.e., either the phenotypic differences have no genetic basis or they result from mutations that are not reflected in AFLP profiles).

Materials and Methods

Sampling

Tubers collected during the ethnotaxonomic survey were used in this AFLP study so that genotypes as distinguished by AFLP data could be compared with the ethnotaxonomy of oca folk cultivars. However, because of the variation between farmers in knowledge of oca varietal names and whether they were applied consistently, a comparison of AFLP data with the names given by each individual farmer would conflate potential genotypic variation within cultivars with inconsistency in the use of names. Therefore, the names supplied by each farmer were compared with a separate grouping based on morphological traits. In this chapter, I report on a comparison of the AFLP data with the tuber morphotypes based on my own visual assessment in which I grouped together tubers that looked similar enough that they might belong to the same clonal genotype. I then called each morphotype group by the name (or names, if recognized as synonyms) that was applied most often to that morphotype by knowledgeable farmers. In most but not all cases, the group to which I independently assigned the tuber agreed with the name given by the farmer (or a variant or synonym of that name). Future stages of this project will incorporate information from the cases of disagreement between the names to which the farmers and I assigned the tuber.

Some of the tubers did not seem to belong definitely with any of the other morphotypes (hereafter called mismatch tubers). In these cases the color and other characteristics of the tuber were noted, and they were either designated as of uncertain identification or tentatively identified as the cultivars they most resembled. The first samples for AFLP included only tubers for which the farmers and I agreed on the cultivar group to which the tuber belonged, whereas later sampling included some of the question-able matches.

AFLP data were generated for 95 tubers collected in the three communi-ties in Pisac district. In addition to *O. tuberosa* accessions, one plant each of three wild tuber-bearing taxa was sampled to compare with the cultivated oca samples. Two of these, *O. picchensis* and the unnamed taxon of Bolivia, were identified by previous results as possible progenitors of octoploid oca (Emshwiller and Doyle, 2002). The third wild taxon, *O. chicligastensis* of northwestern Argentina, is another candidate as a putative progenitor, based on unpublished ncpGS sequence data. An additional 30 oca samples from other areas in Peru and Bolivia were included in the assessment of AFLP polymorphism but were not part of the ethnotaxonomic comparison.

DNA Isolation and Fluorescent AFLP Procedure

DNA was isolated from silica gel dried leaves using DNeasy Plant Kits (Qiagen, Carlsbad, CA, USA). DNA template was prepared by restriction with *Eco*RI and *Mse*I and ligation with T4 DNA ligase (from New England Biolabs, Beverly, MA, USA) of adapters supplied with the Applied Biosystems AFLP Plant Mapping Kit (for Small Plant Genomes) according to the man-ufacturer's instructions (except that templates were diluted by only 1/5, not 1/20, at each step). The labeled amplification products were separated by electrophoresis through LongRanger acrylamide gels in an ABI Prism 377 automated DNA sequencer and visualized using GeneScan software. GeneScan-500 (ROX) size standards permitted automatic sizing of frag-ments. Data were scored using GeneScan and GenoTyper software (PE Applied Biosystems, Foster City, CA) to create the binary matrix, which was then edited by hand. Repeatability was assessed by including some replicate samples, including separate DNA isolations from the same plant prepared for AFLP and run either on the same gel or on separate gels, dif-ferent restriction–ligation reactions prepared from the same DNA sample, template from one preselective amplification that was amplified twice with the same selective primer combination but on separate dates and run on separate gels, and the same selective amplification product run on more than one gel. Here I report results with a single AFLP primer combination, *Eco*RI-AC/*Mse*I-CAC, which was chosen based on good amplification and polymorphism detection. The primer pair is designated here in abbrevi-ated form as "ac/cac" (based on the two and three selective bases of the *Eco*RI and *Mse*I primers, respectively).

Data Analyses

In order to explore the relationship between the genotypes distinguished by AFLP and the morphological groups recognized in the folk taxonomy in Pisac, AFLP data were analyzed with several ordination and clustering methods based on genetic distance and similarity. These included principal component analysis (PCA), principal coordinate analysis (PCOA) using the Gower general similarity index, unweighted pair group method with arithmetic mean (UPGMA, using Jaccard's or Nei and Li distance measures), and minimum variance as implemented in the Multivariate Statistical Package (MVSP; KCS, 2003) (see Appendix II for discussion of analytical methods). Neighbor-joining (NJ) analyses (using Nei and Li distance measures) were conducted using PAUP* (Swofford, 1998). Although they varied in details, these different analyses gave similar results with respect to the points discussed later, so only the NJ results are shown (figures 14.3 and 14.4).

Results and Discussion

AFLP Polymorphism and Reproducibility

The data matrix for primer combination ac/cac included 116 peaks of 95–505 bp (smaller fragments were excluded as being mostly monomorphic or not unambiguously scorable). Polymorphism was assessed not only among the oca accessions from Pisac and the three wild *Oxalis* taxa but also the 30 oca samples from other areas. Among this larger sample, data from ac/cac included 86 peaks that were polymorphic in oca, 7 monomorphic in all samples, 13 monomorphic in oca but absent in at least one wild tuber-bearing taxon, and 10 absent in oca but present in at least one wild tuber-bearing taxon.

Replicate samples run on the same gels had profiles that were remarkably similar, not only in identical presence of bands, but even in their shapes and relative sizes. Duplicate samples run on different gels were less similar in shapes of profiles and were not necessarily identical in band presence (up to 4.3% difference, especially if reaction strength varied; see table 14.2). Unreliable bands were detected and eliminated from the data matrix based on the replicate samples, which to date have been run for about 10% of accessions. Additional replicates are a high priority for very divergent samples because their differences might possibly result from weak reactions or degraded or contaminated DNA templates (see Dyer and Leonard,

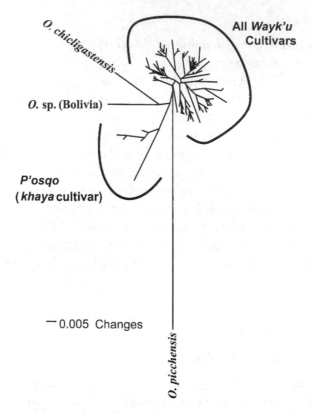

FIGURE 14.3 Results of neighbor-joining analysis of AFLP data (primer combination ac/cac) using Nei and Li distance measure, displayed as an unrooted network. Three wild tuber-bearing *Oxalis* taxa are included, in addition to the cultivated oca accessions. Note the separation of all of the *wayk'u* cultivars from the *p'osqo* tubers (used exclusively for processing into *khaya*). In this unrooted network the three wild *Oxalis* taxa join the branches between the *khaya* and *wayk'u* use categories. Individual oca sample numbers are removed for clarity.

2000). As has been observed by others and is discussed later, replicate AFLP profiles are not always identical (Douhovnikoff and Dodd, 2003).

Correspondence of AFLP Data with Use Categories

The AFLP data agree with the classification of oca by Quechua farmers in Pisac District into two use categories. That is, the different oca use categories were particularly distinct from each other in their preliminary AFLP data, as revealed in both ordination (PCOA and PCA) and clustering (UPGMA and NJ) analyses. *P'osqo*, the cultivar usually used in Pisac for processing into *khaya*,

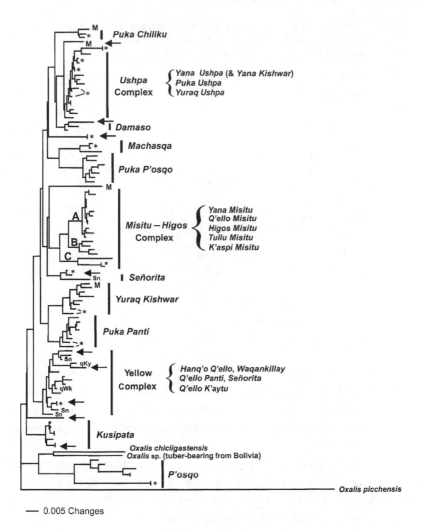

— 0.005 Changes

FIGURE 14.4 Results of the same neighbor-joining analysis as in figure 14.3, here displayed as a phenogram rooted with the most divergent wild taxon, *O. picchensis*. Bars on the right indicate the morphotypes and folk cultivars to which the tubers were assigned, based on tuber morphology. The three complex cultivars list some of the subcultivars included in them, but not all *misitu* subcultivars listed were sampled. Subclusters A, B, and C in the *misitu* complex are discussed in the text. Four of the tubers (marked with an *M*) were purchased from a market in Cusco; 2 of these do not match any of the genotypes from Pisac communities. Asterisks indicate replicate samples from the same tuber. Arrows point to the "mismatch" tubers, which could not be unambiguously assigned to a cultivar group, as discussed in the text. Samples in the yellow complex are *hanq'o q'ello* (both yellower and redder variants) unless indicated as "mismatch" tubers (*arrow*), *señorita (Sn)*, *q'ello k'aytu (qKy)*, or *q'ello waqankillay (qWk)*.

Table 14.2 Maximum pairwise distances between replicate samples or between samples within the same cluster. The simple cultivars listed are those that had at least seven samples.

Groups Compared	Standard Distance	Nei and Li Distance
Replicates (13 pairs)	0.043	0.0083
Simple Cultivars		
Puka p'osqo	0.087	0.0146
Yuraq panti	0.094	0.0185
Puka panti	0.060	0.0131
Kusipata	0.035	0.0065
P'osqo	0.181	0.0450
Complex Cultivars		
Ushpa	0.077	0.0150
Yellow group	0.112	0.0245
Misitu within A	0.068	0.0117
Misitu A to B	0.137	0.0243
Misitu C to A or B	0.224	0.0424

is separated from all the *wayk'u* cultivars in the results of NJ analyses, and the three wild tuber-bearing taxa are found between the two clusters in unrooted NJ networks when using Nei and Li distances (figure 14.3). Results with different distance measures (e.g., standard distance or simple matching) or different algorithms (e.g., UPGMA) differ somewhat in the branch lengths and in the arrangements between the wild species and the two use categories, but they are consistent in the distinct separation of the two use categories from each other. These results suggest the interesting hypothesis that these use categories may have different evolutionary histories.

Correspondence of AFLP Data with Simple Folk Cultivars

Overall there was a good correspondence between the AFLP data and the morphological groups recognized by Quechua farmers in the three Pisac communities. With only a few exceptions, each morphotype that is generally recognized by farmers forms a separate cluster in the results of NJ analysis (figure 14.4). Tubers of the same morphotype clustered together regardless of the distance measure used, although there were differences in branch lengths and some rearrangements in relationships between the clusters, as well as within them, in different analyses. Thus, data from a single AFLP primer pair were able to distinguish all simple folk cultivars recognized by most farmers. In some cases, the AFLP data could clearly distinguish between genotypes that were so similar in color that they are easily confused when distinguished on visually observable tuber traits alone (e.g., the pink tubers of *kusipata* and *puka panti* or the white tubers of *yuraq kishwar* and *yuraq ushpa*). In most cases tubers of the same folk cultivar (e.g., *puka panti, kusipata, puka p'osqo, yuraq kishwar*) had similar but not necessarily identical AFLP profiles (figure 14.4 and table 14.2), indicating that they probably are members of the same clonal lineage (genet). Different tubers of the same cultivar often had a few differences, which might reflect either real differences between them (somatic mutations) or experimental error (AFLP artifacts or scoring ambiguities). Replicate AFLP profiles often are not exactly identical, and Douhovnikoff and Dodd (2003) determined that real differences between samples from different ramets or even different leaves from the same stem may be more numerous than those from experimental error. Evidence of somatic mutation in clonal lineages has also been documented for other marker types (e.g., variable number of tandem repeats; Rogstad et al., 2002). Data from additional primer combinations may distinguish more genotypes from within these clusters and may possibly show that some of these clusters do not represent a single genet (i.e., that some of the tubers in the group are separated by at least one sexual generation). Nonetheless, even if they are not all of the same clonal lineage, their close similarities suggest that they are probably closely related genotypes (e.g., siblings or parent–offspring).

A few rare morphotypes were encountered very infrequently in the household stores of oca tubers. Some of these were among cultivars that farmers had mentioned, in response to inquiries, as cultivars that were disappearing or had disappeared. These cultivars could be distinguished from the others on the basis of AFLP data but do not form a cluster in the NJ network because only one or two of each has been included in this sample (e.g., *Damaso, machasqa,* and *puka chiliku*). The AFLPs were also helpful in the case of the mismatch tubers that could not be identified unambiguously with any of the other morphotypes. Although the AFLP data indicated that a few of these tubers did belong with one of the known cultivar groups, in most cases the AFLP data confirmed these tubers as being distinct from any of the cultivar clusters. Thus at least some of these mismatch tubers are indeed different clonal genotypes than the predominant ones in the named cultivars. The presence of both the low-frequency named cultivars and the mismatch tubers indicate that the genotypic diversity in these communities would be underestimated by a cursory survey of morphotypes.

Correspondence of AFLP Data with Complex Folk Cultivars

Each of the complex cultivars (heterogeneous cultivar groups or cultivars that include subcultivars) is discussed separately because they differ with respect to their correspondence with the AFLP data.

The *Ushpa* Group

The AFLP data confirmed the farmers' classification of the *ushpa* group despite their wide range of tuber pigmentation. All *ushpa* tubers had very similar AFLP profiles, diverging no more than samples within simple cultivars (table 14.2). The color variants were scattered within the *ushpa* cluster, so the data from primer combination ac/cac do not clearly distinguish between them. However, additional AFLP data may distinguish between the different color shades. On the other hand, the color differences might not be reflected in differences in AFLP at all, as might be the case if they are the result of somatic mutations or especially if the differences are developmental. Therefore this group represents a case in which the farmers' classification based on the prized floury texture is a better clue to the genetic similarity of these tubers than is their color variation (although the mottled or splotched patterning of the pigmentation is an important similarity). This underscores the importance of the farmers' close familiarity with their cultivars, through not only growing but also eating them.

The *Misitu* Group

The situation in the *ushpa* group contrasts with the *misitu* group, in which the tubers have enough AFLP differences that they do not appear to be a single clone. Some of the tubers had very similar (or even identical) AFLP profiles with ac/cac, and these are probably clone mates (see cluster A in figure 14.4; table 14.2). However, other *misitu* tubers differ in several markers. The divergence between *misitu* clusters A and B (up to 0.137 standard distance, 0.0243 Nei and Li distance; see table 14.2) is in the same range as divergence between samples of different morphotypes (e.g., between *puka panti* and *yuraq kishwar*), suggesting that these different *misitu* subgroups probably are separated by at least one cycle of sexual recombination. *Misitu* group C tubers were still more divergent (table 14.2), but their differences must be confirmed by additional replicate samples. These AFLP differences validate the discrimination, by at least a few farmers, of different subtypes of *misitu* on the basis of tuber shape and secondarily by shade of coloration. These latter distinctions reflect real genetic differences, as the *misitu* group appears, on the basis of this single primer combination, to be polyclonal. In the results of NJ analysis of this single AFLP primer combination (figure 14.4), these putative separate clonal genotypes all join a single cluster, so they appear to be closely related rather than having converged independently on the same streaked color pattern.

The Yellow Group

Finally, the cluster designated as the yellow group (figure 14.4) is heterogeneous both molecularly and morphologically. Their tubers have yellow as the primary color, with varied patterns of secondary red pigment in some cultivars. I initially saw them as comprising at least three different morphotypes, and the farmers also gave them different names. *Q'ello k'aytu* tubers are yellow with red eyes, whereas the morphotype called either *q'ello panti* or *señorita* is evenly pale yellow, without markings. The morphotype called *hanq'o q'ello* or *waqankillay* grades from a yellow base to variable degrees of red at the apex. Although I expected that some of these morphotypes might comprise multiple genotypes, I did not anticipate that morphotypes would be intermingled within a single heterogeneous cluster (figure 14.4). Divergences in this cluster overall are greater than within simple *wayk'u* cultivars (see table 14.2), suggesting they are not a single clone. The four sampled *q'ellu panti* or *señorita* tubers separate from each other and group in several places in the network, some among the *hanq'o q'ellu* tubers and others outside the yellow cluster. Increased sampling and data from additional primer combinations will be necessary to determine how many different genotypes make up this complex yellow group.

Insights from AFLP About *P'osqo*, the Single Cultivar for Making *Khaya*

Ethnobotanically and morphologically there seemed to be a single homogeneous cultivar, known by a single Quechua name, *p'osqo* (meaning sour, tart, bitter, fermented; Morató Peña and Morató Lara 1995), that was exclusively used for processing into *khaya* (although some farmers also used the firm cultivar *kusipata*, or indeed any undersized tubers, for processing into *khaya* as well). Despite their morphological similarity, however, the AFLP data of the six *p'osqo* tubers had surprisingly high divergence (figures 14.3 and 14.4). Indeed, variation within *p'osqo* (up to 0.181 standard distance, 0.0450 Nei and Li distance; table 14.2) is greater than within any *wayk'u* cultivars except *misitu* and greater than many comparisons between different *wayk'u* cultivars.

In addition to the divergence between the two use categories discussed earlier, the AFLP data also provide a hint that the *p'osqo* tubers might differ in ploidy level from the *wayk'u* cultivars. Like the three wild tuber-bearing taxa, the six *p'osqo* accessions amplified a smaller number of bands than most of the *wayk'u* cultivars. The number of peaks scored per plant in the ac/cac profiles ranged from 32 to 62 in the samples overall. Most *wayk'u* samples amplified 48–62 peaks, but the wild taxa and the *p'osqo* accessions amplified only 32–47 peaks. Whereas most studies have found cultivated *O. tuberosa* to be octoploid, *O. picchensis* is tetraploid (Emshwiller, 2002b). The other wild tuber-bearing taxa probably are polyploid as well because they consistently have multiple sequence types for ncpGS (i.e., they show fixed heterozygosity, one of the criteria of allopolyploidy; Emshwiller and Doyle, 2002, and unpublished data). The divergence of AFLP data and the smaller number of peaks amplified by the *p'osqo* tubers both led to the speculation that the *p'osqo* genotypes might have a lower ploidy level than the dominant octoploid level in most oca studied to date. A similar situation has been found in potato, in which species of lower ploidy level amplify a smaller number of peaks for most primer pairs than species with higher ploidy levels (Kardolus et al., 1998). Thus it may be that oca is similar to the situation in Andean native potatoes, in which the several use categories comprise different species of *Solanum* of several ploidy levels (Brush et al., 1981; Zimmerer, 1991b). This would also be consistent with the clustering analyses in that the *p'osqo* accessions grouped with two of the wild tuber-bearing taxa (figure 14.3) or in other analyses were more distant from *wayk'u* oca cultivars than were those two wild taxa.

More molecular, morphological, and cytological data are clearly needed to confirm this difference in ploidy level and to investigate the relationship between use categories. If *p'osqo* has a lower ploidy level, then the question arises as to its relationship to the more common octoploid cultivars of oca. Some possibilities include that *p'osqo* might represent a surviving line from the progenitor of octoploid oca (meaning that oca was initially domesticated at a lower ploidy level) or, alternatively, that *wayk'u* oca derives from a different progenitor and different origin of polyploidy, and perhaps that the two use categories have entirely separate origins of domestication. In either scenario, it is likely that there has been little or no gene flow between the two use categories for a long time.

Conclusions and Research Needs

These preliminary data from a single AFLP primer pair indicate that the named folk cultivars, at least when applied by knowledgeable farmers, usually designate either individual clonal genotypes or groups of genetically similar genotypes. The AFLP data also indicate that the classification of oca into two use categories by farmers in Pisac reflects a fundamental biological difference. Nonetheless, several aspects of these results indicate that the genotypic diversity of oca in Pisac is underestimated by the number of named cultivars. Many mismatch tubers were confirmed as genotypes that did not belong to any of the primary clusters. A few cultivars were found at very low frequency (and probably would be missed in a brief germplasm collecting visit). Complex cultivars such as *misitu* apparently include more than a single clonal genotype, but only a few knowledgeable farmers distinguished them with separate names. Additional AFLP data may uncover other differences within these clusters. Thus a cursory look at the number of morphologically different tuber types would substantially underestimate the genetic diversity present in these communities.

Others studies have also found that folk taxonomy corresponds well overall but provides a net underestimate of genotypic diversity compared with molecular data. Such was the case in the pioneering research in native Andean potatoes by Quiros et al. (1990) and other studies in potato and cassava that found that individual cultivar names are applied to more than one genotype (e.g., Zimmerer and Douches, 1991; Elias et al., 2000a, 2001b).

Turnover in the composition of clones cultivated over time has been noted in temporal studies of oca (Ramírez, 2002). It is still unknown whether the infrequent genotypes sampled herein reflect such genotypic turnover and, if so, whether these genotypes are coming or going (i.e., whether they represent new recombined genotypes that are not yet at high frequency and are new introductions to the community or, alternatively, whether some of them are in decline). The possibility that some of the uncommon cultivars such as *machasqa* and *Damaso* are disappearing was suggested by the recollections of older farmers that they had been more abundant in the past. Interestingly, the few tubers found of these and some other rare cultivars usually were small, suggesting that they may be declining because of increasing viral load, mutational load, or perhaps clonal senescence. There is a need for more temporal studies of oca and of other clonal crops (e.g., Hamlin and Salick, 2003) to elucidate the causes of genotypic turnover. Understanding of spatial structure of genetic diversity at various scales is also crucial for vegetatively propagated crops. Although there is agreement that the relationship between in situ and ex situ conservation should be complementary, there is little understanding of how this can be accomplished. Data on the evolution of clonal crops in traditional agricultural systems are paramount for this goal.

Acknowledgments

I especially thank the people of the communities of Amaru, Sacaca, and Viacha, without whose help this study would have been impossible. I thank El Centro para el Desarrollo de los Pueblos Ayllu (CEDEP Ayllu) for logistical support in Pisac district, Dr. Alfredo Grau for collaboration in collection of *O. chicligastensis,* Dr. Handojo Kusumo for help getting started with AFLP, Josh Crea, Sofía Lopez, and Anna Mullenneaux for help with DNA isolations, and the Pritzker Laboratory of Molecular Systematics and Evolution at the Field Museum for facilities. Funding for fieldwork was provided by the Fulbright Commission of Peru, with additional support for initial laboratory research from Abbott Laboratories to the Field Museum. I thank Dr. Fabian A. Michelangeli and the book editors for comments that helped improve the manuscript.

References

Alarcón Avendaño, M. F. 1976. *Ritmo de Tuberización en Cinco Clones Seleccionados de Oca (*Oxalis tuberosa *Mol.).* Thesis, Universidad Nacional de San Antonio Abad del Cusco, Cusco, Peru.

Apaza Apaza, S. 1980. *Fuentes de Resistencia en 147 Clones de Oca (*Oxalis tuberosa *Mol.) al Ataque del Nemátodo Dorado (*Globodera *ssp.).* Thesis, Universidad Nacional Técnica del Altiplano, Puno, Peru.

Arbizu, C., R. Blas, M. Holle, F. Vivanco, and M. Ghislain. 1997. Advances in the morphological characterization of oca, ulluco, mashua, and arracacha collections. In *International Potato Center: Program Report 1995–1996,* 110–118. International Potato Center, Lima, Peru.

Arbizu, C. and E. Robles. 1986. *Catálogo de los Recursos Genéticos de Raíces y Tubérculos Andinos.* Universidad Nacional de San Cristóbal de Huamanga, Ayacucho, Peru.

Boster, J. S. 1984. Classification, cultivation, and selection of Aguaruna cultivars of *Manihot esculenta* (Euphorbiaceae). *Advances in Economic Botany* 1: 34–47.

Boster, J. S. 1985. Selection for perceptual distinctiveness: Evidence from Aguaruna cultivars of *Manihot esculenta. Economic Botany* 39: 310–325.

Boster, J. S. 1986. Exchange of varieties and information between Aguaruna manioc cultivators. *American Anthropologist* 88: 428–436.

Brush, S. B., H. J. Carney, and Z. Huamán. 1981. Dynamics of Andean potato agriculture. *Economic Botany* 35: 70–88.

Brush, S., R. Kesseli, R. Ortega, P. Cisneros, K. Zimmerer, and C. Quiros. 1995. Potato diversity in the Andean center of crop domestication. *Conservation Biology* 9: 1189–1198.

Brush, S. B. and J. E. Taylor. 1992. Diversidad biológica en el cultivo de papa. In E. Mayer (ed.), *La Chacra de Papa: Economía y Ecología*, 215–260. Centro Peruano de Estudios Sociales, Lima, Peru.

Bustinza López, T. 1979. *Rendimiento en Almidón, Harina y Materia Seca en 50 Clones de la Colección Ocas Cusco*. Thesis, Universidad Nacional de San Antonio Abad del Cusco, Cusco, Peru.

Castillo Peña, G. 1974. *Estudio de la Variabilidad Morfológica de 142 Clones de la Colección de Ocas del Cusco*. Thesis, Universidad Nacional de San Antonio Abad del Cusco, Cusco, Peru.

de Azkue, D. and A. Martínez. 1990. Chromosome number of the *Oxalis tuberosa* alliance (Oxalidaceae). *Plant Systematics and Evolution* 169: 25–29.

del Río, A. H. 1990. *Análisis de la Variación Isoenzimática de* Oxalis tuberosa Molina *"Oca" y Su Distribución Geográfica*. Thesis, Universidad Ricardo Palma, Lima, Peru.

Douhovnikoff, V. and R. S. Dodd. 2003. Intra-clonal variation and a similarity threshold for identification of clones: Application to *Salix exigua* using AFLP molecular markers. *Theoretical and Applied Genetics* 106: 1307–1315.

Dyer, A. T. and K. J. Leonard. 2000. Contamination, error, and nonspecific molecular tools. *Phytopathology* 90: 565–567.

Elias, M. and D. McKey. 2000. The unmanaged reproductive ecology of domesticated plants in traditional agroecosystems: An example involving cassava and a call for data. *Acta Oecologica* 21: 223–230.

Elias, M., D. McKey, O. Panaud, M. C. Anstett, and T. Robert. 2001a. Traditional management of cassava morphological and genetic diversity by the Makushi Amerindians (Guyana, South America): Perspectives for on-farm conservation of crop genetic resources. *Euphytica* 120: 143–157.

Elias, M., O. Panaud, and T. Robert. 2000a. Assessment of genetic variability in a traditional cassava (*Manihot esculenta* Crantz) farming system, using AFLP markers. *Heredity* 85: 219–230.

Elias, M., L. Penet, P. Vindry, D. McKey, O. Panaud, and T. Robert. 2001b. Unmanaged sexual reproduction and the dynamics of genetic diversity of a vegetatively propagated crop plant, cassava (*Manihot esculenta* Crantz), in a traditional farming system. *Molecular Ecology* 10: 1895–1907.

Elias, M., L. Rival, and D. McKey. 2000b. Perception and management of cassava (*Manihot esculenta* Crantz) diversity among Makushi Amerindians of Guyana (South America). *Journal of Ethnobiology* 20: 239–265.

Emshwiller, E. 1998. Ethnobotanical study of factors affecting genetic erosion in cultivated *Oxalis tuberosa* in Peru. *American Journal of Botany* 85: 55–56 (abstract); www.ou.edu/cas/botany-micro/bsa-abst/section4/abstracts/3.shtml.

Emshwiller, E. 2002a. Biogeography of the *Oxalis tuberosa* alliance. *Botanical Review* 68: 128–152.

Emshwiller, E. 2002b. Ploidy levels among species in the "*Oxalis tuberosa* alliance" as inferred by flow cytometry. *Annals of Botany* 89: 741–753.

Emshwiller, E. and J. J. Doyle. 1998. Origins of domestication and polyploidy in oca (*Oxalis tuberosa*: Oxalidaceae): nrDNA ITS data. *American Journal of Botany* 85: 975–985.

Emshwiller, E. and J. J. Doyle. 2002. Origins of domestication and polyploidy in oca (*Oxalis tuberosa*: Oxalidaceae). 2. Chloroplast-expressed glutamine synthetase data. *American Journal of Botany* 89: 1042–1056.

Guamán, S. 1997. *Conservación In Situ Caracterización y Evaluación de la Biodiversidad de Oca (*Oxalis tuberosa*) y Papalisa (*Ullucus tuberosus*) en Candelaria (Chapare) y Pocanche (Ayopaya)*. Thesis, Universidad Mayor de San Simón, Cochabamba, Bolivia.

Hamlin, C. C. and J. Salick. 2003. Yanesha agriculture in the upper Peruvian Amazon: Persistence and change fifteen years down the "road." *Economic Botany* 57: 163–180.

IPGRI. 2001. *On Farm Management of Crop Genetic Diversity and the Convention on Biological Diversity's Programme of Work on Agricultural Biodiversity.* International Plant Genetic Resources Institute, Rome, Italy; www.biodiv.org/doc/meetings/sbstta/sbstta-07/information/sbstta-07-inf-07-en.pdf.

IPGRI/CIP. 2001. *Descriptores de Oca* Oxalis tuberosa *Mol.* International Plant Genetic Resources Institute (IPGRI), Rome, Italy. www.ipgri.cgiar.org/publications/pdf/143.pdf.

Jackson, M. T., J. G. Hawkes, and P. R. Rowe. 1980. An ethnobotanical field study of primitive potato varieties in Peru. *Euphytica* 29: 107–113.

Kardolus, J. P., H. J. van Eck, and R. G. van den Berg. 1998. The potential of AFLPs in biosystematics: A first application in *Solanum* taxonomy (Solanaceae). *Plant Systematics and Evolution* 210: 87–103.

KCS. 2003. *MVSP (Multivariate Statistical Package),* version 3.13g. Kovach Computing Services, Anglesey, Wales.

LaBarre, W. 1947. Potato taxonomy among the Aymara Indians of Bolivia. *Acta Americana* 5: 83–103.

León Salas, J. 1972. *Determinación de la Curva de Tuberización en Cinco Clones de Oca (*Oxalis tuberosa *Mol).* Thesis, Universidad Nacional de San Antonio Abad del Cusco, Cusco, Peru.

Liu, B. and J. F. Wendel. 2003. Epigenetic phenomena and the evolution of plant allopolyploids. *Molecular Phylogenetics and Evolution* 29: 365–379.

Martins, R. 1976. *New Archaeological Techniques for the Study of Ancient Root Crops in Peru.* Ph.D. dissertation, University of Birmingham, Birmingham, UK.

Medina Hinostroza, T. C. 1994. *Contaje Cromosómico de la Oca (*Oxalis tuberosa *Molina) Conservada In Vitro.* Thesis, Universidad Nacional del Centro del Perú, Huancayo, Peru.

Morató Peña, L. and L. Morató Lara. 1995. *Quechua Qosqo–Qollaw–Basic Level: Trilingual Textbook for Classroom Instruction, Grammar and Dictionary.* The Latin American Studies Program, Cornell University, Ithaca, NY, USA.

National Research Council. 1989. *Lost Crops of the Incas: Little-Known Plants of the Andes with Promise for Worldwide Cultivation.* National Academy Press, Washington, DC, USA.

Nazarea, V. 1998. *Cultural Memory and Biodiversity.* University of Arizona Press, Tucson, AZ, USA.

Pearsall, D. M. 1992. The origins of plant cultivation in South America. In C. W. Cowan and P. J. Watson (eds.), *The Origins of Agriculture,* 173–205. Smithsonian, Washington, DC, USA.

Peña Paredes, A. 1978. *Determinación de Gravedad Específica, Materia Seca y Almidón en 95 Clones de Oca (*Oxalis tuberosa *Mol).* Thesis, Universidad Nacional de San Antonio Abad del Cusco, Cusco, Peru.

PLEC. 2001. International Symposium on Managing Biodiversity in Agricultural Ecosystems, Montreal, Canada, November 8–10, 2001. www.unu.edu/env/plec/cbd/abstracts/compiled-abstracts.pdf.

Prain, G. and D. Campilan. 1999. Farmer maintenance of sweetpotato diversity in Asia: Dominant cultivars and implications for in situ conservation. *Impact on a Changing World. Program Report 1997–98,* 317–327. CIP, Lima, Peru.

Prain, G. D., I. G. Mok, T. Sawor, P. Chadikun, E. Atmadjo, and E. Relwary Sitmorang. 1995. Interdisciplinary collecting of *Ipomoea batatas* germplasm and associated indigenous knowledge in Irian Jaya. In L. Guarino, V. R. Rao, and R. Reid (eds.), *Collecting Plant Genetic Diversity,* 695–711. CAB International, Oxon, UK.

Quiros, C. F., S. B. Brush, D. S. Douches, K. S. Zimmerer, and G. Huestis. 1990. Biochemical and folk assessment of variability of Andean cultivated potatoes. *Economic Botany* 44: 254–266.

Ramírez, M. 2002. On farm conservation of minor tubers in Peru: The dynamics of oca (*Oxalis tuberosa*) landrace management in a peasant community. *Plant Genetic Resources Newsletter* 132: 1–9.

Rea, J. and D. Morales. 1980. *Catálogo de Tubérculos Andinos.* Ministerio de Asuntos Campesinos y Agropecuaria. Instituto Boliviano de Tecnologia Agropecuaria (IBTA) Programa de Cultivos Andinos, La Paz, Bolivia.

Rivero Gonzáles, C. E. 1973. *Determinación Fitoquímica del Contenido de Almidón en 100 Clones de Oca de la Colección de Ocas del Cusco.* Thesis, Universidad Nacional de San Antonio Abad del Cusco, Cusco, Peru.

Rogstad, S. H., B. Keane, and J. Beresh. 2002. Genetic variation across VNTR loci in central North American *Taraxacum* surveyed at different spatial scales. *Plant Ecology* 161: 111–121.

Salick, J., N. Cellinese, and S. Knapp. 1997. Indigenous diversity of cassava: Generation, maintenance, use and loss among the Amuesha, Peruvian Upper Amazon. *Economic Botany* 51: 6–19.

Seminario, J. and I. Rimarachín. 1995. *Universidad y Biodiversidad Regional.* Instituto de Estudios Andinos de la Universidad Nacional de Cajamarca, Cajamarca, Peru.

Shah, A. A., H. Stegemann, and M. Galvez. 1993. The Andean tuber crops mashua, oca and ulluco: Optimizing the discrimination between varieties by electrophoresis and some characters of the tuber proteins. *Plant Varieties and Seeds* 6: 97–108.

Shigeta, M. 1996. Creating landrace diversity: The case of the Ari people and ensete (*Ensete ventricosum*) in Ethiopia. In R. Ellen and K. Fukui (eds.), *Redefining Nature: Ecology, Culture and Domestication,* 233–268. Berg, Oxford, UK.

Soltis, D. E. and P. S. Soltis. 1999. Polyploidy: Recurrent formation and genome evolution. *Trends in Ecology and Evolution* 14: 348–352.

Stegemann, H., S. Majino, and P. Schmiediche. 1988. Biochemical differentiation of clones of oca (*Oxalis tuberosa,* Oxalidaceae) by their tuber proteins and the properties of these proteins. *Economic Botany* 42: 37–44.

Swofford, D. L. 1998. *PAUP*: Phylogenetic Analysis Using Parsimony (*and Other Methods),* version 4.0b10. Sinauer Associates, Sunderland, MA, USA.

Terrazas, F. 1996. Identificación y estudio de la dinámica de microcentros de biodiversidad de RTAS y manejo, evaluación y utilización de la biodiversidad de raíces y tubérculos andinos. In M. Holle (ed.), *Programa Colaborativo Biodiversidad de Raíces y Tubérculos Andinos,* 15–36. Centro Internacional de la Papa, Lima, Peru.

Trognitz, B. R., M. Hermann, and S. Carrión. 1998. Germplasm conservation of oca (*Oxalis tuberosa* Mol.) through botanical seed. Seed formation under a system of polymorphic incompatibility. *Euphytica* 101: 133–141.

Valladolid, A., C. Arbizu, and D. Talledo. 1994. Niveles de ploidía de la oca (*Oxalis tuberosa* Mol.) y sus parientes silvestres. *Agro Sur (Universidad Austral de Chile, Facultad de Ciencias Agrarias)* 22 (número especial): 11–12.

Vallenas Ramirez, M. 1997. Refrescamiento de las accesiones del banco de germoplasma de oca conservada por semilla botánica. In *IX Congreso Internacional de Cultivos Andinos "Oscar Blanco Galdos": Libro de Resumenes Curso Pre-congreso,* 29–30 (abstract). Centro de Investigación en Cultivos Andinos (CICA), Universidad Nacional de San Antonio Abad del Cusco and Asociación ARARIWA para la Promoción Técnico Cultural Andina, Cusco, Peru.

Wendel, J. F. 2000. Genome evolution in polyploids. *Plant Molecular Biology* 42: 225–249.

Zimmerer, K. S. 1991a. Managing diversity in potato and maize fields of the Peruvian Andes. *Journal of Ethnobiology* 11: 23–49.

Zimmerer, K. S. 1991b. The regional biogeography of native potato cultivars in highland Peru. *Journal of Biogeography* 18: 165–178.

Zimmerer, K. S. 1998. The ecogeography of Andean potatoes: Versatility in farm regions and fields can aid sustainable development. *BioScience* 48: 445–454.

Zimmerer, K. S. and D. S. Douches. 1991. Geographical approaches to crop conservation: The partitioning of genetic diversity in Andean potatoes. *Economic Botany* 45: 176–189.

Kenneth Birnbaum

CHAPTER 15

Crop Genetics on Modern Farms
Gene Flow Between Crop Populations

The Green Revolution and other modern farming practices dramatically changed the composition of farmers' fields. In early assessments, a few modern varieties bred to produce high yields in very specific conditions were found to be rapidly replacing traditional varieties, which were bred and selected by farmers over millennia (Frankel and Hawkes, 1975; Frankel et al., 1995). This apparent abandonment of traditional varieties was cause for concern because crop breeders often used these cultivars as a source for resistance traits to combat devastating crop epidemics (Frankel and Hawkes, 1975; Frankel et al., 1995). However, careful fieldwork later demonstrated that traditional crops were not doomed, especially in marginal farming conditions. Several studies in different regions of the world showed that farmers often maintained traditional varieties even while adopting modern cultivars (Brush, 1992, 1995, 2000; Bellon and Brush, 1994; Maxted et al., 1997). Thus, a tenuous coexistence appears to have developed decades after the Green Revolution. In this chapter I focus on the genetic implications of that coexistence, paying particularly close attention to gene flow between modern and traditional crops.

The central issue is that modern crop populations are typically large and genetically homogeneous. They can swamp out a smaller population when interbreeding occurs, causing rapid losses of genetic diversity in traditional diverse populations (Ryman et al., 1995). The first part of this chapter addresses the scope of such crop-to-crop hybridization, explores the population genetics involved in this type of gene flow, and focuses on critical parameters to quantify the loss of genetic diversity in traditional populations due to gene flow.

Another intriguing issue is the fate of traditional populations when gene flow is low enough to allow them to survive at some level. On a population level, their genetic structure probably will change through gene flow from modern varieties. The post–Green Revolution represents an opportunity for farmers to dramatically reshape their genetic resources. In the second part of this chapter I focus on some examples of how farmers orchestrate genetic change. Even in the age of transgenics, farmers may still be the ultimate engineers of crop genetics, mixing modern varieties with traditional ones.

The critical task will be distinguishing when the level of gene flow shifts from potentially beneficial to detrimental, causing rapid losses of genetic diversity. The critical level of gene flow is crop specific, depending on factors such as population size and life history traits. Here I discuss some methods to assess the effects of gene flow to help determine its effect on genetic diversity in crop populations.

Definitions: Modern Versus Traditional Crops

It is important to clarify the meaning of *modern crop varieties.* The term is used here to mean any crop variety that is planted in large numbers ranging from hundreds to even tens of thousands of individuals. Such modern varieties may be inbred lines or hybrid lines derived from many generations of breeding. Alternatively, they may be varieties considered to be landraces or traditional varieties that have been singled out for large-scale cultivation. This latter definition purposefully blurs the distinction between modern and traditional varieties. In this chapter, traditional varieties are considered direct descendants of a diverse population from which any individual genotype is propagated on a small scale. Thus, the concern with hybridization between modern and traditional crop populations may include the introduction of exotic alleles into a population with the threat of outbreeding depression (Brown, 2000; Allendorf et al., 2001). When such gene flow occurs between two species, the phenomenon has been called extinction by hybridization (Rhymer and Simberloff, 1996). Alternatively, populations can also be affected by gene flow from genotypes within the same population, in which case the main threat is the loss of genetic diversity caused by highly skewed breeding success. Because the allelic invasion comes from within the same population or a closely related one, the latter case can be called death by dilution.

Population Genetic Issues

One of the primary issues in the conservation of crop plants is maintaining the high levels of genetic diversity that are typical of traditional crop populations (Brown, 2000; Maxted et al., 1997). The reasoning is that this diversity may be useful in the future. In general, smaller populations undergo more genetic drift and tend to lose allelic diversity, or the genetic diversity associated with genes. Thus the census size of a crop population is important, but it is not the only factor determining population size for the sake of retaining allelic diversity. For example, a population in which breeding success is skewed to a few individuals can suffer a significant decrease in effective population size. The drift effects caused by breeding disparities (above random noise) can be quantified by equating them to an equivalent population size that has the same degree of genetic drift, $N_{e(v)}$, the variance effective population size, hereafter called effective population size (Crow and Kimura, 1970).

In wild populations, disparities in breeding success can lead to $N_{e(v)}/N$ ratios of about 0.5 (Nunney and Elam, 1994). However, in crop populations, proliferation of grafted or inbred lines has the potential to greatly skew the breeding success of a few individuals, with dramatic effects on effective population size. An analogous conservation problem has been studied in captive breeding programs and fishery management, where a highly prolific stock population feeds progeny into a source population (Ryman et al., 1995). As an example of the severity of the problem in plants, a population of 1000 crop plants in which 999 plants mate randomly but a single variety contributes 10% of all gametes has an effective size of only 100.

Modern crop varieties are cultivated in ways that can dramatically increase the breeding success of a few varieties. As mentioned earlier, crop varieties often are derived from inbred lines to create genetically uniform seed stocks. Many tree varieties are grafted to form orchards of genetic clones. The collective breeding success of genetically identical individuals is, in effect, the breeding success of a single individual. Thus, some of the critical parameters in measuring the effects of gene flow in crop populations are the size of genetically uniform populations, the number of different varieties within the largely uniform populations, and the breeding success of each of the genetically uniform varieties. In addition, simulations have shown that the longevity of the uniform varieties can have a critical impact on long-term genetic diversity (Birnbaum et al., 2002).

Crop Hybridization Is Widespread

Hybridization between domesticated crops and their wild relatives has been studied more extensively than crop-to-crop hybridization (Lee and Snow, 1998; Ellstrand et al., 1999; Jenczewski et al., 1999; Burke et al., 2002; Montes-Hernandez and Eguiarte, 2002). It can be used to gauge the potential for crop-to-crop hybridization because analogously it involves spontaneous gene flow between closely related plants that occur in the same habitat.

In one survey of the 13 most important food crops, there was evidence of cross-hybridization between crops and wild relatives in 12 of 13 cases that were examined (Ellstrand et al., 1999). Additional evidence for hybridization has been found in sunflower (Burke et al., 2002), squash (Montes-Hernandez and Eguiarte, 2002), radish (Lee and Snow, 1998), and clover (Jenczewski et al., 1999). The level of gene flow into wild populations appears to be high enough in some cases to threaten those populations. For example, there was evidence that wild relatives of rice and cotton could face extinction through hybridization with crop populations (Ellstrand et al., 1999). Researchers have also noted that hybrid populations could be more prone to population crashes caused by outbreeding depression (Brown, 2000; Allendorf et al., 2001).

All the conditions that lead to gene flow between crops and their wild relatives exist between modern and traditional crop populations. Indeed, modern and traditional crops typically are more closely related than crops and their wild relatives, a factor that should lower the barrier to gene flow. In addition, different varieties of crop plants often are planted on the same farm. Thus the opportunity for hybridization between modern and traditional crops probably is greater than between crop plants and their wild relatives.

Hybridization in the Field

Despite the potential for loss of genetic diversity through cross-pollination, few studies have examined the effects of gene flow between varieties in crop plants. The studies that do exist suggest that gene flow is likely to be common (Jenczewski et al., 1999; Louette, 2000; Birnbaum et al., 2003). However, one intriguing theme in these studies is that farmers appear to create barriers to gene flow. For example, Louette (2000) showed evidence of gene flow from exotic maize varieties into a population of Mexican cultivars. Most of the gene flow was limited to short distances (within a few meters), but planting practices that placed different varieties side by side created conditions in which gene flow was likely to be widespread. However, in experiments that measured quantitative traits over two generations, farmers appeared to select seed in order to maintain specific characteristics of the original varieties. This showed how farming practices had the potential to oppose gene flow from contaminating varieties to maintain quantitative traits. Whether these practices could limit gene flow enough to maintain genetic diversity is not known.

A similar scenario was found in a study of *Medicago sativa* in Spain (Jenczewski et al., 1999). The authors studied gene flow from wild relatives into domesticated varieties of *Medicago* using quantitative traits, which are presumably under farmer selection, and allozymes, which are considered neutral (unselected) markers. First, genetic markers indicated that hybridization was common in some populations. Interestingly, in some hybrid populations the neutral allozyme markers provided evidence of a high level of gene flow, whereas quantitative traits remained distinct. The authors concluded that selection for specific traits by farmers was a likely explanation for why quantitative traits between the two populations remained distinct.

The level of gene flow was specifically examined in a case study of avocados (*Persea Americana* var. *americana*) in Central America (Birnbaum et al., 2003). This tree population provided a good system to study the effects of gene flow because of the crop's life history traits and recent cultivation history. The study area, on the Pacific Coast of Costa Rica, included the putative region where the West Indian variety of avocados was domesticated and remains a center of diversity for this crop species. The use of grafting techniques in the past 20 years created orchards of genetic clones. This meant that some cohorts in the population arose before the large-scale planting of grafts dramatically changed the genetic composition of orchards. These cohorts could be compared with more recent cohorts in which gene flow from grafts occurred. In addition, avocado has a high rate of outcrossing, making it a good system to study the effects of gene flow from a highly uniform population, which consisted of grafted varieties in this case. To assess the level of gene flow between the populations, DNA microsatellite molecular markers were used in a parentage analysis.

Although only five grafted varieties made up 40% of the population, their collective rate of gene flow into the population was only 14.5%. That is, among all the gametes that made up the next generation of avocado seedlings, the grafted varieties contributed only 14.5%, which was significantly less than expected under rules of random mating. Computer modeling showed that the observed level of gene flow led to minor losses of genetic diversity over 150 years. Some clues to what prevented higher gene flow came from the farmers. Although they typically planted both grafted and traditional varieties on the same farm, they tended to separate them physically, apparently slowing the rate of gene flow by cross-pollination. They also avoided planting the seed of grafted avocados, largely preventing maternal gene flow from the genetically uniform subpopulation. Although other factors may have contributed to the moderate gene flow between the populations, farming practices appeared to present significant barriers. Thus the intriguing result was that farmers had a strong hand in conserving the genetic diversity of their traditional population.

Hybridization and Allele Frequency Shifts

The aforementioned studies raise the possibility that gene flow from modern or select varieties may not cause the extinction of traditional populations but rather push them toward a new evolutionary trajectory. Is there any evidence of such evolutionary shifts?

The Costa Rican avocado population offered an opportunity to examine the effect of gene flow from grafted varieties over time (Birnbaum et al., 2003). Because the widespread grafting of few varieties started only about 20 years before the study, older trees represented a sample of the population before the onset of gene flow from graft varieties. To address whether graft gene flow caused changes in population structure, 10 DNA microsatellite markers were used to assess allele frequency changes in the avocado population in San Jerónimo, Costa Rica. DNA fingerprint patterns from leaf samples of 20- to 25-year-old trees ($n = 56$), which were established before the arrival of grafted trees, were compared with 0- to 5-year-old trees ($n = 88$), which were established two to three generations after grafted trees began interbreeding with the population.

Microsatellites themselves are largely considered neutral markers, having no selective advantage or disadvantage (Birnbaum and Rosenbaum, 2002), but they serve as markers for changes in segments of the genome. A unique microsatellite allele indicates the presence of a potentially unique set of functional alleles at loci that are physically linked to it. One way to summarize the genetic composition of a sample population for a given set of markers is the distribution of allele frequencies, which is a plot of the relative frequency of each allele found in the sample (figure 15.1). Statistical tests can then assess whether two allele frequency distributions are different (accounting for possible sampling noise). Results from

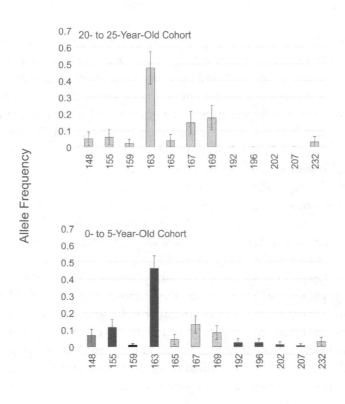

Microsatellite Alleles (base pairs)

FIGURE 15.1 Allele frequency distributions at one locus showing the introduction of graft variety alleles into the population. Different microsatellite alleles, of different sizes (*x*-axis), are shown in the top panel in a sample of 56 individuals that were 20–25 years old. In the lower panel, the frequency of microsatellite alleles among younger individuals (age cohort 0–5 years old, *n* = 88) are shown. In the young cohort, black bars indicate alleles that are present in the varieties that were frequently grafted. Thus, some of the graft alleles were already present in the older cohort (alleles 148, 155, 163), whereas other alleles appear to be more recent and unique to the grafted varieties (alleles 192, 196, 202, 207). These unique graft alleles provide support for graft varieties interbreeding with the local population in the last 20 years. The locus depicted is AVO 102 (see Birnbaum et al., 2003).

many loci can be combined to make an overall statement about genetic change.

A comparison of the avocado population, using the exact test for population differentiation (Goudet et al., 1996; Raymond and Rousset, 1995), shows a significant difference in the allele frequency distribution between the 20- to 25-year-old cohort and 0- to 5-year-old cohort ($p < .0015$), meaning that the composition of genetic markers in the avocado population has changed since the arrival of grafts.

Three observations indicate that graft introgression is the likely cause of the allele frequency shifts. First, a similar set of cohort samples taken in a control sampling site, Londres, where no mature grafts were present, showed no significant change in allele frequencies over the same cohorts (exact test for population differentiation, $p > .18$). Second, the difference in allele frequency distributions over cohorts in San Jerónimo was no longer significant when 16 trees determined to be graft progeny were excluded from the comparison (exact test for population differentiation, $p < .08$). Finally, graft alleles consistently increased in frequency in the most recent cohort. For example, several alleles at the microsatellite locus AVO102 that were present in graft varieties were not detectable in samples of the older cohorts (20–25 years old) but were found in samples of the younger cohorts (0–5 years old; figure 15.1). Similar patterns were evident at nine other loci examined. Thus, grafts caused a directed genetic change in the population as measured by the microsatellite markers. This shift in population structure was caused in part by gene flow from varieties that were exotic to the population examined. However, many graft alleles that increased in frequency were from varieties native to the population. Overall, this shows how new microevolutionary trajectories can be driven by the changes in the patterns of gene flow within a crop population.

The Farmer's Hand: Creating New Varieties

The role of hybridization as a force of genetic change in a population also raises an intriguing question about specific effects farmers have in guiding evolutionary trajectories. At a minimum, farmers have a significant impact on population change by controlling the level of gene flow; for example, they appear to decrease gene flow between two subpopulations by increasing the spacing between them. It also seems likely that farmers play a more active role in shaping what genetic traits from graft populations increase in frequency in the traditional population. We lack a body of research that examines how farmers control the population structure of crops over time, although some work has been done on population size during domestication (for example, see Eyre-Walker et al., 1998). However, case studies offer some insights into the ways in which farmers may control gene flow between populations to shape crop genetics.

From the avocado study in Jerónimo, Costa Rica, parentage analysis with molecular markers illustrated how experimentation creates a phase of cultivation in which farmers can screen for favorable genetic traits to enter the traditional population. The competition in the domestic avocado market in Costa Rica led to a market-driven atmosphere in which farmers were constantly searching for new varieties for which buyers would pay higher prices. As a result, farmers continually sampled new varieties, and some farmers planted the seeds of favorable varieties in addition to cultivation by grafting. Farmers know by experience that such experiments usually are doomed to failure because progeny from a desirable variety rarely bear fruit similar to that of the parent. Fruit characteristics are controlled by multiple loci in

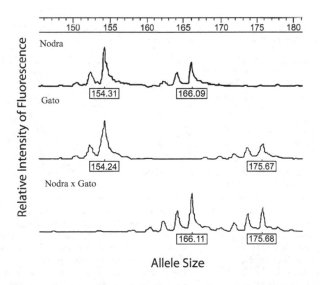

Allele Size

FIGURE 15.2   The Mendelian inheritance of microsatellite alleles is used in a parentage analysis to assess gene flow from grafted varieties. Peaks, which are shown in the Genotyper program, represent the size of microsatellite alleles (in base pairs) as they migrate on a polyacrylamide gel. Allele sizes within a range of one base pair are assumed to be the same allele. The individual at the bottom panel was determined to be the F₁ offspring of the 2 individuals above, inheriting the 166 allele from the upper individual, Nodra, and the 176 allele from the individual in the middle panel, Gato.

this highly diverse population, and open pollination invariably reshuffles a fortuitous combination of alleles. However, these experiments occasionally do succeed.

In one case, a farmer planted the most widely grafted avocado in town, called Gato, next to the second most popular grafted variety, Nodra. The farmer's wife planted out several seedlings of Gato tree as an experiment. As suspected, parentage analysis using DNA microsatellite fingerprinting showed that several of the experimental seedlings were Nodra × Gato hybrids (figure 15.2). Thus, two highly marketable varieties were hybridized to create a new genotype in three of the six progeny examined on the farm (table 15.1). Any of these trees possessing favorable fruits probably will be widely grafted, greatly expanding its reproductive potential. Thus the example shows how farmers incorporate and amplify graft alleles in the population after testing on the farm.

In a study of barley in the Fertile Crescent, Ceccarelli and Grando (2000) measured important agronomic traits of crosses between traditional

Table 15.1 Parentage Analysis of 6 Avocado Saplings from a San Jerónimo Farm Showing How Farmers Selectively Amplify and Combine Specific Genotypes

	Parent #1 —Gato			Parent #2 —Nodra		
Seedling	Shared Alleles	Inclusionary Confidence of Parentage[a]	LOD Score[b]	Shared Alleles	Inclusionary Confidence of Parentage[a]	LOD Score[b]
445	10/10	99%	7.1	2/9	<1%	<0
446	10/10	99%	5.3	9/9	99%	7.2
447	9/9	99%	4.8	9/9	99%	8.0
448	9/9	99%	10.9	3/9	<1%	<0
449	9/9	99%	4.4	10/10	99%	9.4
450	9/9	97%	9.2	4/9	<1%	<0

For each candidate progeny, 9 or 10 loci were used for the parentage analysis. Parentage was determined by first identifying individuals that share an allele from a parental graft at each locus (genotyping errors are permitted at 1 locus in LOD score analysis). The frequency of the allele in the general population determines the confidence with which a parentage assessment can be made. In this case, 6 seedlings were determined with high confidence to be the progeny of Gato, and 3 of the 6 (*shaded rows*) also had Nodra as the likely other parent. The result showed how one farmer combined alleles from 2 popular grafted varieties despite the fact that these varieties typically are propagated clonally. The results from 2 types of analysis are presented: inclusionary analysis (Westneat and Webster, 1994; Dow and Ashley, 1996) and a log-likelihood test (LOD) (Meagher, 1986; Marshall et al., 1998).

[a]The probability that only the parent in question could have passed on the shared alleles in a population of 1500 separate genotypes.

[b]How many times more likely the variety in question is to be the parent of the seedling than a randomly chosen genotype in the population.

landraces and improved varieties. They found that under stress conditions, crosses between landraces and improved varieties possessed beneficial characteristics of both parents. For example, these hybrids retained higher yields associated with landraces under stressful conditions but superior to either parent for plant height, a factor important to local farmers for ease of harvesting. The authors stated that such crosses are likely to occur naturally on farms, demonstrating how farmers can combine the resources generated by plant breeders with those of traditional farming systems.

Conclusions

Crop domestication is a period of rapid genetic change resulting from population bottlenecks and intense selection by farmers. However, crop evolution and the domestication process continue into the present. Several lines of evidence suggest that we are undergoing rapid changes in the population structure of domesticated crops. How the rate of change compares with previous periods since domestication remains an important question.

Much attention has been paid to gene flow from genetically modified crops into open pollinated populations. The effects of genetic modifications at the species and community level and on human health are important issues. However, changes in the way farms are managed raise another important issue concerning gene flow from one crop population to another. In the regions that harbor crop genetic diversity, modern farming practices, even on traditional farms, often create two distinct subpopulations. One is highly homogeneous, carrying little genetic diversity. The other often is a genetically diverse traditional crop population, typically the result of a long history of cultivation.

The problem is not necessarily the particular alleles that move from one population to another but rather the quantity. That is, the homogeneous subpopulations often are large, and their high rate of gene flow into the diverse population has the potential to swamp the allelic diversity of the traditional population. The traditional population may technically survive, but its diversity of alleles may decrease dramatically. Valuable traits could be lost.

Is there any evidence that extensive gene flow from a genetically narrow population has led to a loss of diversity in a traditional population? At present, there is good evidence that hybridization between different crop populations is common. However, there is little evidence to say conclusively whether levels of diversity, as measured by either molecular markers or quantitative traits, have been lost. It may still be too early to tell because such genetic erosion can occur over many generations. In the avocado study presented, some methods have been presented that will help assess the effects of crop gene flow on genetic diversity.

Interestingly, farmers appear to be an important factor in the gene flow equation. Several studies show that farming practices limit gene flow from homogeneous populations into diverse ones. This may slow the loss of genetic diversity and at least buy time in the efforts to conserve crop genetic resources. A more optimistic scenario is that farmers are simply navigating crop populations through a new period of crop evolution, adding favorable alleles to the gene pool while maintaining diversity.

The role of introgression as an evolutionary force raises important questions about the nature of its effects. Are neutral and quantitative traits becoming more or less diverse in traditional crop populations? Does gene flow from genetically homogenous individuals decrease or increase the fitness or yield of a crop population? Does gene flow lead to changes in important, complex agronomic traits? How does farmer selection reinforce or oppose allele frequency changes due to gene flow?

Research is needed on crop populations that can be monitored as introgression occurs on farms with supplemental trials in controlled experiments to measure genetic changes in complex traits. Such experiments will work best in a crop with well-developed genetic tools and short generation times, such as an annual crop (e.g., corn, soybean, or rice). In addition, more work is needed to determine the effects of introgression on genetic diversity in crops with different breeding systems, life history traits, and management regimes (Wolf et al., 2001). In annual crops where age cohorts are not available, historic collections may help us compare changes in populations before and after the use of genetically homogeneous populations.

References

Allendorf, F. W., R. B. Leary, P. Spruell, and J. K. Wenberg. 2001. The problems with hybrids: Setting conservation guidelines. *Trends in Ecology and Evolution* 16: 102–108.

Bellon, M. R. and S. B. Brush. 1994. Keepers of maize in Chiapas, Mexico. *Economic Botany* 48: 196–209.

Birnbaum, K., P. N. Benfey, C. M. Peters, and R. Desalle. 2002. ManagedPop: A computer simulation to project allelic diversity in managed populations with overlapping generations. *Molecular Ecology Notes* 2: 615–617.

Birnbaum, K., R. Desalle, C. M. Peters, and P. N. Benfey. 2003. Integrating gene flow, crop biology and farm management in the on-farm conservation of avocado (*Persea americana,* Lauraceae). *American Journal of Botany* 90: 1619–1627.

Birnbaum, K. and H.C. Rosenbaum. 2002. A practical guide to microsatellites. In R. DeSalle, G. Giribet, and W. Wheeler (eds.), *Techniques in Molecular Systematics and Evolution*, 351–364. Birkhäuser Verlag, Basel, Switzerland.

Brown, A.D.H. 2000. The genetic structure of crop landraces and the challenge to conserve them in situ on farms. In S.B. Brush (ed.), *Genes in the Field: On Farm Conservation of Crop Diversity*, 29–48. Lewis Publishers, Boca Raton, LA, USA.

Brush, S.B. 1992. Reconsidering the green revolution: Diversity and stability in cradle areas of crop domestication. *Human Ecology* 20: 145–167.

Brush, S.B. 1995. In situ conservation of landraces in centers of crop diversity. *Crop Science* 35: 346–354.

Brush, S.B. 2000. *Genes in the Field: On-Farm Conservation of Crop Diversity*. Lewis Publishers, Boca Raton, LA, USA.

Burke, J.M., K.A. Gardner, and L.H. Rieseberg. 2002. The potential for gene flow between cultivated and wild sunflower (*Helianthus annuus*) in the United States. *American Journal of Botany* 89: 1550–1552.

Ceccarelli, S. and S. Grando. 2000. Barley landraces in the Fertile Crescent: A lesson for plant breeders. In S.B. Brush (ed.), *Genes in the Field: On Farm Conservation of Crop Diversity*, 51–76. Lewis Publishers, Boca Raton, LA, USA.

Crow, J.F. and M. Kimura. 1970. *An Introduction to Population Genetics Theory*. Harper & Row, New York, NY, USA.

Dow, B.D. and M.V. Ashley. 1996. Microsatellite analysis of seed dispersal and parentage of saplings in bur oak, *Quercus macrocarpa*. *Molecular Ecology* 5: 615–627.

Ellstrand, N.C., H.C. Prentice, and J.F. Hancock. 1999. Gene flow and introgression from domesticated plants into their wild relatives. *Annual Review of Ecology and Systematics* 30: 539–563.

Eyre-Walker, A., R.L. Gaut, B. Hilton, D.L. Feldman, and B.S. Gaut. 1998. Investigation of the bottleneck leading to the domestication of maize. *Proceedings of the National Academy of Sciences* 95: 4441–4446.

Frankel, O.H., A.H.D. Brown, and J.J. Burdon. 1995. *The Conservation of Plant Biodiversity*. Cambridge University Press, Cambridge, UK.

Frankel, O.H. and J.G. Hawkes. 1975. *Crop Genetic Resources for Today and Tomorrow*. Cambridge University Press, Cambridge, UK.

Goudet, J., M. Raymond, T. de Meeus, and F. Rousset. 1996. Testing differentiation in diploid populations. *Genetics* 144: 1933–1940.

Jenczewski, E., J.M. Prosperi, and J. Ronfort. 1999. Evidence for gene flow between wild and cultivated *Medicago sativa* (Leguminosae) based on allozyme markers and quantitative traits. *American Journal of Botany* 86: 677–687.

Lee, T.N. and A. Snow. 1998. Pollinator preferences and the persistence of crop genes in wild radish populations (*Raphanus raphanistrum*, Brassicaceae). *American Journal of Botany* 85: 333–339.

Louette, D. 2000. Traditional management of seed and genetic diversity: What is a landrace? In S.B. Brush (ed.), *Genes in the Field: On-Farm Conservation of Crop Diversity*, 109–142. Lewis Publishers, Boca Raton, LA, USA.

Marshall, T.C., J. Slate, L.E.B Kruuk, and J.M. Pemberton. 1998. Statistical confidence for likelihood-based paternity inference in natural populations. *Molecular Ecology* 7: 639–655.

Maxted, N., B.V. Ford-Lloyd, and J.G. Hawkes. 1997. Complementary conservation strategies. In N. Maxted, B.V. Ford-Lloyd, and J.G. Hawkes (eds.), *Plant Genetic Conservation: The In Situ Approach*, 15–39. Chapman & Hall, London, UK.

Meagher, T. R. 1986. Analysis of paternity within a natural population of *Chamaelirium leteum*. 1. Identification of the most likely male parents. *American Naturalist* 128: 199–215.

Montes-Hernandez, S. and L. E. Eguiarte. 2002. Genetic structure and indirect estimates of gene flow in three taxa of *Cucurbita* (Cucurbitaceae) in western Mexico. *American Journal of Botany* 89: 1156–1163.

Nunney, L. and D. Elam. 1994. Estimating the effective population size of conserved populations. *Conservation Biology* 8: 175–184.

Raymond, M. and F. Rousset. 1995. An exact test for population differentiation. *Evolution* 49: 1280–1283.

Rhymer, J. M. and D. Simberloff. 1996. Extinction by hybridization and introgression. *Annual Review of Ecology and Systematics* 27: 83–109.

Ryman, N., P. E. Jorde, and L. Laikre. 1995. Supportive breeding and variance effective population size. *Conservation Biology* 9: 1619–1628.

Westneat, D. F. and M. S. Webster. 1994. Molecular analysis of kinship in birds: Interesting questions and useful techniques. In B. Schierwater, B. Steit, G. P. Wagner, and R. DeSalle (eds.), *Molecular Ecology and Evolution: Approaches and Applications*, 91–126. Birkhäuser Verlag, Basel, Switzerland.

Wolf, D. E., N. Takebayashi, and L. H. Rieseberg. 2001. Predicting the risk of extinction through hybridization. *Conservation Biology* 15: 1039–1053.

Sarah M. Ward

Molecular Marker and Sequencing
Methods and Related Terms

What Is a Marker?

The basic function of all genetic markers is the detection of genotypic variation between individuals. Molecular marker techniques are used to detect the presence of specific DNA sequences in the nuclear or organelle genomes of a plant. Most molecular marker systems use some form of electrophoresis to separate either different DNA sequences or the proteins they encode. Before the development of electrophoresis few genetic markers were available in plants. Researchers relied on variation in phenotypic traits such as flower color or seed type, preferably controlled at a single locus. Markers based on phenotypic differences provide some information on individual genotypes and on the levels of genetic variation in plant populations but are limited in scope and availability. Other disadvantages associated with phenotypic markers are that they may interact with environmental factors that affect the observed phenotype, they may not be selectively neutral (differing fitness levels associated with different phenotypes may result in generational changes in genetic variation measured at a given locus), and inferring variation at the DNA level from phenotypic observation presumes gene expression. Silent (i.e., nonexpressed) alleles cannot be detected, leading to underestimation of the genotypic variation actually present in the population.

Development of protein electrophoresis techniques in the 1950s, and later the use of gel electrophoresis to separate DNA fragments generated either by restriction enzymes or by the polymerase chain reaction (PCR), opened the way for a growing array of molecular markers. Although phenotypic markers are still used, development of molecular marker systems has made possible a wide range of applications. These include using molecular markers to determine whether a plant carries a particular allele, investigating the composition and structure of plant genomes, and investigating phylogenetic or taxonomic relationships between plants by comparing differences in DNA sequences. In this appendix the most widely used current molecular marker systems are described, together with a brief overview of the laboratory techniques that make them possible.

Key Laboratory Techniques Used in Molecular Marker Systems

Gel Electrophoresis

Electrophoresis separates molecules such as proteins and nucleic acids based on differences in their size, shape, and electrical charge. **Starch gels** are used to separate proteins such as allozymes, and **agarose** and **polyacrylamide** (PAGE) **gels** that can achieve higher resolution are more commonly used to separate polymorphic DNA fragments. When direct current is applied to the gel to create an electrical field, the preloaded molecules migrate through the pores in the gel, reaching different locations depending on their rate of movement. Those that are smaller or have a higher charge density will move faster and further. The gel is removed from the electrophoresis chamber and stained or probed to visualize the relative positions of the separated molecules. Commonly used stains include **Coomassie blue dye** for proteins and **ethidium bromide** or **silver nitrate** for DNA. **Probes** are molecules that identify and attach to a specific subset of the separated molecules; for example, an individual DNA sequence can be located on the gel by first denaturing it and then probing with a complementary oligonucleotide that will anneal to the exposed single-stranded base sequence. Probes typically are labeled with radioisotopes or chemoluminescent dyes so they can be tracked.

Other electrophoresis techniques used in marker work include **two-dimensional electrophoresis** (2DE) and **capillary electrophoresis** (CE). In 2DE, separation of very similar molecules is achieved through two consecutive electrophoresis runs. During the first run, molecules are separated along a pH gradient: A molecule stops migrating when it reaches the **isoelectric point (pI)** in the gradient where its net charge is zero. During the second run, electrical current is applied at right angles to the direction of the first run, and molecules are further separated on the basis of molecular weight. Two-dimensional electrophoresis is especially powerful for separating similar proteins and has been widely used in proteomics.

In capillary electrophoresis a sample of the mixture to be separated is loaded into a small tube of fused silica. The tube is filled with buffer, and a high-voltage current is applied. Molecules in the mixture move at different speeds in the resulting electrical field depending on their size and charge, passing through a detection system based on their absorbance of ultraviolet or other short-wave light beamed through the tube. Different mixture components are recorded on a graph as peaks. Concentration of each component can be quantified from the peak area, and light absorbance, migration time, charge, and size allow identification of different molecules. Capillary electrophoresis is increasingly used for sequence-based DNA markers such as single nucleotide polymorphisms (SNPs) because it can be automated for high-throughput systems and uses very small sample quantities.

Restriction

Restriction is the targeted cutting of DNA using enzymes to break the phosphodiester bonds in the sugar–phosphate backbone of the DNA strand. Hundreds of different enzymes capable of this targeted cutting, known as **restriction endonucleases,** have been isolated from bacteria, where they defend the cell against viral invasion by digesting foreign DNA. Restriction endonucleases cut DNA only at specific locations known as **restriction sites.** Type I and Type III restriction endonucleases have one subunit for target site recognition and another for restriction. Consequently the actual cutting of the DNA by these enzymes may take place up to several hundred bases distant from the recognition site, and the sequence actually cleaved is not always specific. By contrast, Type II restriction endonucleases are highly targeted in their mode of action. The restriction site for a Type II endonuclease typically consists of a palindromic nucleotide sequence (reading the same forwards and backwards on opposite strands) four or six bases long and specific to one enzyme. For example, the recognition and target site for the widely used *EcoRI* Type II restriction endonuclease is

5' GAATTC 3'
3' CTTAAG 5'

EcoRI cleaves DNA only where it finds this target sequence, and it always breaks the phosphodiester bond between 5' GA 3'. DNA from any organism can be restricted, not only bacterial.

Targeted DNA restriction using Type II endonucleases is the basis of recombinant DNA technology because sequences cleaved using the same enzyme have compatible cut ends that can be ligated even if the restricted DNA is from different species. Some molecular marker systems such as restriction fragment length polymorphism (RFLP) and amplified fragment length polymorphism (AFLP) use DNA restriction. These markers exploit the fact that genotypically distinct individuals have target sequences at different locations in their genomes, so restriction with the same endonuclease generates DNA fragments of different lengths. The RFLP and AFLP marker systems are described in more detail later in this appendix.

Polymerase Chain Reaction

The term *polymerase chain reaction* and the use of this technique to amplify single copy DNA sequences using site-specific primers was first described in two key articles by Saiki et al. (1985) and Mullis et al. (1986), based on in vitro DNA replication protocols earlier proposed by Panet and Khorana (1974). At its most basic PCR synthesizes multiple copies of a DNA segment lying between two known sequences. This entails first denaturing the DNA to be copied (the template DNA) to expose the base sequence, and then adding two single-stranded DNA primers, each up to approximately 30 bases long and complementary to at least part of the known flanking sequences. The primers anneal to the exposed flanking sequences in a 5'→3' direction on opposite strands of the template DNA. A DNA polymerase (typically a thermostable Taq polymerase) then copies both strands of the DNA lying between the annealed primers by adding nucleotides to the 3' end of each primer. This completes one cycle of the PCR reaction. The next cycle is initiated by reheating the reaction mixture to denature the original DNA template and the newly formed copies. More primers then anneal to the exposed flanking bases, and copying of the intervening DNA is repeated, this time generating twice as much product as in the first cycle because the number of available priming sites and DNA template sequences has doubled. With each subsequent reaction cycle the amount of amplified DNA continues to double, so after 25–30 cycles thousands of copies have been made of the targeted DNA segment located between the priming sites. The ends of each

DNA copy are defined by the 5' termini of the primers, and the length of each amplified DNA fragment depends on the distance between the priming sites. A unique segment of a genome that has been amplified by a pair of PCR primers in this way is called a **sequence tagged site**.

Theoretically, any DNA segment can be amplified with a high degree of specificity and fidelity using this PCR protocol, provided suitable flanking sequences to serve as priming sites can be identified on opposite DNA strands, each oriented in a 5'→3' direction. Numerous adaptations of the basic PCR method have been developed since Mullis and his co-workers first published their original description. An important modification of PCR for molecular marker applications is random priming PCR, which does not require prior knowledge of the nucleotide sequences flanking the DNA to be amplified. This is described in more detail later in this appendix. Other widely used PCR adaptations include reverse transcriptase PCR (RT-PCR), nested PCR, and real time PCR, each of which is described briefly here.

Reverse Transcriptase PCR

Whereas conventional PCR uses a DNA polymerase to makes multiple DNA copies of sequences from a DNA template, RT-PCR uses the enzyme reverse transcriptase to produce cDNA copies from an RNA template. One application of this technique is to use appropriate primers to detect and investigate rare mRNA transcripts occurring at low frequencies in the cell. After the initial PCR cycle, additional cDNA copies of the rare mRNA sequence are available as templates in the reaction mixture, so the final cDNA amplification product is sufficiently abundant to allow sequencing of the rare RNA.

Nested PCR

In nested PCR, two pairs of primers are used sequentially to amplify the same locus. The second pair of primers anneal within the amplified PCR product produced by the first pair of primers. This results in a final PCR product shorter than that generated by the first primers. Nested PCR greatly increases accuracy of amplification. For example, if the template DNA contains paralogs (different genes with similar sequences, often coding for related products), it is possible that the first pair of primers will anneal at more than one site. If nested PCR is used, however, the probability of the second primer pair also amplifying the incorrect sequence is very low.

Real Time PCR

In real time PCR (Belgrader et al. 1998) the amount of amplification product present is automatically monitored at the end of each cycle, for example by incorporating fluorescent dye into the newly synthesized DNA and measuring the amount and wavelength of light emitted. Amplification output thus measured is recorded on a graph instead of the final PCR product being run on a gel for visualization. Real time PCR allows very rapid and sensitive detection of the nucleic acid sequences targeted by the primers used; detection times can be as low as a few minutes as opposed to hours when the PCR product must be visualized on a gel. Simultaneous detection of more than one DNA sequence is made possible by **multiplexing**: using more than one primer pair in a single reaction tube, each pair labeled with a different color fluorescent dye.

Combinations of PCR methods can also be used, depending on the desired outcome. For example, RT-PCR can be nested, or inverse PCR output can be monitored in real time.

DNA Sequencing

Analyzing genotypic differences between individuals by direct comparison of nucleotide base sequences at selected loci is now feasible as DNA sequencing technology becomes faster, more accurate, and less expensive. The most widely used sequencing technology is based on the dideoxy method, also known as the Sanger sequencing technique after its inventor, Fred Sanger. Dideoxy sequencing is based on replicating DNA strands in vitro and halting the replication process at random points in the sequence by incorporating a color-labeled artificial base. Denatured DNA is first mixed with the components needed for in vitro replication: DNA polymerase, a primer, and deoxyribonucleotide triphosphates (dNTPs) of all four bases: dATP, dGTP, dCTP, and dTTP. This mixture is then divided between four reaction tubes, to each of which is added a small quantity of one of four different dideoxyribonucleotide triphosphates ddNTPs: ddATP, ddGTP, ddCTP, or ddTTP. A ddNTP resembles the equivalent dNTP closely enough to be incorporated into the newly replicated DNA strand, but unlike a normal dNTP it has no 3' hydroxyl group. The absence of the hydroxyl group prevents the DNA polymerase from adding the next nucleotide triphosphate as it normally would, so replication of an individual DNA strand is terminated every time a ddNTP is added. Each of the four reaction tubes in the sequencer has just one of the four ddNTPs, so DNA strands replicated in that tube all terminate at different points in the sequence but with the same base. For example, in the tube to which ddCTP was added, all the strands eventually terminate with the base cytosine. Incorporation of the ddNTP occurs randomly as replication proceeds, so in this tube some DNA strands terminate with cytosine after just a few bases have been replicated, whereas in others replication proceeds further before a ddCTP is added. The replicated contents of each tube are then loaded onto individual lanes on a sequencing gel, and the different lengths of replicated DNA are separated. The final base sequence is determined by reading of the gel from the bottom up. The smallest DNA fragment that was the first to incorporate a ddNTP migrates the furthest, and the tube this smallest fragment was produced in (ddATP, ddCTP, ddGTP, or ddTTP) reveals which was the first base at the 5' end of the DNA to be sequenced.

Scoring DNA sequencing gels by hand is slow and tedious, and automated scanning techniques have been developed that greatly improve sequencing speed and accuracy. Dramatic improvements in DNA sequencing technology have evolved from projects to sequence the entire genomes of different organisms, especially the Human Genome Project. The latest generation of automated DNA sequencers dispenses altogether with gel separation of the replicated DNA fragments in favor of capillary electrophoresis combined with fluorescent dye labeling of the ddNTPs. Using laser scanning and a different dye color for each of the four ddNTPs, modern capillary sequencers produce sequence readout as a four-color chromatogram with a different colored peak for each base. Fluorescence-based sequencing can also be performed using PCR instead of conventional in vitro replication to generate the DNA fragments. **Cycle sequencing** of this type enables fast and highly automated sequencing of very small amounts of DNA. One form of PCR-based sequencing used for reading short pieces of DNA (e.g., to detect single base changes at a key locus) is **pyrosequencing.** This technique reveals the sequence of a single-stranded DNA fragment by using it as template to synthesize a new complementary strand. Each new dNTP is added to a special PCR mix one at time; if it is complementary to the next base in the sequence, it will be incorporated, and this reaction releases an inorganic phosphate molecule (PPi). The enzyme ATP sulfurylase in the PCR mix converts each PPi to ATP, which in turn provides energy to catalyze the production of oxyluciferin and visible light from luciferin molecules also present in the PCR mix. The amount of visible light produced is proportional to the number of dNTPs incorporated and is detected by a small camera. Another enzyme, apyrase, then degrades any unincorporated dNTPs and excess ATP remaining in the PCR mix, and the next dNTP is added. Pyrosequencing can provide 30–50 bases of sequence information in 45 minutes or less, and automated pyrosequencers can run multiple samples simultaneously. Cycle sequencing techniques such as this have led to the increasing availability of directly read DNA sequence variations such as single nucleotide polymorphisms SNPs as markers. SNPs are described in more detail in the next section.

Commonly Used Molecular Marker Systems

Isozymes and Allozymes

These earliest molecular markers do not target plant DNA directly but instead rely on variation in the electrophoretic mobility of the gene protein products to indicate differences in DNA sequence. Proteins of different molecular weight or net charge can be extracted from plant tissue, separated on a gel, and seen as spots or bands when the gel is stained. Variant forms of enzymes in plants have been widely used as molecular markers in this way. **Isozymes** are enzymes that catalyze the same reaction in the cell but are coded for by separate genes at different loci. **Allozymes** are distinct versions of the same enzyme produced by different alleles at a single locus. Allozyme markers in particular have been used extensively in plant population analyses and genetic diversity studies. They have the advantage of being cheap and easy to produce, and as codominant markers they can distinguish between heterozygote and homozygote genotypes. Typically, individuals homozygous for one of two alleles at a given allozyme locus each generate a single protein band of slightly different size and position on the gel, whereas the allozyme profile for a heterozygous individual contains both bands. Markers based on enzyme variants have several limitations, however, because they are gene products rather than the actual DNA. Allozyme and isozyme markers usually fail to detect very small genetic differences, and proteins varying in amino acid sequence but similar in size and charge often comigrate to the same location on the gel and are not recognized as distinct. Isozyme and allozyme markers also depend on gene expression. Consequently alleles not transcribed and translated at the time the tissue was sampled are not detected, and neither are silent alleles where expression of the gene product is permanently suppressed.

Seed Storage Proteins

Seed storage proteins represent another category of gene protein products that have been used as genetic markers in plant phylogenetic and population diversity studies. Like isozymes and allozymes, seed proteins can be extracted and electrophoresis used to separate proteins that have different molecular weights. The process is inexpensive and simple but subject to the many of the same limitations associated with enzyme markers. The extent of detectable variation between seed storage proteins varies widely with species and often is too limited for many of the genetic analyses possible with molecular markers.

DNA-Based Marker Systems

Markers based on differences in DNA sequence can be used to analyze mtDNA, cpDNA, and nuclear DNA (Table A.1). DNA sequence variations occurring at the same locus in different individuals are known as **polymorphisms.** Such variations can be detected by the following methods, either singly or in combination: direct sequencing of the DNA bases, digesting the DNA with restriction enzymes, or amplifying selected parts of the DNA using primers and PCR.

Restriction Fragment Length Polymorphisms

This was the first widely used DNA marker. To generate RFLPs, DNA is digested with a restriction endonuclease to produce fragments that are separated by electrophoresis on an agarose gel. A thin membrane made of a material such as nylon or nitrocellulose is then pressed onto the gel so the fragments are transferred to it. The transferred fragments on the membrane are known as a **Southern blot.** The membrane is washed with a strong alkali to denature the DNA and is then incubated in a solution containing a **probe.** This is a piece of single-stranded DNA of known sequence prelabeled with a chemical dye or radioisotope that allows it to be tracked. The probe hybridizes to any DNA fragments on the membrane that contain a complementary base sequence. The tracking dye or isotope in the probe labels the DNA fragments to which the probe hybridized. Photographing the blotted membrane reveals the labeled DNA fragments as bands on the final RFLP image.

Polymorphisms in the RFLP marker system arise for two reasons. First, variations in the DNA base sequence between individuals cause target sites

Table A1.1 Comparison of Commonly Used Molecular Markers

Marker	Polymorphisms	Codominant	Prior Sequence Knowledge Needed	Random Sampling of Anonymous Loci	Technical Complexity and Cost
Isozymes	Low	Yes	No	No	Low
RFLP	Medium	Yes	Yes	Yes	High
RAPD	Medium	No	No	Yes	Medium
ISSR	High	No	No	Yes	Medium
AFLP	High	No	No	Yes	High
SSR	High	Yes	Yes	No	High
SNP	High	Yes	Yes	No	High
EST	High	Yes	No	No	High

for the restriction endonuclease to be at different places in the genome, so the digested DNA fragments are of varying lengths and migrate to different points in the gel during electrophoresis. Second, variations in the DNA sequence between target sites may result in the probe hybridizing to a matching sequence on some fragments but not on others. Polymorphic RFLP markers are seen as individual bands that are present or absent, creating distinctive patterns associated with specific plant genotypes.

Like many allozymes, RFLP markers have the advantage of being codominant. This made them valuable as markers in early genetic linkage maps, for which they have been widely used. However, a major disadvantage of the RFLP technique is that the need for suitable probes necessitates prior sequence knowledge of the plant genome under investigation. Sometimes probes already developed for other plant species can be used if the target sequence for hybridization is similar; this is often true for cpDNA and mtDNA, where nucleotide sequences are more likely to be conserved between species. RFLPs are also slower, more expensive, and require larger amounts of DNA than markers using PCR. For these reasons they are increasingly being replaced by PCR-based techniques.

Randomly Amplified Polymorphic DNA

Randomly amplified polymorphic DNA (RAPD) was the first widely used marker in plant biotechnology using PCR to amplify DNA sequences at multiple genome locations simultaneously. The RAPD technique, first described by Williams et al. (1990), uses oligonucleotide (usually 10-base) single-stranded DNA primers with an arbitrary sequence to amplify DNA fragments at random. This was a departure from standard PCR procedures that use two different primers with sequences chosen to complement the DNA flanking a specific fragment for amplification as a sequence-tagged site. The DNA fragments amplified by the RAPD procedure are those in which two base sequences complementary to the primer sequence occur in opposite orientation and on opposite strands of the DNA not more than approximately 2000 bases apart. In a typical RAPD reaction one primer anneals at several sites that meet this requirement, scattered at random through the genome. Consequently, several different PCR products are amplified that are then separated on an agarose gel and visualized using a DNA-specific stain such as ethidium bromide. Different distances between the paired priming sites at different locations in the genome result in amplified DNA fragments of varying size. The pattern of DNA bands seen on the gel when the amplified fragments are separated and stained is called a DNA profile or DNA fingerprint and is characteristic of any individual primer–genome combination. Variations on the basic RAPD concept include using even shorter random primers such as five-base sequences. Shorter primers find more annealing sites because there are fewer complementary bases to match, so even more potentially polymorphic bands will be generated.

The RAPD marker technique has proved extremely popular for a wide range of genetic diversity and phylogenetic studies. It is fast, inexpensive, and simple. Unlike RFLPs or site-specific PCR markers such as simple sequence repeats (SSRS), RAPDs do not require specific probes or primers, so they can be used without any previous knowledge of the genome sequence. RAPD markers also have a number of disadvantages, however. First, they are usually dominant. This is because a priming site associated with allele *A* is present in both the homozygote *AA* and the heterozygote *Aa*. Unless the other allele *a* has the same priming sites but at a different distance from each other, thus amplifying a different-sized PCR product from the same primer, an identical single RAPD band will be produced from that locus both by *AA* and *Aa*. Second, RAPD priming sites may not sample all parts of the genome with equal probability: Some researchers have reported a tendency for RAPD primers to preferentially amplify repeat DNA. Even where this is not the case, RAPDs are anonymous amplified sequences, and it cannot be assumed that they represent transcribed DNA unless the RAPD product is actually sequenced and identified. Third, the RAPD technique is extremely sensitive to small differences in the PCR conditions. This lack of reliable reproducibility is a significant problem with this marker system, especially where lack of amplification and consequent absence of an expected band is caused by reaction failure rather than sequence difference. These false negatives can make individual genotypes appear more diverse than they really are, inflating estimates of phylogenetic distance or genotypic diversity.

Inter–simple sequence repeats (ISSRs) now provide a more reliably reproducible alternative to RAPD in situations in which randomly primed PCR markers are needed. This system is described in more detail later in this appendix.

Simple Sequence Repeat

Like RAPDs, SSR markers are generated using PCR. Unlike RAPD primers, however, SSR primers target a particular kind of base sequence within the genome. An SSR locus consists of a short nucleotide sequence or **motif** repeated in the same orientation multiple times (e.g., AGGAGGAGGAGGAGG . . .). This kind of nucleotide pattern is known as a **tandem repeat.** SSRs composed of short motifs—up to 6 nucleotides—are called **microsatellites,** and those based on longer motif sequences—up to 60 nucleotides—are known as **minisatellites.** The number of repeats present can vary from two or three to several dozen for microsatellites, whereas minisatellites can contain up to several hundred copies of the motif. SSRs are a common feature of eukaryotes and are widely dispersed through plant and animal genomes. Although they are usually found in noncoding DNA, some SSRs have been found to play a role in the regulation of gene expression when, for example, they are present in transcription factor binding regions. Longer repeat sequences such as **variable number tandem repeats** (VNTRs) are more characteristic of animal genomes and are widely used as markers in human genetic studies. The most common tandem repeats in most plant genomes are short stretches of DNA, usually 10–20 bp long, consisting of a repeat of a single base (a mononucleotide repeat such as AAAAAA . . .), two bases (a dinucleotide repeat such as AGAGAGAG . . .), or three bases (a trinucleotide repeat such as GACGACGAC . . .). Mononucleotide repeats are found in chloroplast DNA, whereas dinucleotide and trinucleotide repeats are more characteristic of the nuclear genome. Tandem repeats usually are flanked by conserved DNA sequences that do not vary between individuals within the same species. If these conserved flanking sequences are known, it possible to design PCR primers that will pick them out and anneal to them. This allows selective amplification of the stretch of repeated nucleotide motifs lying between the primers. The amplified DNA is visualized as bands, usually on a PAGE gel to provide the high resolution needed. The number of repeated motifs at an SSR locus can be highly variable between individuals in a population and also varies between species, so using PCR to amplify microsatellite sequences at the same loci in different individuals generates bands of different molecular weights, producing distinct DNA banding profiles on a gel.

SSRs have several advantages as markers. Like other PCR-based systems, they require small quantities of template DNA and can be generated quickly once the priming sequences are known. They are codominant: Different numbers of repeated nucleotide motifs at the same SSR locus on homologous chromosomes can be visualized as separate bands on the gel, so heterozygotes can be distinguished from homozygotes. SSR loci tend to be hypervariable: These regions of the chromosome rapidly accumulate different numbers of repeated nucleotide motifs. One reason for this is that SSRs usually are in nonexpressed regions of the DNA, so changes in sequence do not have deleterious effects subject to selection, as would be more likely with sequence alteration within an active exon.

The fact that SSR markers are based on nonexpressed DNA is also a potential disadvantage. Microsatellites are extensively used to characterize and quantify overall genetic variation in populations, but they are not always sufficiently closely linked to actively expressed genes to be useful as markers to tag such genes. However, the greatest disadvantage associated with SSRs is that knowledge of the flanking DNA sequences is necessary before suitable primers can be developed. Obtaining this sequence information is expensive and time-consuming because it entails the generation of short-fragment genomic DNA libraries that must be screened for SSR regions, followed by sequencing to identify the flanking nucleotides. So far this has limited SSR use to crops of major economic importance such as soybean, rice, wheat, and maize. However, new techniques for rapid identification of potential SSR regions in cloned DNA fragments are accelerating primer development for additional plant species. Researchers have also found that nucleotide sequences flanking SSR loci sometimes are sufficiently well conserved between related species that SSR primers developed for one species can be used in another.

Inter–Simple Sequence Repeats

The ISSR marker system is also sometimes called randomly amplified microsatellite polymorphism (RAMP). Like SSRs, this marker system targets the tandem repeat regions found scattered throughout plant genomes. However, there are two key differences between SSR and ISSR. First, SSR primers are used in forward-and-reverse pairs to target specific predetermined microsatellite or minisatellite loci for amplification. On the other hand, ISSR uses single primers that typically consist of a few dinucleotide or trinucleotide sequence repeats complementary to a microsatellite sequence, plus one to three anchoring bases. Second, whereas in SSR marker systems the repeat sequences themselves are amplified, in ISSR the amplification product is the intervening nucleotide sequence lying between two microsatellites. For example, the ISSR primer 5' $(CA)_8$ AG 3' anneals to sites on the genome where the complementary base sequence 3' $(GT)_8$ TC 5' is present on each strand in opposite orientation, and it amplifies the interrepeat stretch of DNA lying between these annealing sites. As a number of such paired microsatellite sites within amplifiable distance of each other often are present in the genome, one ISSR primer commonly generates several bands of different size. This polymorphism reflects the varying distances between the paired tandem repeats. Sometimes extraction of an ISSR band from the gel followed by endonuclease restriction of the amplified DNA it contains reveals additional polymorphisms, not because of different distances between the paired repeats but because of variations in the base sequence lying between them. This procedure is known as ISSR-RFLP.

ISSR has some of the same advantages as RAPD. No previous sequence knowledge is needed, it is often possible to generate multiple bands with one primer, and the amplified products can be separated and visualized cheaply on low-resolution agarose gels using ethidium bromide, although more bands can be scored with the higher resolution of a PAGE gel and silver staining. The ISSR system was first described by Zietkiewicz et al. (1994), and is used for various marker applications in an increasing number of plant species. ISSR has proved especially useful for the detection and analysis of genetic diversity in nondomesticated plants for which genome sequence data have not been developed. Researchers using ISSR markers report that they are more stable and repeatable than RAPD, that segregation at ISSR loci follows Mendelian patterns of inheritance, and that ISSR primers tend to generate more bands than RAPD primers, thus sampling more points on the genome simultaneously and providing more information. For these reasons ISSR is increasingly replacing RAPD where a random priming PCR-based marker is needed. One disadvantage of the ISSR technique is that, as in RAPD and AFLP, the markers produced usually are dominant, so heterozygotes cannot readily be distinguished from homozygotes using this system.

Amplified Fragment Length Polymorphisms

AFLP marker technology combines DNA restriction with a version of randomly primed PCR, simultaneously sampling as many as 60 loci scattered throughout the genome. The technique was first described by Vos et al. (1995) and has been used in many plant species for analysis of genetic diversity and evolutionary or phylogenetic relationships and for taxonomic studies. AFLPs are generated by first cutting genomic DNA into fragments using two restriction enzymes simultaneously. Short single-stranded pieces of synthesized DNA known as adapters are then ligated to the cut ends of the fragments. The sequences of the adapters are known, allowing a subset of the genomic DNA fragments to be amplified via PCR using primers complementary to the adapters plus an extension of one to three additional bases at the 3' end. Only the genomic DNA fragments with bases at each end complementary to the adapter extensions will be amplified. Therefore, AFLPs can distinguish between allelic sequences at a locus differing by as little as a single base pair. Polymorphic amplified fragment lengths that can be used as markers result from changes in the nucleotide sequence that add or eliminate endonuclease restriction sites, sequence variations at the ends of the restricted DNA that determine which fragments will be included in the amplified subset, or insertions or deletions within the amplified DNA fragments affecting the size of the final PCR product. Simultaneous exploitation of these multiple sources of sequence variation often enables the AFLP system to distinguish between very similar genotypes that cannot be differentiated using other markers.

Like RAPDs, AFLPs have the advantages of sampling loci at random throughout the genome and of requiring no previous knowledge of nucleotide sequence. This has made the AFLP system popular with researchers working with plant species for which little or no genomic information is available, including many regional or minor crops and nondomesticated plants. Because AFLPs sample many more loci at one time than RAPDs, more scorable bands are generated in one amplification run, typically 50–60 for AFLPs as opposed to 5–6 for RAPDs. AFLPs also have good repeatability, reliably generating identical marker profiles from the same DNA and enzyme–primer combinations on different occasions. The chief disadvantage of the AFLP system is that it requires more sophisticated equipment than RAPD or ISSR, especially because PAGE gels with silver staining or fluorescent dyes are needed to separate the large numbers of amplified fragments produced. Automated scanning of AFLP gels is also desirable because scoring the numerous bands manually is tedious and error-prone. As with RAPDs and ISSRs, another potential drawback for some applications is that AFLP markers usually are dominant.

Single Nucleotide Polymorphisms

SNPs occur where the nucleotide sequence at the same locus in a genome differs by a single base between individuals. This is the most common form of DNA sequence variation, occurring in both expressed and nonexpressed parts of the genome. SNP frequency is estimated at 1 every 10,000 bases in eukaryotes, unevenly distributed throughout the genome because of varying rates of mutation, recombination, and selection at different loci. SNPs often are closely grouped together on the same DNA strand so they are co-inherited as a block, a process called linkage disequilibrium. A group of linked SNPs on a single DNA strand is described as forming a haploid genotype, or **haplotype.**

Single nucleotide differences in the DNA sequence often are the basis of other marker system polymorphisms. For example, a single base difference could mean that one individual has a RAPD priming site or an AFLP restriction site that another individual lacks, resulting in different RAPD or AFLP profiles that differentiate the two genotypes. Thus a polymorphic RAPD, AFLP, or other marker may be an indirect indication of an SNP in the DNA. High-throughput automated nucleotide sequencing, such as the Sanger sequencing or pyrosequencing described earlier, enables direct detection of SNPs on a large scale, and this approach is increasingly used in SNP work. Comparing accumulated SNPs within the base sequence at the same locus provides an alternative approach to estimating genetic distance when examining evolutionary and phylogenetic relationships between individuals. The underlying assumption is that individuals with similar versions of a given sequence are more closely related than individuals with more dissimilar versions. The more dissimilar sequences are presumed to be the result of greater accumulation of SNPs over time. For example genotype I (AATGCCGA) is considered closer to genotype II (AATGGCGA) than to genotype III (AATTGCGA). Tracing the accumulation of SNPs by sequencing the same locus in multiple individuals allows visual representation of phylogenetic relationships via construction of a **gene tree:**

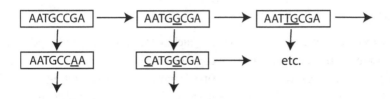

The reconstruction of phylogenetic and evolutionary relationships via SNPs and gene trees should be approached with caution, however, especially if a single gene tree based on one locus is used. The base substitutions that generate SNPs may not occur at a constant rate, and repeated substitution at the same base position over a longer period of time cannot be readily distinguished from a single base change. SNP saturation in short hypervariable DNA sequences, when substitution has occurred at all base positions at least once, can result in highly misleading gene trees. Sampling multiple SNP loci and combining SNP data with other measures of similarity help avoid these pitfalls.

Several loci in the nuclear, chloroplast, and mitochondrial genomes of plants are routinely sampled as sources of sequence variation that can be used to interpret phylogenetic relationships. Such variation may take the form of SNPs or larger rearrangements involving more than one base, such as indels (insertions or deletions of short sequences). Selection of the appropriate loci for sequencing depends in part on the level of taxa being examined. Noncoding nuclear loci tend to accumulate sequence changes most rapidly and therefore are used most often for analysis of closely related taxa such as species. For example, polymorphisms in the internal and external transcribed spacer (ITS and ETS) sequences that form part of the nuclear ribosomal DNA repeat have been widely used to track lineages within genera. The ETS locus lies just upstream of the ribosomal 18s subunit sequence, and the two ITS regions lie between the 18s and 5.8s subunits (ITS 1) and the 5.8s and 26s subunits (ITS 2). These DNA spacer sequences are transcribed with the ribosomal RNA but not translated; consequently, they can accumulate polymorphic sequence changes such as base substitutions and indels at a rapid rate without affecting gene function. More conserved loci that do not accumulate sequence changes as quickly are sampled for polymorphisms distinguishing between higher taxa such as families. Examples of widely used polymorphic sequences at this level are the chloroplast *rbc*L gene and the *nad1* and *cox* genes in the mitochondrion.

Expressed Sequence Tag

An expressed sequence tag (EST) is a nucleotide sequence obtained from a cDNA clone that in turn was derived from an mRNA. Therefore ESTs can be considered randomly sequenced DNA fragments corresponding to partial gene transcripts. Expressed sequence tags were first used in the early 1990s as part of the Human Genome Project, where they provided researchers with a rapid technique for identifying and locating expressed genes. The use of ESTs has been rapidly extended to other species, and as of 2003 more than 3 million sequences from approximately 200 plant species have been deposited in public EST databases.

To generate plant ESTs, RNA is first extracted from tissue in which genes of interest are likely to be expressed. For example, a researcher interested in identifying genes activated in response to heat might extract RNA from a leaf that has been stressed at high temperatures. To increase the overall yield of processed mRNA as opposed to other types of RNA, the extracted RNA molecules are screened, and those without a polyadenine sequence at the 3' end are eliminated.

cDNA is then transcribed in vitro from the mRNA using reverse transcriptase. This produces a DNA version of the mRNA sequence. The cDNA is unlikely to be identical to the gene that actually produced the mRNA because the gene (and the first raw RNA gene transcript it produces) contains introns that are spliced out to generate the processed mRNA for translation. Nevertheless, sequence similarity between the cDNA copy and the original gene often is sufficiently close that cDNA can be used to identify and even physically locate the original gene.

Next, the cDNA sequences are placed in plasmid vectors ("cloned") and inserted into *E. coli* bacterial cells ("transformed") to create a cDNA library. A randomly selected subset of the cDNA sequences in the library is released by endonuclease digestion, and the released DNA fragments are partially sequenced from the 5' or 3' end. This creates the expressed sequence tags. Researchers have debated the relative merits of sequencing the cDNA fragments from the 5' as opposed to the 3' end. The 5' sequences are more likely to provide protein coding information because the 3' end often contains untranslated regions (UTRs). However, features of the 3' end (including the UTRs) are more likely to be unique to a gene. This makes 3' end

ESTs more useful for distinguishing between genes of similar sequence. As sequencing techniques improve it is increasingly feasible to generate both 5' and 3' end sequences from the same cDNA fragment.

Finally, the sequences are compiled to create an EST database. ESTs can then be compared with known gene sequences for possible matches. Large EST databases can be systematically analyzed for recurrent motifs or patterns that might provide useful information about gene identity and function, a process known as **data mining.**

Production of expressed sequence tags can be automated to a high degree in the laboratory. However, generating a set of ESTs does not immediately provide markers that can be used to detect genotypic variation between individuals. There are several methods for creating markers from ESTs. If different genotypes are represented in the EST collection, it is possible to look for sequence variations such as SNPs within the ESTs themselves. Another possibility is to use the EST sequence to design matching PCR primers that will amplify either the 5' or 3' region immediately flanking the gene from which the EST was derived. There is a better chance of finding polymorphisms in this flanking region because it is not expressed DNA and therefore may not be as highly conserved as the gene itself. EST databases can also be searched for repeat motifs embedded in the sequence tags that can be used to create SSR markers.

Markers derived from ESTs have a number of potential advantages because they are based on expressed gene sequences rather than randomly sampled and anonymous genomic DNA. ESTs differentially expressed in different tissues or under certain environmental conditions are useful for identifying key genes. Likewise, because it is based on expressed DNA, there is a good chance that an EST-derived marker linked with a desirable phenotypic trait will map directly to a gene associated with that trait. Another advantage of EST-based markers is that they are more likely to transfer between species than markers based on random genome sampling such as RAPDs or AFLPs. This is because the coding DNA from which ESTs are obtained tends to be more highly conserved than noncoding parts of the genome. ESTs are also independent of genome size or structure, so they are useful in dealing with polyploids or other large plant genomes with high levels of repetitive DNA.

The most obvious limitation associated with ESTs is that the mRNA extracted from a cell (and the cDNA library produced from it) represents only the genes actively transcribed at the time of sampling. Genes expressed at a low level will be underrepresented in the cDNA library, and genes not transcribed when the mRNA was collected will be missing from the cDNA altogether. Most current EST collections do not cover the entire functional genome and often are biased towards highly expressed genes, but as more ESTs are developed and the sequences made available the databases will become more comprehensive.

Mapping Markers onto the Genome

Positioning molecular marker loci within the genome, in relation to each other and to other loci containing genes that code for traits of interest, entails the generation of a mapping population. A good mapping population consists of many individuals within a generation segregating for different combinations of marker genotypes and readily identifiable morphological or physiological phenotypes. Phenotypes and marker genotypes of these individuals are recorded, and the data are analyzed for association between the presence of different alleles at marker loci and particular phenotypes. If one marker frequently occurs in combination with another or is consistently found in plants that express a certain trait, cosegregation is said to be occurring. Cosegregation of two markers or of a marker and a trait indicates that the relevant loci are syntenic (on the same chromosome) and situated sufficiently close together that there is a lower probability of recombination between them at meiosis; in other words, they are linked. The smaller the distance between the linked loci, the rarer recombination between them becomes and the more often alleles at those loci are seen together in individuals in the mapping population. Calculating the percentage of individuals in which recombination is seen provides an estimate of genetic distance between two linked loci: 1% recombination is equivalent to 1 centimorgan (cM). In a simple example, if 90% of plants in a mapping population with black seeds also have a particular AFLP band, and only 10% of black-seeded plants lack this AFLP band, the distance between the AFLP locus and the gene coding for black seeds can be estimated at 10 cM. Repeating this analysis for many different markers and traits allows construction of a genetic map giving the relative positions of the loci in the genome.

Mapping individual markers and loci controlling single-gene traits is fairly straightforward. However, many traits of interest in crop plants are quantitative: They exist as a range of phenotypes with no clear distinction between classes and often are controlled by multiple genes. Mapping these quantitative trait loci (QTLs) onto the genome is a more complex procedure. Segregation for different phenotypic values of the quantitative trait in a mapping population is recorded, together with the presence or absence of a marker. All individuals with a particular marker genotype are grouped as a class, and the phenotypic variance for the quantitative trait among classes is analyzed. If a significant portion of the variance is associated with a particular marker, it is likely that the marker is linked to a QTL contributing to the trait.

Analysis of cosegregation in mapping populations allows marker and trait loci to be positioned in the genome relative to each other, a procedure known as genetic mapping. Physical mapping using in situ hybridization techniques provides a more direct method of locating markers or other key DNA sequences within the plant genome. In situ hybridization is similar to the RFLP technique already described, using single-stranded nucleic acid probes labeled with a dye to anneal to complementary sequences in the DNA under examination. The difference is that whereas RFLPs are generated by probing of digested DNA fragments separated on a gel, with in situ hybridization the probe sequences anneal to entire denatured chromosomes on a slide, and the final result is examined under a microscope. Fluorescence in situ hybridization (FISH) uses labeled probes containing parts of the genome, such as a key gene or gene fragment. Allowing the probe to hybridize to a set of mitotic chromosomes on a slide reveals the location of the complementary gene sequence: Where the probe anneals, a section of the chromosome is lit up by the colored fluorescent dye. Multiple probes with different-colored labels can be used simultaneously to reveal the physical locations of various DNA sequences within the genome, a process sometimes called chromosome painting. It is also possible to probe denatured mitotic chromosomes with labeled total genomic DNA, typically from another plant species; this procedure is known as genomic in situ hybridization (GISH). *Arabidopsis* genomic DNA has been used as a probe in this way to physically locate in other plant genomes the common DNA sequences conserved between species. GISH also enables the genomes of related species to be compared: Using the total genomic DNA of one species to probe the chromosomes of another reveals the location and extent of sequences similar enough to allow hybridization between the two genomes.

Final Note: Are Molecular Markers Good Substitutes for Direct Observation of Phenotypic Traits?

Molecular markers are used by plant researchers in many fields: genomics, plant breeding, population genetics, germplasm conservation, taxonomy, and evolutionary and phylogenetic studies. Although markers have become a widely accepted tool, the underlying assumption that a strong relationship always exists between molecular marker data and population characteristics such as phenotypic trait variance or evolutionary history is rarely questioned. Sometimes the correlation is good: There are numerous published examples of taxonomic groups based on marker information echoing those derived from more traditional phenotypic observations and of marker-derived similarity coefficients accurately reflecting known geographic or evolutionary relationships (e.g., Tranel and Wassom, 2001; Fernandez et al., 2002; Mignouna et al., 2003). In other situations the surrogate value of molecular marker data has been questioned (Patterson et al., 1993; Reed and Frankham, 2001). More research is needed in this area. Meanwhile, we should bear in mind that molecular markers provide a great deal of information about individual genotypes and population gene pools, but they do not tell the whole story.

References

Belgrader, P., W. Bennett, D. Hadley, G. Long, R. Mariella, F. Milanovich, S. Nasarabadi, W. Nelson, J. Richards, and P. Stratton. 1998. Rapid pathogen detection using a microchip PCR array instrument. *Clinical Chemistry* 44: 2191–2194.

Fernandez, M. E., A. M. Figueiras, and C. Benito. 2002. The use of ISSR and RAPD markers for detecting DNA polymorphism, genotype identification and genetic diversity among barley cultivars with known origin. *Theoretical and Applied Genetics* 104: 845–851.

Mignouna, H. D., M. M. Abang, and S. A. Fagbemi. 2003. A comparative assessment of molecular marker assays (AFLP, RAPD and SSR) for white yam (*Dioscorea rotunda*) germplasm characterization. *Annals of Applied Biology* 142: 269–276.

Mullis, K., F. Faloona, S. Scharf, R. Saiki, G. Horn, and H. Erlich. 1986. Specific enzymatic amplification of DNA in vitro: The polymerase chain reaction. *Cold Spring Harbor Symposia on Quantitative Biology* 51: 263–273.

Panet, A. and H. G. Khorana. 1974. Studies on polynucleotides. The linkage of deoxyribonucleic templates to cellulose and its use in their replication. *Journal of Biological Chemistry* 249: 5213–5221.

Patterson, C., D. M. Williams, and C. J. Humphries. 1993. Congruence between molecular and morphological phylogenies. *Annual Review of Ecology and Systematics* 24: 153–188.

Reed, D. H. and R. Frankham. 2001. How closely correlated are molecular and quantitative measures of genetic variation? A meta-analysis. *Evolution* 55: 1095–1103.

Saiki, R. K., S. Scharf, F. Faloona, K. B. Mullis, G. T. Horn, H. A. Erlich, and N. Arnheim. 1985. Enzymatic amplification of beta-globin genomic sequences and restriction site analysis for diagnosis of sickle cell anemia. *Science* 230: 1350–1354.

Tranel, P. J. and J. J. Wassom. 2001. Genetic relationships of common cocklebur accessions from the United States. *Weed Science* 49: 318–325.

Vos, P., R. Hogers, M. Bleeker, M. Reijans, T. van de Lee, M. Hornes, A. Frijters, J. Pot, J. Peleman, M. Kuiper, and M. Zabeau. 1995. AFLP A new technique for DNA fingerprinting. *Nucleic Acids Research* 23: 4407–4414.

Williams, J.G.K., A.L. Kubelik, K.J. Livak, J.A. Rafalski, and S.V. Tingey. 1990. DNA polymorphisms amplified by arbitrary primers are useful as genetic markers. *Nucleic Acids Research* 18: 6531–6535.

Zietkiewicz, E., A. Rafalski, and D. Labuda. 1994. Genome fingerprinting by simple sequence repeat (SSR)-anchored polymerase chain reaction amplification. *Genomics* 20: 176–183.

Timothy J. Motley, Hugh Cross,
Nyree Zerega, and Mallikarjuna K. Aradhya

Molecular Analyses

The study of crop evolution, origins, and conservation entails the assessment of genetic variability within and between populations and species at different genetic, evolutionary, and taxonomic hierarchical levels. Molecular biology has greatly increased the amount of data and computational intensity of population genetic and phylogenetic systematic analyses. Numerous methods of analysis are available, several of which are used in the studies presented in this volume. This appendix is not meant to be exhaustive, but rather the following glossary is meant to serve as a quick reference for common population genetic and phylogenetic terms and methods of analysis that are found in this volume and other works. For those who would like more information or would like to further explore these methods, a list of suggested reading (basic and advanced) and some Internet links are provided.

Glossary

accelerated transformation (ACCTRAN): An optimality criterion for resolving ambiguous character state optimization; homoplasious characters are treated as reversals to the plesiomorphic condition, and initial transformations are placed as near to the root of the tree as possible.

alignment: The juxtaposition of amino acids or nucleotides in homologous molecules to maximize similarity or minimize the number of inferred changes in the sequences. Alignment is used to infer positional homology before or concurrent with phylogenetic analyses.

analysis of molecular variance (AMOVA): A hierarchical partitioning of genetic diversity of a population for haplotypic data (Excoffier et al., 1992). Data are partitioned into diversity between groups of populations, between the populations within groups, and between the individuals within a population.

apomorphy: A character state derived by evolution from an ancestral state (**plesiomorphy**). A novel evolutionary trait.

autapomorphy: A derived character state unique to a particular taxon. A uniquely derived character state. A type of **apomorphy** that is unique to a single terminal taxon. Compare to **synapomorphy**.

Bayesian analysis: Similar to **maximum likelihood** methods, Bayesian analysis is based on using a probabilistic model of how the observed data are produced. Bayesian inferences are based on the posterior probability of a hypothesis, which is the probability derived after taking into account observed data. For phylogenetic inference using Bayesian analysis it uses the posterior probability distribution of a phylogenetic tree, conditioned on the observed matrix of aligned sequences, and integrates all parameters (models of evolution). However, with so many possible parameters, the posterior probability distribution of trees is impossible to calculate analytically; therefore, for phylogenetic analyses a simulation technique called the Markov chain Monte Carlo (MCMC) method is used to approximate the posterior probabilities of trees.

bootstrap consensus tree: A consensus tree formed from the clades that have received at least 50% bootstrap support.

bootstrap support: A statistical method used to assess support for the relationships resulting from a phylogenetic analysis, based on repeated random sampling with replacement from an original sample to provide a collection of new pseudoreplicate samples, from which sampling variance can be estimated. The results of this method are presented as percentages and can be interpreted for a given node as the percentage of pseudoreplicates in which the given node was found.

Bremer support or decay index: A method to assess support for the relationships in a phylogeny that is based on length differences in parsimony analyses with and without a particular clade. If a cladogram is five steps longer without a given clade, then the support for that clade would be five.

character: Any heritable attribute of an organism that can be used for recognizing, differentiating, or classifying a taxon. Characters are used from morphological, behavioral, developmental, and molecular data and are usually described in terms of their states, which can be binary (e.g., "spines present" vs. "spines absent," where "spines" is the character and "present" and "absent" are its states) or multistate (e.g., the character "fruit shape" can have many states, including "ovoid," "pyriform," and "obovate").

chronogram: A branching diagram (tree) in which the branch lengths represent units of time estimated from a molecular clock.

clade: A branch on a cladogram (composed of a monophyletic group; see **monophyly**) identified by at least one **synapomorphy**. Clade is from the Greek word *klados,* meaning branch or twig.

cladistics: A general term that refers to a method for inferring phylogenies in which parsimony is preferred, and all lineages (**clades**) in this analysis are defined by shared, derived characters (**synapomorphies**). The only assumptions in a cladistic analysis are that organisms are related by common descent and that relationships between them are best represented in a hierarchical, bifurcating pattern (**dendrogram** or **cladogram**).

cladogram: A branching diagram (tree) resulting from a cladistic analysis, assumed to be an estimate of a **phylogeny** (see also **dendrogram**).

cluster analyses: Analyses of multivariate genetic relationships within and between populations and species based on pairwise genetic similarity or difference coefficients. Cluster analysis encompasses numerous different algorithms that classify individuals into groups, or clusters, so that the degree of association is stronger between members of the same cluster than between members of different clusters.

convergent evolution: A character similarity that has evolved independently in two or more organisms and that are not inherited from a common ancestor. This is a specific type of **homoplasy** in which the similarities are a result of adaptation for similar function in both organisms.

decay index: See **Bremer support**.

delayed transformation (DELTRAN): An optimality criterion for resolving ambiguous character state optimization in a phylogenetic analysis. Homoplasious characters are treated as independent gains, and initial transformations are placed as far from the root of the tree as possible.

dendrogram: Any branching diagram (or tree) (including **chronogram**, **cladogram**, **phylogram**, **phenogram**). The points of branching in a dendrogram are called nodes.

distance: Usually treated as a measure of evolutionary divergence, that is, phylogenetic distance increases with increasing evolutionary divergence. Distances usually are expressed pairwise among the **terminal taxa** and can be calculated based on a specified evolutionary model; the model specifies the probabilities of character state changes through evolutionary time.

fixation index (Wright's *F* statistics): Fixation is increased homozygosity resulting from inbreeding. *F* statistics measure the difference between the mean heterozygosity among the subdivisions in a population and the potential frequency of heterozygotes under random mating. Fixation indices can be determined for different hierarchical levels of a population structure to indicate the degree of differentiation of individuals within the subpopulation (F_{IS}), of subpopulations within the total population (F_{ST}), and of individuals within the total population (F_{IT}) (Wright, 1978). These statistics are related as follows: $(1 - F_{IS})(1 - F_{ST}) = 1 - F_{IT}$. Another measure of genetic differentiation between subpopulations (G_{ST}) allows for more than two alleles at a locus.

haplotype networks: A method of representing relationships of populations within a species by the number of changes between haplotypes. It is preferred over using branching trees (dendrograms) for demonstrating infraspecific relationships because networks allow for recombination among the individuals.

heterozygosity: Possessing different alleles at the same locus. Levels of heterozygosity often are used as a measure of genetic diversity.

heuristic method: Any method of analysis that does not guarantee finding the optimal solution to a problem and that involves computationally efficient strategies that should produce a solution at least close to the optimal one. They are usually used to obtain a large increase in speed over exact methods (e.g., branch and bound and exhaustive methods in phylogenetic analyses).

homology: Similarity caused by common evolutionary origin. Two structures are considered homologous when they are derived from the same structure or trait inherited from a common ancestor, although it may have been modified through descent.

homoplasy: Similarity caused by independent evolutionary change. Thus homoplasy is a false **homology** and can be misleading in phylogenetic analyses. Homoplasy is caused by **convergent** or parallel evolution (although some distinguish between these two terms), which is an independent gain or loss of a character.

Hudson–Kreitman–Aquade (HKA) test: A neutrality test used to compare rates of divergence between species with the levels of polymorphism within species (Hudson et al., 1987).

incongruence: In phylogenetic analyses of the same group of organisms, when trees produced from different data show different topologies they are said to be incongruent. Incongruence can arise from several causes, including lineage sorting (random changes in the lineage before speciation), hybridization, and paralogy (gene duplication) (see Johnson and Soltis, 1998).

incongruence length difference (ILD) test: Measures the proportion of inferred homoplasy attributed to the combined data sets. The test compares the sum of the tree lengths with a null distribution (generated by random character permutation among partitions) to detect areas of hard incongruence (i.e., strongly supported character conflict) (Farris et al., 1994, 1995).

ingroup: In a phylogenetic analysis, the set of taxa that are hypothesized to be more closely related to each other than any are to the outgroup, generally the study group whose phylogeny is being reconstructed.

jackknife support: An estimate of tree branching reliability achieved through data resampling based on elimination of a portion of the original data as an average measure of pseudoreplicate variance from the original sample.

Kimura 2-parameter distance: A model of evolution used in many methods of inferring phylogenies, such as likelihood. This model assumes that all transitions and all transversions are equally likely (Kimura, 1980).

Kishino–Hasegawa (KHT) test: A statistical test for comparing two phylogenetic trees using differences in the support provided by individual characters to determine whether the topologies of the two trees are significantly different from each other (Kishino and Hasegawa, 1989).

likelihood ratio test: A method for testing alternative hypotheses of molecular evolution by comparing the likelihood score of the alternative hypothesis to a null hypothesis to test whether they are significantly different from what is expected by random fluctuation.

majority rule consensus tree: A consensus tree formed from the clades that occur in at least 50% of the original cladograms.

maximum likelihood (ML): A criterion for estimating a parameter from observed data under an explicit model. In phylogenetic analysis, the optimal tree under the maximum likelihood criterion is the tree that is the most likely to have occurred given the observed data and the assumed model of evolution. The optimal tree is the one that maximizes the statistical likelihood that the specified evolutionary model produced the observed character state data; the models specify the probabilities of character state changes through evolutionary time.

maximum parsimony (MP): A method for inferring phylogenies based on the principle of minimizing the number of events needed to explain the data. In phylogenetic analysis, the optimal tree under the maximum parsimony criterion is the tree that entails the fewest number of character state changes. The method is also called **parsimony.**

minimum evolution (ME): A distance-based method that allows selection of the shortest evolutionary tree from all possible topologies. Branch lengths for the ME trees are generally estimated from the observed pairwise distances using either linear programming or least square methods. Neighbor-joining or UPGMA algorithms are used to build the ME trees.

molecular clock: The assumption that for neutral (not under selection) genes the mutation rate will be constant over time and all lineages will evolve in a clocklike manner. With this assumption times of species divergence can be estimated by comparing their gene sequences.

monophyly: A group of organisms that has a unique origin from a single ancestral taxon and includes the ancestor and all its descendants. Contrast with **paraphyly** and **polyphyly**.

neighbor joining: A simple method of stepwise tree construction that starts by grouping the two individuals with the smallest distance and then progressively adds more distant individuals and new groups. With each step the distance matrix is adjusted so each pair is the average divergence from all other groups. Neighbor-joining trees can be constructed so observed distances between individuals are equal to the sum of the branch lengths connecting them (Saitou and Nei, 1987).

nonparametric rate smoothing (NPRS): This is a "rates of evolution" model that relaxes the assumptions of a molecular clock by estimating local rates of evolution for each node of the tree and then minimizing, or smoothing, the differences in those rates from ancestral to descendent lineages (Sanderson, 1997; Johnson and Soltis, 1998). This rate smoothing is accomplished using an optimality criterion that is a sum of squared differences in local rate estimates compared from branch to neighboring branch.

outgroup: A taxon (or taxa) that is not part of the ingroup but is included in a phylogenetic analysis in order to provide a root for the ingroup and to help differentiate between **apomorphies** and **plesiomorphies** in the ingroup. See also **polarity.** An outgroup should be closely related to the ingroup.

paraphyly: A group of organisms that includes their most recent common ancestor and some but not all of its descendants. This is very similar to **polyphyly,** and the two are sometimes used interchangeably. Contrast with **monophyly.**

parsimony: A principle for choosing between scientific theories that states that the simplest explanation that accounts for the greatest number of observations is preferred to more complex explanations. In other words, among competing hypotheses, one should always choose the simplest explanation of a phenomenon that takes the fewest leaps of logic (ad hoc assumptions). Also known as Occam's razor. In phylogenetic analyses this means that the most parsimonious tree is the one that takes the fewest evolutionary steps or character state changes (see **maximum parsimony**).

Pearson chi-square statistic: To assess the significance of interpopulation heterogeneity in allele frequencies using an MN contingency table, $(M-1)(N-1)$ df, where M = number of population and N = number of alleles.

penalized likelihood: A semiparametric approach for estimating rates of evolution in cases when lineages are not evolving in a clocklike manner. It combines a parametric model having a different substitution rate on every branch with a nonparametric roughness penalty that discourages rates from changing too much. The optimality criterion is the log likelihood minus the roughness penalty. The relative contribution of the two components is determined by a smoothing parameter. The optimal value of smoothing is chosen through an empirical cross-validation method (Sanderson, 1997, 2002).

percentage of polymorphic loci: The proportion of the number of variable loci to the total number of loci. This is a measure of genetic diversity that can be applied to a variety of molecular marker data.

phenogram: A branching diagram (tree) that links entities by estimates of overall similarity.

phylogenetics: Field of biology that involves the study of evolutionary relationships between organisms. It includes the discovery of these relationships and the study of the causes that result in these patterns.

phylogeny: A hypothesized set of evolutionary relationships between organisms usually represented as a bifurcating (branching) tree (see **dendrogram**).

phylogeography: The study of the biogeography of populations or species as revealed by a comparison of their estimated phylogenies with their geographic distributions.

phylogram: A branching diagram (tree) that depicts inferred historical relationships between organisms and in which the branches are drawn proportional to the amount of inferred character change.

plesiomorphy: An ancestral character state for the taxa under consideration. Plesiomorphies were acquired by an ancestor deeper in the phylogeny than the most recent common ancestor of the taxa under consideration. Contrast with **apomorphy.**

polarity: Evolutionary ordering of character states for the taxa under consideration, to determine which states were acquired by an ancestor deeper in the phylogeny than the most recent common ancestor (**plesiomorphy**) and which states were derived within the taxa under consideration (**apomorphy**).

polyphyly: A group of organisms that does not include their most recent common ancestor. The group has multiple evolutionary origins, and members of this group appear on different branches of a phylogenetic tree, and the branch that includes the most recent common ancestor of the group includes other groups. See **paraphyly.**

principal component analysis (PCA): An ordination technique using Eigen analysis that effectively reduces complex multidimensional data into two or three meaningful linear orthogonal vectors, explaining as much as possible the variation in the original data. The orthogonal vectors are projected on to two- or three-dimensional plots to visualize the groups or clusters so that the variance of the pairwise distances within clusters is minimized and between clusters is maximized.

principal coordinate analysis (PCOA): This method is similar to PCA and plots the data in dimensional plots, but ordination is based on distance and dissimilarity measures rather than linear correlations.

Shannon diversity index (H): An index to characterize the diversity of species (and the diversity of crop varieties). It accounts for both abundance and evenness of the species present. The proportion of species i relative to the total number of species (pi) is calculated and then multiplied by the natural logarithm of this proportion ($\ln pi$). The resulting product is summed across species and multiplied by -1.

Shimodaira–Hasegawa test: A test for comparing statistical differences between trees, similar to the Kishino–Hasegawa test, but with allowances for comparing between multiple trees simultaneously (Shimodaira and Hasegawa, 1999).

sister groups (or sister taxa): The two groups resulting from the splitting of a single lineage and that are most closely related to one another.

strict consensus tree: A consensus tree formed from clades shared by all the original cladograms.

synapomorphy: A shared derived character state (**apomorphy**). A novel evolutionary trait that is shared by two or more groups descending from an immediate common ancestor. These are used to define a clade or monophyletic group in a phylogenetic analysis.

Tajima's test: This tests the neutral theory of molecular evolution by calculating D statistics (two-tailed test) under the assumption that the neutral model estimates of the number of segregating sites and of the average number of nucleotide differences are correlated (Tajima, 1989).

terminal taxon: The taxon or named group at the tips of the branches of a tree. Terminals may represent almost any kind of group, including higher taxa (e.g., genera, families, species, populations, individuals, and genes).

topology: The branching sequence of a tree.

unweighted pair group method with arithmetic means (UPGMA): A simple method of stepwise tree construction using a sequential clustering algorithm in which distance values are assigned equal weights and similarity is used to create the cluster relationships and order the tree.

Recommended Reading

Avise, J. C. 2004. *Molecular Markers, Natural History, and Evolution,* 2nd ed. Sinauer Associates, Sunderland, MA, USA.

Felsenstein, J. 2004. *Inferring Phylogenies.* Sinauer Associates, Sunderland, MA, USA.

Hall, B. G. 2004. *Phylogenetic Trees Made Easy,* 2nd ed. Sinauer Associates, Sunderland, MA, USA.

Hartl, D. L. and A. G. Clark. 1997. *Principles of Population Genetics*. Sinauer Associates, Sunderland, MA, USA.

Hillis, D. M., C. Moritz, and B. K. Mable. 1996. *Molecular Systematics*, 2nd ed. Sinauer Associates, Sunderland, MA, USA.

Page, R. D. M. and E. C. Holmes. 1998. *Molecular Evolution, A Phylogenetic Approach*. Blackwell Science Ltd., Oxford, UK.

Some Useful Links

Extensive guide to phylogeny methods and programs by John Felsenstein: evolution.gs.washington.edu/phylip/software.html

Links from the Willi Hennig Society: www.cladistics.org/education.html

University of Oxford Evolutionary Biology Group Software page: evolve.zoo.ox.ac.uk/software.html

References

Excoffier, L., P. E. Smouse, and J. M. Quattro. 1992. Analysis of molecular variance inferred from metric distances among DNA haplotypes: Application to human mitochondrial DNA restriction data. *Genetics* 131: 479–491.

Farris, J. S., M. Kallersjo, A. G. Kluge, and C. Bult. 1994. Testing significance of incongruence. *Cladistics* 10: 315–319.

Farris, J. S., M. Kallersjo, A. G. Kluge, and C. Bult. 1995. Constructing a significance test for incongruence. *Systematic Biology* 44: 570–572.

Hudson, R. R., M. Kreitman, and M. Aquadé. 1987. A test of neutral molecular evolution based on nucleotide data. *Genetics* 116: 153–159.

Johnson, L. A. and D. E. Soltis. 1998. Assessing congruence: Examples from empirical data. In P. S. Soltis, D. E. Soltis, and J. J. Doyle (eds.), *Molecular Systematics of Plants*, Vol. 2, 297–343. Chapman & Hall, New York, NY, USA.

Kimura, M. 1980. A simple method for estimating evolutionary rates of base substitutions through comparative studies of nucleotide sequences. *Journal of Molecular Evolution* 16: 111–120.

Kishino, H. and M. Hasegawa. 1989. Evaluation of the maximum likelihood estimate of the evolutionary tree topologies from DNA sequence data, and the branching order in Hominoidea. *Journal of Molecular Evolution* 31: 151–160.

Saitou, N. and M. Nei. 1987. The neighbor-joining method: A new method for reconstructing phylogenetic trees. *Molecular Biology and Evolution* 4: 406–425.

Sanderson, M. J. 1997. A nonparametric approach to estimating divergence times in the absence of rate consistency. *Molecular Biology and Evolution* 14: 1218–1231.

Sanderson, M. J. 2002. Estimating absolute rates of molecular evolution and divergence times: A penalized likelihood approach. *Molecular Biology and Evolution* 19: 101–109.

Shimodaira, H. and M. Hasegawa. 1999. Multiple comparisons of log-likelihoods with applications to phylogenetic inference. *Molecular Biology and Evolution* 16: 1114–1116.

Tajima, F. 1989. Statistical method for testing the neutral mutation hypothesis by DNA polymorphism. *Genetics* 123: 585–595.

Wright, S. 1978. *Evolution and the Genetics of Populations*, Vol. 4. *Variability Within and Among Natural Populations*. University of Chicago Press, Chicago, IL, USA.

Printed in the USA
CPSIA information can be obtained
at www.ICGtesting.com
JSHW011506221024
72173JS00005B/1220

9 780231 133166